Praise for John Vaillant's

FIRE WEATHER

"A gripping narrative and a loud wake-up call. . . . Impossible to stop [reading]."   —*The Washington Post*

"Riveting. . . . A minute-by-minute disaster-movie narrative of the inferno."   —*The Guardian*

"Gripping. . . . Vaillant takes readers back into the deep history of the boreal forests before thrusting us into the Beast's fiery heart. *Fire Weather* is a report from the front lines of environmental cataclysm and a prediction of what more will surely come."   —NPR

"Epic. . . . A tale of terror from a climate change front line. . . . *Fire Weather* includes a lot about the science of fire and weather. But it is also a book about the cognitive dissonance in climate change discourse."   —*Financial Times*

"A refreshingly clear explanation of this hazy, uncanny moment in the earth's history. . . . Vaillant is the type of journalist who picks a single narrative and monomaniacally researches it, plunging himself deeper and deeper into the murky details, and then emerges, many years later, with a small universe cupped in his hands. . . . By turns heart-racing and horrifying."   —*New York*

"Vaillant writes so vividly that he can make subjects like the mining of bituminous sand . . . fascinating. . . . A timely warning of more smoke to come."   —*Slate*

"Tortuously timely. . . . Vaillant's book offers vital context for how the world's forests became more flammable."   —*WIRED*

"A gripping book that brings readers to the front lines of a major forest fire, while also exploring the intertwined history of oil and gas development and the study of climate change. Its lessons should not be soon forgotten."

—*Science*

"No book feels timelier. . . . A deeply reported narrative of one of Canada's most destructive recent wildfires. . . . A strongly argued polemic on the culpability of the petrochemical industry in a hotter, increasingly flammable world. . . . Vaillant's description of the fire rips along, an adrenaline-soaked nightmare that is impossible to put down."

—*Air Mail*

"A compulsively readable journey into our fiery times. At the center, Vaillant gives us fire itself as a character—fast, hungry, and evolving to shape the warming decades to come. You might never hear an engine or watch a bonfire the same way again."

—Bathsheba Demuth,
author of *Floating Coast*

"A glimpse into to a climate apocalypse. . . . We aren't done producing and using fossil fuels, and our world is heating up. Those two trends are inevitably going to bang into each other again, and Vaillant's book is a useful look at how that might unfold."

—*The New York Times* DealBook

"Vaillant is one of the great poetic chroniclers of the natural world, and here he captures the majesty and horror of one of its great disasters—and what made it tragically possible."

—David Wallace-Wells,
#1 bestselling author of *The Uninhabitable Earth*

"Mixes a beat-by-beat account of a wildfire with the history of oil extraction, climate change. . . . Masterfully done."

—*PBS NewsHour*

JOHN VAILLANT
FIRE WEATHER

John Vaillant's acclaimed, award-winning nonfiction books, *The Golden Spruce* and *The Tiger*, were national bestsellers. His debut novel, *The Jaguar's Children*, was a finalist for the Rogers Writers' Trust Fiction Prize and the International Dublin Literary Award. In 2023, Vaillant won the Baillie Gifford Prize and was chosen as a finalist for the National Book Award. He has also received the Governor General's Literary Award, the Windham-Campbell Literature Prize, and the Pearson Writers' Trust Prize for Nonfiction. Vaillant has written for, among others, *The New Yorker*, *The Atlantic*, *National Geographic*, and *The Guardian*. He lives in Vancouver.

ALSO BY JOHN VAILLANT

The Jaguar's Children
The Tiger
The Golden Spruce

FIRE WEATHER

FIRE WEATHER

///

*ON THE FRONT LINES
OF A BURNING WORLD*

///

JOHN VAILLANT

Vintage Books
A Division of Penguin Random House LLC
New York

TO THE SCIENTISTS AND VISIONARIES

CONTENTS

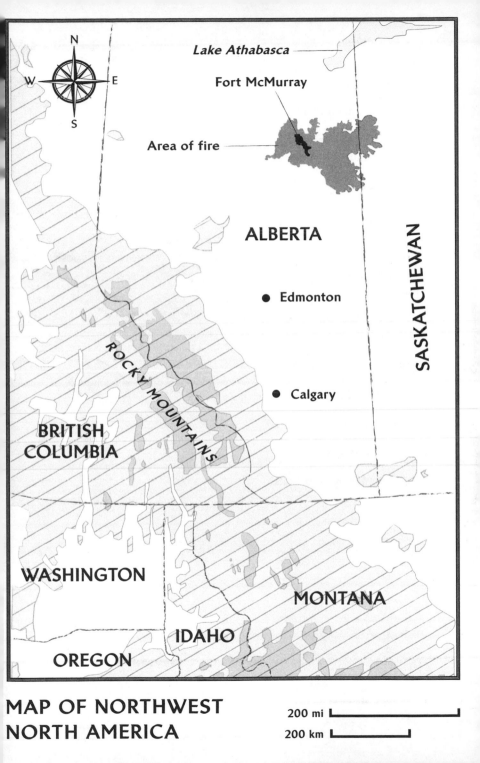

MAP OF NORTHWEST
NORTH AMERICA

200 mi
200 km

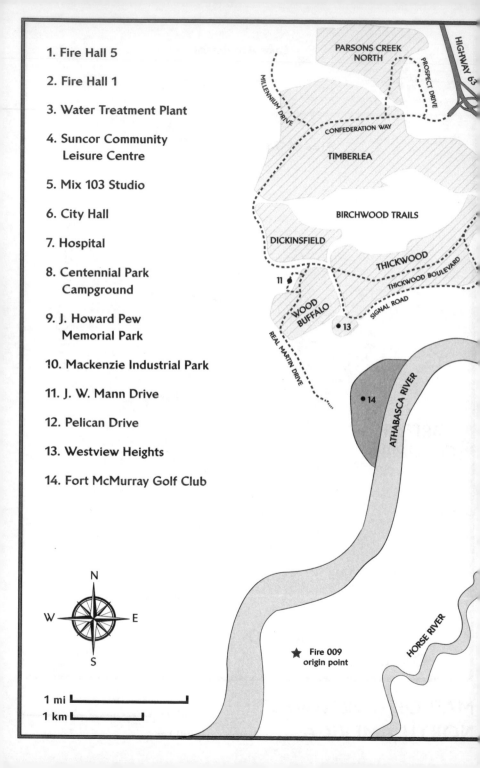

1. Fire Hall 5

2. Fire Hall 1

3. Water Treatment Plant

4. Suncor Community Leisure Centre

5. Mix 103 Studio

6. City Hall

7. Hospital

8. Centennial Park Campground

9. J. Howard Pew Memorial Park

10. Mackenzie Industrial Park

11. J. W. Mann Drive

12. Pelican Drive

13. Westview Heights

14. Fort McMurray Golf Club

PARSONS CREEK NORTH

HIGHWAY 63

PROSPECT DRIVE

MILLENNIUM DRIVE

CONFEDERATION WAY

TIMBERLEA

BIRCHWOOD TRAILS

DICKINSFIELD

THICKWOOD

THICKWOOD BOULEVARD

11

WOOD BUFFALO

SIGNAL ROAD

● 13

REAL MARTIN DRIVE

● 14

ATHABASCA RIVER

HORSE RIVER

N
W E
S

★ Fire 009 origin point

1 mi

1 km

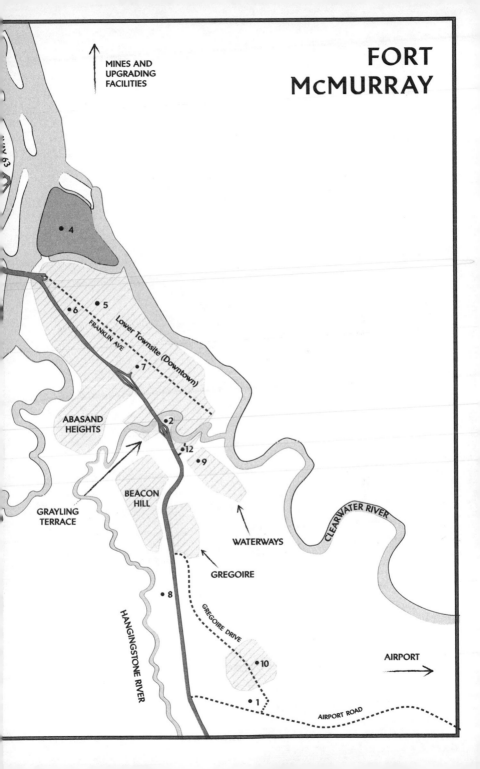

PART ONE

ORIGIN STORIES

The National Weather Service issues a "Fire Weather Watch"
when weather and fuel conditions may lead to rapid or dramatic
increases in wildfire activity.

In this great chain of causes and effects, no single fact
can be considered in isolation.

—ALEXANDER VON HUMBOLDT

PROLOGUE

On a hot afternoon in May 2016, five miles outside the young petro-city of Fort McMurray, Alberta, a small wildfire flickered and ventilated, rapidly expanding its territory through a mixed forest that hadn't seen fire in decades. This fire, farther off than the others, had started out doing what most human-caused wildfires do in their first hours of life: working its way tentatively from the point of ignition through grass, forest duff, and dead leaves—a fire's equivalent to baby food. These fuels, in combination with the weather, would determine what kind of fire this one was going to be: a creeping, ground-level smolder doomed to smother in the heavy dew of a cool and windless spring night, or something bigger, more durable, and dynamic—a fire that could turn night into day and day into night, that could, unchecked and all-consuming, bend the world to its will.

It was early in the season for wildfires, but crews from the Wildfire Division of Alberta's Ministry of Forestry and Agriculture were on alert. As soon as smoke was spotted, wildland firefighters were dispatched, supported by a helicopter and water bombers. First responders were shocked by what they saw: by the time a helicopter with a water bucket got over it, the smoke was already black and seething, a sign of unusual intensity. Despite the firefighters' timely intervention, the fire grew from 4 acres to 150 in two hours. Wildfires usually settle down overnight, as the air cools and the dew falls, but by noon the following day this one had expanded to nearly 2,000 acres. Its rapid growth coincided with a rash of broken temperature records across the North American subarctic that peaked at 90°F on May 3 in a place where temperatures are typically in the 60s. On that day, Tuesday, a smoke- and wind-suppressing inversion lifted, winds whipped up to twenty knots, and a monster leaped across the Athabasca River.

Within hours, Fort McMurray was overtaken by a regional apocalypse that drove serial firestorms through the city from end to end—for days. Entire neighborhoods burned to their foundations beneath a towering pyrocumulus cloud typically found over erupting volcanoes. So huge and energetic was this fire-driven weather system that it generated hurricane-force winds and lightning that ignited still more fires many miles away. Nearly 100,000 people were forced to flee in what remains the largest, most rapid single-day evacuation in the history of modern fire. All afternoon, cell phones and dashcams captured citizens cursing, praying, and weeping as they tried to escape a suddenly annihilating world where fists of heat pounded on the windows, the sky rained fire, and the air came alive in roaring flame. Choices that day were stark and few: there was Now, and there was Never.

A week later, the fire's toll conjured images of a nuclear blast: there was not just "damage," there was total obliteration. Trying to articulate what she saw during a tour of the fire's aftermath, one official said, "You go to a place where there was a house and what do you see on the ground? Nails. Piles and piles of nails." More than 2,500 homes and other structures were destroyed, and thousands more were damaged; 2,300 square miles of forest were burned. By the time the first photos were released, the fire had already belched 100 million tons of carbon dioxide into the atmosphere, much of it from burning cars and houses. The Fort McMurray Fire, destined to become the most expensive natural disaster in Canadian history, continued to burn, not for days, but for months. It would not be declared fully extinguished until August of the following year.

Wildfires live and die by the weather, but "the weather" doesn't mean the same thing it did in 1990, or even a decade ago, and the reason the Fort McMurray Fire trended on newsfeeds around the world in May 2016 was not only because of its terrifying size and ferocity, but also because it was a direct hit—like Hurricane Katrina on New Orleans—on the epicenter of Canada's multibillion-dollar petroleum industry. That industry and this fire represent supercharged expressions of two trends that have been marching in lockstep for the past century and a half. Together, they embody the spiraling synergy between the headlong rush to exploit hydrocarbons at all costs and the corresponding increase in heat-trapping greenhouse gases that is altering our atmosphere in real time. In the spring of 2016, halfway through the hottest year of the hottest decade in recorded history, a new kind of fire introduced itself to the world.

"No one's ever seen anything like this," Fort McMurray's exhausted and grieving fire chief said on national TV. "The way this thing happened, the way it traveled, the way it behaved—this is rewriting the book."

1

If a tree burns in the forest and nobody sees it . . .

In Canada, this is more than a philosophical question. Canada contains 10 percent of the world's forests, vast tracts of which are uninhabited. But "vast" is an ineffective descriptor when it comes to Canada, its forests, or its fires. One way to grasp the magnitude of this country is to get in a car in Great Falls, Montana, and head up I-15 to Sweetgrass, on the Canadian border. Once you've crossed into Coutts, Alberta, reset your odometer and point your car north. Then, settle into your seat for a couple of days. With the Rocky Mountains on your immediate left, this route takes you up the western edge of the Prairies, through Lethbridge, Calgary, and Red Deer—wheat and cattle country. Once past the northern metropolis of Edmonton, you will find yourself increasingly alone on the road, surrounded by broad expanses of hardscrabble subarctic prairie—fields frozen solid or half drowned and barely fit for cattle feed.

On the main road, now no wider than a residential street, hamlets with one blinking light and a gas station slide by and not another for fifty miles. To the east and west, gravel range roads run out to the vanishing point, and man-made structures appear more and more as intermittent novelties. Here, a schoolhouse-sized Ukrainian church with its tin-sheathed onion dome stands alone against a windswept loneliness so profound it suggests the Russian steppe. There, a barn collapses asymmetrically beneath the weight of a hundred heavy years, fully half of them spent clenched in the fist of winter, the people long gone. Farther on, a ten-acre lake so startlingly blue that mere reflection, even of Alberta's sky, seems insufficient to explain it. Somewhere along the way, you will cross an unmarked divide where deer give way to moose, crows give way to ravens, and coyotes give way to wolves. By the time you get to North Star, the wide-open spaces for which Alberta is famous will be filled in by low, mixed forest and bogland that bears a strong resemblance to Siberia. By the time you stop for coffee in a lonely place called Indian Cabins, it will be tomorrow and your odometer will be approaching one thousand miles, but you will still be in Alberta.

Up here, in the landlocked subarctic, things seem to occur in outsized dimensions: lakes can be the size of inland seas and the trout inhabiting them can weigh a hundred pounds; large wild animals, including the continent's biggest bison, outnumber people. In Wood Buffalo National Park, the second-largest national park in the world, is the world's largest known beaver dam. Spotted in 2007 with the aid of a satellite, it is more than twice as long as the Hoover Dam, and it appears to be growing. In 2010, an adventurous man from New Jersey named Rob Mark set out to visit it. He was allegedly the first person to do so, and it was hard going. "The foliage is so thick," Mark told the CBC, "you can't see very far . . . then it turns into muskeg, which is incredibly difficult to walk on. And then it goes out to complete bog swamp." It explains why so few outsiders frequent this place in the warmer months, and why winter is the preferred season for cross-country travel. "The mosquitoes," added Mark, "are absolutely horrific."

One exception to the general gigantism can be found in the trees, which seldom exceed sixty feet in height or a hundred years in age. These woods, a shifting mix of pine, spruce, aspen, poplar, and birch, are known collectively as the boreal forest,[*] and whatever they may lack in individual size, they compensate for in sheer numbers. Girdling the Northern Hemisphere in a circumpolar band, the boreal forest is the largest terrestrial ecosystem, comprising almost a third of the planet's total forest area (more than 6 million square miles—

---

[*] "Boreal," which means "northern," is derived from Boreas, the Greek god of the north wind.

larger than all fifty U.S. states). Fully a third of Canada is covered by boreal forest, including half of Alberta. Continuing west, over the Rocky Mountains, through British Columbia, the Yukon, Alaska, and across the Bering Strait into Russia (where it is known as the taiga), the boreal forest stretches all the way to Scandinavia and then, undeterred by the Atlantic Ocean, makes landfall on Iceland before picking up again in Newfoundland and continuing westward to complete the circle, a green wreath crowning the globe.

As densely wooded as the boreal might appear from the roadside, it is, within, something far more amphibious, containing more sources of fresh water than any other biome. In this sense, the circumboreal forest resembles a kind of hemispheric sponge that happens to be covered in trees, their billions of miles of roots weaving the continents together in a subterranean warp and weft. While not as openly fluid as Florida's Everglades, the boreal's countless lakes, ponds, bogs, rivers, and creeks serve a similar function of gathering, storing, filtering, and flushing fresh water. Billions of birds, representing hundreds of species, live in and migrate through this ecosystem.

One reason the trees never get very big or very old is because, in spite of all that water, they burn down on a regular basis. They're designed to. In this way, the circumboreal is truly a phoenix among ecosystems: literally reborn in fire, it must incinerate in order to regenerate, and it does so, in its random patchwork fashion, every fifty to a hundred years. This colossal biome stores as much, if not more, carbon than all tropical forests combined and, when it burns, it goes off like a carbon bomb. In North America, the epicenter for these stratospheric explosions is northern Alberta. Because of this, every town up here, big or small, faces the same dilemma: where the houses end, the forest begins. There are bears, wolves, moose, and even bison in there, but the most dangerous thing hiding in those woods is fire. Under the right conditions, a big boreal fire can come on like the end of the world, roaring and unstoppable. These are fires that can burn thousands of square miles of forest along with everything in it and still be out of control.

Virtually unknown and, at the time, unseen by all but a handful of people, is the Chinchaga Fire of 1950, the largest fire ever recorded in North America. Igniting on the border of British Columbia and

Alberta in June of that year, it burned eastward across northern Alberta for more than four months, impacting approximately 4 million acres, or 6,400 square miles, of forest (roughly, the combined area of Connecticut and Rhode Island, or three times the size of Prince Edward Island). The fire generated a smoke plume so large it came to be known as the Great Smoke Pall of 1950. Rising forty thousand feet into the stratosphere, the plume's colossal umbra lowered average temperatures by several degrees, caused birds to roost at midday, and created weird visual effects as it circled the Northern Hemisphere, including widespread reports of lavender suns and blue moons. Prior to the Chinchaga Fire, the last time such effects had been reported on this scale was following the eruption of Krakatoa in 1883. Carl Sagan was sufficiently impressed by the effects of the Chinchaga Fire to wonder if they might resemble those of a nuclear winter.

⁓⁓⁓

Every year, the National Oceanic and Atmospheric Administration (NOAA), in cooperation with fire scientists from Canada and Mexico, issues a document called the *North American Seasonal Fire Assessment and Outlook*, which attempts to predict the likelihood of wildfires across the continent. The *Outlook* includes maps for each month of fire season, and they are color-coded, with red indicating a likelihood of increased fire activity and green indicating a decrease. Like 2015 before them, the monthly maps for 2016 showed a lot more red than green, and the map for May showed more red than all the others: in addition to large swaths of Mexico, the American Midwest, and all of Hawai'i, red covered much of southern Canada— from the Great Lakes all the way to the Rocky Mountains. It was an enormous area and included most of Alberta's active petroleum fields. In the middle of that hot zone, in the middle of the forest, sat Fort McMurray.

Fort McMurray is an anomaly in North America. Located six hundred miles north of the U.S. border and six hundred miles south of the Arctic Circle, the city is an island of industry in an ocean of trees. Without the lure of petroleum, this part of Alberta would resemble Siberia in even more ways than it already does: sparsely populated; its rivers spun like compass needles toward the Arctic Ocean; its trees low, short-lived, and prone to fire. Here, half a dozen permanent settlements dot a region the size of Kentucky, and only one has a population over 800: in 2016, Fort McMurray and its satellite communities were home to an international population of nearly 90,000 people living in 25,000 houses and buildings ranging from trailer homes and condominiums to McMansions and high-rise concrete apartments. The city's "urban service area"—the area covered by garbage collection and firefighting services—covers sixty square miles (100 sq. km.) of convoluted terrain laced with creeks and ravines that are further fragmented by two major rivers and two tributaries. Together, they surround and entwine the city like the writhing arms of an octopus.

Scattered across the surrounding landscape in semipermanent "man camps" was an additional shadow population of roughly fifty thousand workers whose numbers ebb and flow with the price of crude oil, the pace of development, and routine maintenance cycles at the processing plants. As one longtime resident put it, "We're just a colony of oil companies." Canada is the world's fourth-largest oil producer and the third-largest exporter. Nearly half of all American oil imports—around 4 million barrels per day, come from there—the equivalent of one ultra large crude carrier ship every twenty-four hours. Of this vast quantity, almost 90 percent originates in Fort McMurray.

Despite being virtually unknown outside of Canada and the petroleum industry, Fort McMurray has become, in the past two decades, the fourth-largest city in the North American subarctic after Edmonton, Anchorage, and Fairbanks. In terms of overtime logged and dollars earned, it is, without a doubt, the hardest-working, highest-paid municipality on the continent. In 2016, two years past a decade-long boom that ended with a sudden drop in global crude oil prices, the median household income was still nearly $200,000 a year. Fort McMurray has earned several nicknames over the years, and one of them is Fort McMoney.

May 3, 2016, began differently for everyone, but in Fort McMurray, it ended the same. For Shandra Linder, it began with a rite of spring. Linder was a labor relations adviser who worked for Syncrude (a portmanteau of "synthetic crude oil"), a mainstay of the local economy. Shandra's husband, Corey, an engineer, was employed there, too, and so were many of their friends. Both Linders worked out of the head office at the Mildred Lake complex, a half hour's drive north of town. By 2016, Shandra Linder had called "Fort Mac" home for nearly twenty years; blond, with a pixie cut, Linder is fit and warm and does not suffer fools. It makes sense once you get to know her and what she does, but to an outsider it might be surprising to see some- one so polished—and female—in such a remote, industry-oriented, testosterone-heavy place. At "Site" (the catch-all term for any mine or other petroleum-related workplace around Fort McMurray), the ratio of men to women runs about twenty-five to one. For work, Linder dressed accordingly: minimal makeup, high collars, dark pants, no heels—clothes suitable for climbing in and out of trucks and SUVs, for working in a world of working men. Linder exudes a quiet con- fidence, in part because working full-time for Syncrude, or its larger counterpart, Suncor, confers a blue-chip status on its employees. Working for these companies is the boreal equivalent to working for Exxon or Shell, and the distinction permeates like a pheromone. As one insider put it, "I am Syncrude and you are not." Tradesmen and machine operators wear their company badges like team colors— even to the bars, where, during the last boom, they signaled to avail- able women like so much plumage. Comparable to a stockbroker's platinum card, worn externally, a company badge communicates vol- umes at a glance: six-figure salary, five-figure truck, four-figure party budget, fungible skills. Meanwhile, the company, also known as "the Owner," or "Mother Syncrude"—asks a lot in return: just like Wall Street or Silicon Valley, working late and weekends is simply part of the job. But that's where the money is: in Fort McMurray, the best time is overtime.

Shandra Linder had already seen the smoke plume southwest of town, because everyone had seen it. It had been there for days, mor- phing on the horizon, a windswept cauliflower of billowing grays and browns that appeared to have sprouted, full blown, from the forest on Sunday afternoon. It had been growing since then, but it was still miles away, and it wasn't the only one. Over the weekend, the Linders had hosted friends who had evacuated due to another fire burning near the new Stonecreek development north of downtown. It was almost a lark: on Sunday, May 1, they'd had cocktails on their back deck in Timberlea, one of many hilltop neighborhoods to the north and west of downtown. There, drinks in hand, putting green and pocket fountain at their feet, they took photos of the big plume developing across the river the same way one would a sunset or a rainbow. They ate chicken and rice, and got a convivial buzz on—life was good in Fort McMurray. Their friends went home the next day.

Because Forestry was on it: boots were on the ground, water bombers were in the air. As far as the Linders and their guests were concerned, whatever was out there was being handled. After all, that is what people do in Fort McMurray: they handle things. Not many regions self-select as rigorously as northern Alberta does, and Fort McMurray selects for workers—tough, adventurous team players, highly motivated to do what it takes and prosper. That includes wildfire fighters, and Alberta Forestry's wildfire crews—with a territory ranging from tallgrass prairie and parkland to the Rocky Mountains and the boreal forest—are considered among the best in the world. In private, some members consider themselves *the* best. Certainly, the beginning of May was a little early for fires—there were still car-sized blocks of winter ice on the riverbanks, and some local lakes had yet to thaw—but otherwise, this was nothing new. Fires cloud the horizon every spring and summer; up here, smoke is simply a feature of the boreal landscape. As Shandra and Corey Linder said, practically in chorus, "It happens every year."

Which was true, until it wasn't.

In the forest, out of sight, things were changing. Winter snowfall had been far below average for two years running and, though it was still early spring in the north, leaves and pinecones crackled underfoot as if it were late summer. Given this, the unseasonable heat, and the fact that five separate wildfires ignited around the city that weekend, it is hard to overstate how unconcerned was Fort McMurray's citizenry. But if you were up at dawn on May 3, as Shandra Linder was, and you had seen the sky, so fresh and clear and full of summer promise, as she had, you might understand why. The brilliance of that morning was so exceptional, even for northern Alberta, that after her morning routine of a dog walk, emails with coffee and a cigarette, and a shower, Linder did something she hadn't done in a long time: she pulled out her favorite navy-blue suit with the skirt, picked some medium heels to go with it, and left her socks in the drawer. Thus attired, she headed off to work in Syncrude's head office at Mildred Lake. In the garage, there were a few vehicles to choose from; in keeping with her outfit and her mood, Linder picked the car she calls "the little one"—a black Porsche that hadn't seen daylight in six months. Winters are long and dark in Fort McMurray, but this one was over, spring was here, and Linder felt as beautiful and hopeful as the day.

She had lots of company; over the past few weeks, her neighbors had been emerging, too, unfurling with the spring flowers that had arrived weeks early that year. Coats and boots worn like a second skin since October were being packed away, and yards were being tidied after half a year of neglect. Garages, where a lot of Fort McMurray actively socializes among tool benches, beer fridges, ATVs, and various works in progress, were opening to the air, sun, and visitors. People were smiling to themselves at the bus stop, faces turned skyward like sunflowers, or Russians, as their bodies remembered the foreign sensation of warm sun on bare skin.

2

It is natural that the idea of bituminous sand
development should be accepted with considerable
hesitancy. It does not conform with the established
order of things.

—Karl A. Clark, *The Bituminous Sands of Alberta*, 1929

To begin to understand Alberta, you really have to understand the
power of the sun and the sky, which seem to take up a disproportionate amount of space here. There aren't many places where, in
a single twelve-hour period, you can see the northern lights in full
regalia and a rainbow so huge and vivid that all seven colors appear to
vibrate in concentric neon arcs—each band distinct from the next. So
big and open is the country, and so clear and dominant the sky, that
these phenomena may accompany you for hours, even as you travel at
highway speed.

Alberta lies due north of Montana and due south of the Northwest Territories; it is roughly the size of Texas, and the two have a lot
in common. Like Texas, Alberta is a kind of energy vortex: along with
wide-open spaces and a patriotic allegiance to the petroleum industry, it shares with its American counterpart a painful familiarity with
natural disasters, including tornadoes, hailstorms, floods, and fires.
Hardworking and independent-minded, Alberta, too, prides itself on
a mythic legacy built around cattle, horses, cowboys, and oil, further
energized by deep wells of "get 'er done" can-do-itiveness, evangelical

Christianity, and cantankerous alienation from its national capital. To give an idea just how alienated Alberta has been from the federal government, consider a snow sculpture erected on the campus quad during the University of Alberta's winter festival in 1981: in response to then Prime Minister Pierre Trudeau's program to share Alberta's oil revenue with less-wealthy provinces, Trudeau was depicted on his knees, fellating an oil derrick rising from the crotch of Alberta's premier. In front of this bracing tableau, also carved in snow, was the caption "TRUDEAU WANTS EVERY DROP."

But not as much as Alberta does. Those two men may have melted into memory, but the resentment persists, and the number of oil and gas wells in Alberta has ballooned into the hundreds of thousands. In an effort to locate every drop, the province has also been crisscrossed with exploratory seismic lines. Today, they run from horizon to horizon, like meridians on a globe. These random-seeming pathways to nowhere, carved and blasted by geologists assessing the mineral and hydrocarbon potential below, are interrupted only by rivers, beaver dams, and the occasional road, test pit, or drill site. In some places, including around Fort McMurray, the landscape has been cross-hatched so tightly by these bulldozed corridors that, from the air, the forest has the appearance of a terraformed waffle. Cutting through these gridworks at oblique angles are man-made rivers of oil and gas in the form of pipelines, and these, too, trace the arc of the earth, out of sight and across the continent. If all of Alberta's pipelines were lined up end to end, they would span the gap between Fort McMurray and the moon, with enough leftover to wrap the equator. Some of these pipelines are four feet in diameter, and much of the petroleum flowing through them is extracted using unconventional methods like fracking, steam-assisted gravity drainage, and strip-mining.

Fort McMurray plays a central role in this endeavor, and, to this end, it is a one-industry town. That industry revolves around bitumen recovery, upgrading, and transport. Without an appreciation for bitumen and its provisional status in the fossil fuel hierarchy, there is no way to comprehend the phenomenon of Fort McMurray. Bitumen (pronounced *BITCH*-amin) is a kind of degenerate cousin to crude oil, more commonly known as tar or asphalt. Surrounding Fort McMurray, just below the forest floor, is a bitumen deposit the size

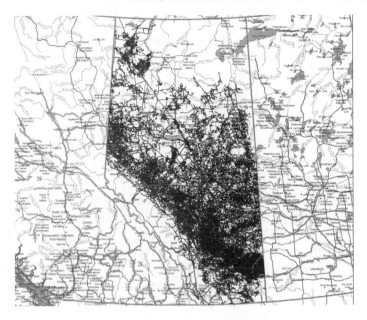

More than 300,000 pipelines serve the fossil fuel industry in Alberta.

of New York State. Sometimes referred to as the Alberta tar sands, or the oil sands, it is one of the biggest known petroleum reserves in the world. In terms of potential barrel volume, it is in a class with Saudi Arabia, Venezuela, and Iran. But as abundant as it is, there's a catch: it's not oil. It is not, strictly speaking, even bitumen; it is what geologists call "bituminous sand." Bituminous sand is to a barrel of oil what a sandbox soaked in molasses is to a bottle of rum. Even after you've dug it out of the ground and separated it from its gritty matrix, you're no closer to a viable form of energy. It's still just bitumen—excellent for tarring roofs and paving roads, but so nonflammable you can put out a campfire with it.

Though it is possible to find deposits of pure bitumen along the Athabasca River, the overwhelming bulk of it occurs in a mineral aggregate comparable to an exit ramp. According to *Oil Sands Magazine*, "A typical oil sands deposit contains about 10% bitumen, 5% water and 85% solids." Those solids are principally quartzite, one of the hardest minerals in the world. Quartzite sand is extraordinarily abrasive, and it is hell on machinery, shovels, dump truck boxes, and pipelines, not to mention the paint job on your truck and your kitchen floor. The process of excavating, separating, and then "upgrading" this pavement-like substance involves elements of strip mining, rock crushing, and steam cleaning—the petrochemical equivalent to squeezing blood from stones. Because of this, there is really no comparison between the petroleum industry in northern Alberta and the petroleum industries in Texas, Saudi Arabia, or any other place, on-or offshore, where oil is drawn from the earth by conventional means.

A bitumen mine is not a place you would let your child play, but it is excavated using equipment familiar to any four-year-old conversant in Tonka technology—and with a similar grandiosity of ambition. In order to access the bitumen, the forest above it must first be removed. In industry parlance, this living material is referred to as "overburden," and the machine used to scrape it off is a Caterpillar D11 bulldozer. The D11 weighs more than a hundred tons, and its blade is twenty-two feet wide; it can plow down a forest like mowing a lawn. But this is entirely in keeping with the scale of things up here; working alongside the D11s are Komatsu D575s, which are even bigger. Once the forest has been removed, enormous electric shovels excavate the bituminous sand in boulder-sized chunks that can weigh a hundred tons and occasionally contain complete dinosaur fossils from the Cretaceous Period. These garage-sized payloads are dumped into a "hauler," and the Caterpillar T797 hauler is one of the biggest dump trucks in the world. It is three stories tall and weighs four hundred tons—unloaded. There are hundreds of machines like this operating in the mines north of Fort McMurray. Far too large for ordinary highways, they must be transported north in pieces. It takes twelve oversized semi loads traveling with escorts to move the component parts of a single hauler. The tires alone are thirteen feet tall and cost $85,000 apiece. When one of them catches on fire—something that happens more often than one might expect, due to the terrific friction their loads generate—it must be deflated from a safe distance with a rifle bullet. Should one of these six-ton tires explode on its own, it will impact its surroundings like a powerful bomb. The hauler's job is to carry the raw bituminous sand over to the "crusher," a kind of mechanical black hole composed of two gigantic, continuously turn-

Hauler trucks at work in a mine north of Fort McMurray

ing studded cylinders. In an effort to articulate the inexorable voracity of this device, one employee explained to me that a crusher "can consume a city bus in three seconds."

The environment these behemoths are currently dismantling is an icebound netherworld as seen through the eyes of Sebastião Salgado, Edward Burtynsky, or J. M. W. Turner: mile upon mile of black and ransacked earth pocked with stadium-swallowing pits and dead, discolored lakes guarded by scarecrows in cast-off rain gear and overseen by flaming stacks and fuming refineries, the whole laced together by circuit board mazes of dirt roads and piping, patrolled by building-sized machines that, enormous as they are, appear dwarfed by the wastelands they have made. The tailings ponds alone cover well over a hundred square miles and contain more than a quarter of a trillion gallons of contaminated water and effluent from the bitumen upgrading process. There is no place for this toxic sludge to go except into the soil, or the air, or, if one of the massive earthen dams should fail, into the Athabasca River. For decades, cancer rates have been abnormally high in the downstream community of Fort Chipewyan. Even people gainfully employed in those mines compare them to Mordor. Grouped near Syncrude's upgrading plant, like an ancient temple complex, are bright yellow ziggurats of solid sulphur bigger than the Pyramids at Giza. Dwarfing even these is Syncrude's six-hundred-foot-tall flare stack; Suncor, just a few miles away, has one too. In 2016, these skyscraping, fire-breathing gnomons were the tallest man-made things for a thousand miles in any direction.

At this scale, humans simply disappear.

Underappreciated by most of us down south is the fact that all of this—man and machine alike—must function year-round, twenty-four hours a day, across some of the world's most extreme temperature fluctuations. Untreated diesel fuel begins to gel at 15°F and bulldozer blades can shatter at -35°F, but Fort McMurray sees temperatures in the -40s every winter and has posted lows in the -60s. The fire department's pumper trucks have built-in heaters to keep the water from freezing en route to a call. And yet, with increasing regularity, the region is seeing summer highs in the 90s. Extremes like this put terrific stress not just on metals, but also on hydraulic hoses, lubricated gears, and any fluid that must be kept flowing at a consistent viscosity. Of course, it is particularly tough on human beings; every plant has dedicated crews who do nothing but build scaffolding, often outside and in all weathers. There, a five-knot zephyr will make -25°F, a routine winter temperature in Fort McMurray, feel like -40°F, a temperature at which exposed flesh and eyeballs freeze in minutes. Shifts are typically ten to twelve hours long, and, in winter, the sun is up for only seven hours a day; even at noon, it sulks on the horizon offering no sensible heat.

But when a stiff north wind carries smoke from the cokers up the river into town, there are few complaints. Most just sniff and say, "Smells like money."

The Alberta government has always worked closely with the fossil fuel industry, to the point where it can be difficult to tell where one ends and the other begins. Together, they have struggled for a century to come up with a brand for all that potential lying just beneath the forest floor. "Nature's Supreme Gift to Industry" was floated in the 1920s, but it didn't catch on. Another worthy effort was the "Magic Sand-Pile" campaign. Aimed at American investors and inventors, the ad ran in a 1947 issue of *Fortune* magazine with the slogan "ALBERTA *has* WHAT YOUR BUSINESS *needs!*" (Low taxes, generous subsidies, and minimal oversight—enticements known today as

THE MAGIC
SAND-PILE

In the north-eastern corner of Alberta, Nature has created the most fabulous "sandpile" in the world. Along the banks of the Athabasca River, 30,000 square miles of land contain billions of tons of oil-soaked sands. Tests have shown that these bituminous sands can yield an amazing number of valuable petroleum products. Your research, your plant methods, could unlock fully the oil reserves in this magic sandpile — authoritatively estimated to contain more than ten times all the oil reserves in the world.

The U. S. Bureau of Mines estimates Alberta's oil-sands to contain 250 billion barrels of oil. 23 per cent of this oil can be made into high-octane aviation gasoline, 17 per cent into high quality motor gasoline. By-products include everything from road-surfacing to roofing products. The oil-recovery content is as high as 25 per cent-- a yield of from 100,000 to 125,000 barrels per acre.

Here, in the tar-sands of Alberta, is an unique opportunity for industry-- an opportunity and a challenge in the free land of free enterprise.

ALBERTA *has* WHAT YOUR BUSINESS *needs!*

WRITE - - - THE INDUSTRIAL
DEVELOPMENT BOARD
Administration Building

GOVERNMENT OF THE
PROVINCE OF ALBERTA
Edmonton, Alberta, Canada

AG-20

the "Alberta Advantage.") The ad copy reads like a call for industrial homesteaders.

Despite its stirring language and double-barreled promise of freedom, the prospect of something bordering on the alchemical buried in a magic sandpile seven hundred miles north of Great Falls was a bit too rich, even for American capitalists and engineers. Besides, there was an awful lot of actual oil to be found down south—not only in Texas, Oklahoma, and California, but in southern Alberta. By 1930, the United States was already veined with more than 100,000 miles of oil pipelines carrying a billion barrels of crude oil annually.

Reinventing and rebranding unwanted things has been a necessary survival strategy in western Canada, but transforming the sow's ear of bituminous sand into the silk purse of Syncrude Sweet Blend is its greatest achievement yet. In language that sounds like a special offer from Starbucks, one oil industry data site describes "this typical light sweet synthetic crude [as] a bottomless blend of hydrotreated naphtha, distillate, and gasoil fractions." And, like rendering something as undrinkable as coffee beans into a Dark Chocolate Melted Truffle Mocha, the process of rendering something as unburnable as tar-coated sand into Syncrude Sweet Blend, Western Canadian Select, or Albian Premium requires a lot of heat and pressure.

Whatever one chooses to call the petroleum residue beneath Fort McMurray, it's not oil—not anymore. Had the industry discovered it 50 million years ago, they could have had more than a trillion barrels of crude oil on their hands, a bonanza by which Saudi Arabia's reserves would pale in comparison. But over thousands of millennia, much of that sweet crude was impacted by inexorable natural forces that caused it to migrate upward and eastward through a huge, geological depression called the Western Canada Sedimentary Basin. This continental reservoir, formerly an inland sea, contains generous deposits of oil, gas, and bitumen that inform much of the geology between the Rocky Mountains and the Precambrian Shield (that long, broad, flat stretch west of the Great Lakes).

By the time this wayward oil found its way into an England-sized sandstone layer known as the McMurray Formation, just beneath the Athabasca Plain, it had already been discovered. Not by the British or the Americans, but by bacteria. Like a plague of mice in a cheese cellar, this nano army raided the continent's greatest petroleum reserve and left behind only wrappers and rinds. These unlikely predators, representing several different genera, are otherworldly creatures that nourish themselves on hydrocarbons, survive without oxygen, and off-gas methane (one of the only characteristics we share besides an appetite for crude oil). Small as they are, they appear to be the masters of a poorly understood domain known to geochemists as the "deep biosphere." Situated at the lower limits of habitability, between the lifeless depths of Earth's crust and the surface realms of sunlight and oxygen, inhabitants of the deep biosphere have been found more than a mile underground and at temperatures that exceed the boiling point. By all accounts, this biome is not only vast but teeming. As Steve Larter, a fellow of the Royal Society who occupies the Canada Research Chair in Geochemistry at the University of Calgary, wrote in 2014, "The oil sands and heavy-oil belts of the world represent the most viable access point to the deep biosphere, which from a cell-balance perspective is the largest biome on the planet."

When Larter tried to calculate how many of these hydrocarbon-eating "extremophiles" inhabited the bitumen deposits surrounding Fort McMurray, the number he arrived at was "well in excess of 10 to the 23rd"—trillions upon quadrillions upon quintillions of hungry creatures thriving in one of our planet's most hostile environments. Given their numbers and their impact, surprisingly little is known about them; their only familiar requirement is water. Nonetheless, they have taken a devastating toll. Approaching this ancient oil with the diligence and discernment of a petroleum engineer, these tiny multitudes cherry-picked the simpler, "sweeter," more marketable hydrocarbons, leaving behind the longer, more complex molecules laden with tarry asphaltenes, resins, salts, heavy metals, and complex sulphur compounds, among other unsavory impurities. Fifty million years on, with the low-hanging fruit largely gone, Alberta has inherited the dregs, and oil refiners don't care for it any more than those intrepid microbes do.

In order to make this hydrocarbon residuum resemble something contemporary oil refineries can actually process, and that foreign markets are willing to buy, it must be artificially restored to its pre-degraded state—in other words, forced back in time. It has been said that helicopters don't fly; they beat the air into submission. The same can be said of the effort required to turn bitumen back into a usable, marketable fuel. It takes two tons of bituminous sand to make a single barrel of bitumen, and, at room temperature, bitumen pours about as well as Nutella. To be capable of ignition, liquid bitumen (the only petroleum product that is heavier than water) must be preheated well past the boiling point. To make it flow through a pipe, it must either be heated or blended with a diluting agent. These diluents are usually natural gas condensates or other industrial thinners, all of which are highly toxic and explosive. Diluted bitumen—"dilbit"—will remain fluid only under sealed conditions in a tank or a pipeline. In the event of a spill or other exposure, the added diluent will evaporate (or burn) off almost immediately, leaving the bitumen to glom on to the nearest firm surface. In the case of a major rupture in Enbridge's Line 6B in 2010, the nearest firm surface was the bottom of the Kalamazoo River. The million-gallon (24,000-barrel) spill impacted forty miles of waterway near Marshall, Michigan, and took five years and more than a billion dollars to mitigate ("clean up" being a misnomer here), making it the most expensive inland petroleum spill in history. The market value of the spilled dilbit was less than 1/1000th of the mitigation cost.[*]

While the end products rendered from bitumen are ultimately useful, taking the form of synthetic crude oil, diesel fuel, and feedstock for other petroleum products, its journey into a form recognizable to a petroleum engineer or a southern oil buyer is an arduous and expensive one, requiring enormous quantities of water, chemicals, and foreign capital. But most of all, it takes brute force, and that brute force is fire. The preferred extraction method, which now accounts for about 80 percent of the region's bitumen production, is steam-assisted gravity drainage, or SAG-D. SAG-D involves layered arrays of wellbores and piping through which steam is injected, in situ, in order to melt the bitumen directly out of its confusion of sand and clay. In both SAG-D and surface mining operations, the steam used

---

[*] In light of our current planetary circumstances, impacts on the natural world represent the truest accounting of the actual "downstream" cost of a petroleum-powered civilization.

to melt the bitumen is heated with natural gas, and the quantities required to do this are astounding.

Natural gas, which is about 80 percent methane, is measured, not in gallons or barrels, but in cubic feet. According to Canada's National Energy Board, the bitumen industry uses more than 2 billion cubic feet of natural gas per day (the energy equivalent of 350,000 barrels of oil), for the sole purpose of separating bitumen from sand. Canada is the fourth-largest producer of natural gas in the world, and in 2017 nearly a third of Canada's total production was devoted to this purpose. Natural gas, it must be said, is an organic fuel that requires minimal refining, yet the result of this colossal energy input is a substance still so dense with impurities that it does not burn. Even after it has been separated from the sand, bitumen requires an additional series of radical, heat-intensive interventions in order to liberate the oil and gas molecules bound up in the tar.

The industry knows this, and it has not been kind. Bitumen and its derivatives are steeply discounted on the open market. One way to measure the value and desirability of a fuel is to calculate its energy return on investment, or EROI. Conventional fuels like crude oil or natural gas have high EROI ratings, generating close to thirty units of energy for every unit of energy used to extract them. Alberta bitumen has an EROI of about six to one for surface mining, and three to one for steam-assisted gravity drainage—ratios so marginal that no conventional petroleum producer would consider them. Given these obstacles, a layperson would be hard pressed to make the business case for bitumen, especially when there is still so much flowing oil in the world.

Alberta has taken these liabilities into account and, in order for the bitumen industry to be even remotely profitable, four conditions must be met: conventional oil must be trading above $50 a barrel; the natural resources needed to produce it (fresh water, natural gas, and the boreal forest ecosystem) must be had for next to nothing; the industry itself must be heavily subsidized; and exploration costs must be nil.* There is a fifth condition, exploited not just by the bitu-

---

* Saudi Arabia, by contrast, can break even at $5 a barrel.

men industry but by the entire burning world: no consequences for emissions. This is what Alberta has built and bet its economy on, with mixed results.

By the early 2000s, these conditions had proved sufficiently favorable to Syncrude and Suncor—and, more recently, to multinationals like ExxonMobil, Chevron, Conoco, Royal Dutch Shell, France's Total, Norway's Statoil, China's Sinopec, and the joint Chinese- and Canadian-owned Husky Energy (among others)—that all of them staked billion-dollar claims in these northern woods, a thousand miles from the nearest tidewater port or major market. Because of these companies, along with their colleagues, competitors, and the generous support of the Canadian and Alberta governments, Fort McMurray has become the center of the largest, most expensive, most energy-intensive hydrocarbon recovery project on Earth. A rough estimate of investment to date is half a trillion dollars.

3

Whereas many incredible miracles occur in the
Babylonian country, there is none such as the great
quantity of bitumen found there. Indeed, there is
so much of it that . . . the people who have gathered
there collect large quantities of it. And although the
multitude is without number, the yield, as with a rich
well, remains inexhaustible.

—Diodorus of Sicily, *Bibliotheca Historica*

Bitumen has been in use for far longer than Alberta has been a province, or Canada a country. There are numerous places in the world,
including Baku in Azerbaijan, Hīt in Iraq, Pitch Lake in Trinidad,
Guanoco Lake in Venezuela, and Sakhalin Island on the east coast of
Russia, where bitumen is found in open deposits, in both solid and
liquid form. During the past six thousand years or so, in most places
and markets where it occurs, bitumen's traditional uses have been
in construction, waterproofing, and paving. At various times, it has
also been used as an adhesive and for embalming the dead. The first
golden age of bitumen began about four thousand years ago in Mesopotamia, between the Tigris and Euphrates rivers, where it was used
to most memorable effect as mortar in massive construction projects.
Until relatively recently, great bergs of solid bitumen would appear
spontaneously on the surface of the Dead Sea, where it would float

due to the water's exceptional buoyancy. This phenomenon occurred so predictably that, in ancient times, the Dead Sea was known as Asphalt Lake.

Prior to European contact, Indigenous peoples in the Athabasca region used bitumen for plugging holes in their canoes and sealing water vessels, much as their counterparts did in the Middle East. To this day, northern Alberta is inhabited by the Dene people, who live widely across the western subarctic. They are members of the Athapaskan language family, and they now share northern Alberta with more recently arrived Cree and several other Indigenous groups, including the Métis, a federally recognized population of mixed Indigenous and European heritage (which, in 2016, included Fort McMurray's mayor).

Historically, the Dene people subsisted on hunting, trapping, fishing, and berry picking. Several different Dene tribes overlapped in the region, and their presence was a determined but tenuous one: the winters were brutal, starvation was commonplace, and, prior to European contact, populations tended to be in the low hundreds, if that. So strong was their desire to trade that the hardiest among them would travel eastward, all the way to the Hudson's Bay Company trading hub of Fort York, in present-day Manitoba, a hazardous journey of nearly seven hundred miles. It was there that the first written reference to Alberta bitumen appeared, in the journal of James Knight, the trading post's chief factor (manager). On June 27, 1715, Knight reported a conversation with Indigenous hunters who told him of a "Great River" that "runs into the Sea on the Back of this Country & they tells us there is a Certain Gum or pitch that runs down the river in Such a bundance [sic] that they cannot land but at certain places."

It would be another sixty years before a European laid eyes on it.

The river those hunters were referring to was the Athabasca— specifically, the area around present-day Fort McMurray, where bitumen forms a black and sandy stratum through the river valley. On particularly hot days, it will trickle down the south- and west-facing cliffs and pool up on the riverbank; in some places it will flow into the river itself, causing temporary slicks. If the weather is hot and the wind is right, you can smell it before you see it. In winter these drips freeze in place, giving the impression that the cliff faces are weeping black tears, but even at twenty below, they carry an odor of roofing tar. Once they arrived, European explorers and traders would note this acrid goo in their journals as one would a hot spring, a waterfall, or any other local curiosity. But they were there to make money, and, while there was a killing to be made in the boreal forest, it wasn't in bitumen, not yet.

Long before Syncrude, Suncor, Exxon, or Shell, there was the Hudson's Bay Company. "The Company," as it came to be known, was the continent's first industrial-scale resource extractor, and it pioneered an approach to business, markets, employees, and the natural world that together could be called "wildfire economics." Using furs as fuel, the European market as fire, and credit as oxygen, the Hudson's Bay Company burned its way across the North American continent, altering it forever while generating extraordinary wealth for a handful of men an ocean away.

The Company was granted a Royal Charter by King Charles II in 1670, three years after John Milton published *Paradise Lost*. In return for actionable intelligence on the Northwest Passage, the king gave this "Company of Adventurers" exclusive trading and mineral rights to the Hudson's Bay watershed, a 2-million-square-mile swath of a continent none of these men had ever seen. This region was appealing because it was known to be rich in fur-bearing mammals, and the beaver, which had been wiped out in Europe, was still in heavy demand for hat making. So lucrative was this trade, and so central was it to the new boreal economy, that beaver pelts became a unit of currency. This "fur standard" was well established by 1778 when Peter Pond, a British loyalist, soldier, and twice-implicated killer from Connecticut, arrived in the Athabasca region—the first recorded European to do so. Pond, who one biographer described as "a fist in the wilderness," was a bare-knuckle entrepreneur and commercial explorer. Operating outside the purview of the Hudson's Bay Company, Pond had been seeking a Northwest Passage himself when he found Dene hunters eager to trade what turned out to be the finest beaver pelts in all of Canada. Nothing has been the same there since. Following Pond's arrival, beaver, along with just about anything else that could grow fur, was packed out of northern Alberta by the ton.*

It is hard to fathom the sheer number of dead animals that passed through Canada's trading posts during the boom years of the fur trade, but the legendary commercial explorer, navigator, and fur trader Alexander Mackenzie kept meticulous records on the subject. Mackenzie, a Scot from the Hebrides, was a partner with Peter Pond in the North West Company, which dominated trade in the region before being absorbed into the Hudson's Bay Company. In 1798, Mackenzie reported a staggering annual take of 106,000 beaver pelts, 32,000 marten, 6,000 lynx, 3,800 wolves, 2,100 bears, and 500 wood bison (among many other species).†

These numbers are all the more astonishing when one considers the fact that each animal had to be killed, collected, and skinned individually, before being transported by foot or canoe to the nearest trading post, which could be a hundred miles away from the hunter's village or killing site. And that was only the beginning of the jour-ney: the Athabasca region is so remote that, during the early days, a full trading cycle—by canoe from the Athabasca eastward across the continent to the port in Montreal and then by ship to England, returning with an equal load of trade goods, again by canoe, on waterways that were frozen half the year and lethally cold and fast the other half—took three years. Slow as it was, the logistics, plan-ning, and brute optimism of the Athabasca fur trade was, for its time and technology, as formidably ambitious and effective as any global enterprise today.

The delivery vans of the day were birchbark canoes, and, delicate as they were, they could match a UPS truck pound for pound. A sig-nal difference between the United States and Canada is that, with the exception of the Rockies, it really is possible to canoe across Canada.

---

* The Peter Pond Mall, a modern-day trading post offering the same easy credit that made economic captives of so many Indigenous hunters, marks the heart of downtown Fort McMurray.

† Mackenzie was the first European to cross the continent overland, arriving at the Pacific coast in 1793, twelve years before Meriwether Lewis and William Clark.

However, journeys through its intricate network of rivers and lakes required occasional portages, some of them miles long, over steep and slippery terrain. In 1800, a common eastbound portage load for one man was two ninety-pound packs of furs, bound into bales, plus his personal gear—and then back for another load, until the entire cargo had been transferred, along with the canoe. These loads, which would give a linebacker pause, were borne barefoot, or in wet moccasins, both of which have negative traction.

Between the rivers, lakes, rain, and snow, it was an amphibious world these men traversed where chronic saturation was simply a condition of existence. One can only imagine the chills, rot, and rash these men endured, to say nothing of the mosquitoes and biting flies. These *voyageurs* were typically French and Iroquois, or combinations of the two. They were paid a pittance for their superhuman labors, and many were injured or killed en route—not by enemies, but by drowning, busted guts, or falls beneath their brutal burdens. Some particularly arduous portages came to look like graveyards, so many were the crosses that lined the way.

Both trappers and traders had things the other desperately wanted, and a fragile peace was maintained beneath a surface tension of trust, fear, and desire. While there was room for some negotiation, credit, and even friendship, a muzzle-loading rifle still cost forty beaver skins, and it was useless without powder and shot—another twenty skins. Then there were tools, traps, flour, blankets, sewing supplies, and gifts to consider. It all added up, and the bottom line was a hard one: anyone who wanted to participate in the modern world of iron, gunpowder, cotton, and alcohol had to play by Company rules. Anyone who didn't was effectively marginalized and, ultimately, doomed. If the technical disadvantages didn't destroy you outright, the public shame for refusing to participate would finish the job. Some things haven't changed; in Alberta's bitumen industry, a man will not be taken seriously if he drives anything less than a Ford 150 pickup, or its equivalent. Tantalizing options are available, just as they were at an eighteenth-century trading post, and a young man can, with a week's wages and a signature, put himself on the hook for $75,000 with alarming ease. It took (and takes) an extraordinarily determined person to resist these pressures. Consider the social and professional costs of not having a cell phone—the muzzle-loader of our day.

At its peak, during the first half of the nineteenth century, the Hudson's Bay Company enjoyed a virtual monopoly, operating hundreds of trading posts from coast to coast, with satellite "offices" as far away as Alaska, Hawai'i, California, and the High Arctic. The Company's vast domain fit perfectly the description used by the Irish philosopher and politician Edmund Burke to describe its southern counterpart, the East India Company: "A state in the guise of a merchant." Encompassing nearly 10 percent of the earth's landmass, the Company's territory was—geographically speaking—the greatest commercial empire the world has ever known, and its network of trading posts effectively staked out the boundaries for what is now the world's second-largest nation after Russia.

The profits once reaped by the Hudson's Bay Company's remote and secretive "governors" in London were spectacular, in part because the Company's practices and policies, dutifully enforced by its Scottish proxies, were so ruthless. As a longtime employee named John M'Lean wrote in 1849:

> Since [1840], the dividends have been on the decline, nor are they ever likely to reach the same amount, for several reasons,— the chief of which is the destruction of the fur-bearing animals. In certain parts of the country, it is the Company's policy to destroy them along the whole frontier; and our general instructions [were] that every effort be made to lay waste the country, so as to offer no inducement to petty traders to encroach on the Company's limits. Those instructions have indeed had the effect of ruining the country, but not of protecting the Company's domains.

As barbaric as this policy might seem today, it is no different, in practice or principle, than the competition-killing tactics used by Standard Oil, Walmart, Amazon, Netflix, or Uber. In this way, corporations and wildfires follow similar growth patterns in that, once they reach a certain size, they are able to dictate their own terms across a landscape—even if it destroys the very ecosystem that enabled them to grow so powerful in the first place.

The Hudson's Bay Company's potent combination of offshore capital (often borrowed) and indebted local labor is how modern Canada—a continental beaver farm and trading company serving Europe's hat industry—came into being. By making beaver skins a standardized unit of currency, and offering irresistibly attractive and useful things in return, the Company and its aggressive competitors turned the inhabitants of the boreal forest, human and animal alike, into a huge, surprisingly efficient profit-making machine—until they exhausted the resource. In so doing, the fur trade shaped Canada's creation myth and set the tone for how extractive industries continue to operate there. Through this lens, Canada in general, and Alberta in particular, could be seen not as "a state in the guise of a merchant," but as a merchant in the guise of a state. This colonial model, which systematically commodifies natural resources and binds local people to the trading post system with company store–style debt, has replicated itself in resource towns across the continent.

In the post–fur trade world, banks, big-box stores, and car dealerships have taken the place of the all-purpose trading post.* Employees in Alberta's bitumen industry are among the highest-paid petroleum workers in the world; nonetheless, heavy debt is rampant, and bankruptcies, layoffs, and foreclosures are common. (In 2019, Canadian household debt, expressed as a percentage of GDP, was the highest in the Group of 7.) This legacy is keenly felt across the boreal, particularly in extractive industries. When it comes to rapidly and radically altering a landscape along with the lives of those who live upon it, only a few things compare to a big boreal fire, and one of them is the profit motive. In May 2016, Fort McMurray was the rare place where one could witness both of these energies unleashing simultaneously.

---

* The Hudson's Bay Company, which exists to this day, got out of the fur business in 1987.

Before the multinational giants and their armies of workers move in, the visionaries must pave the way. One of those intrepids was Sidney Ells, a bronze bust of a man who, in 1913, dug, drilled, and blasted nine tons of bitumen samples in the vicinity of Fort McMurray. In those days, the all-but forgotten trading post was a backwater consisting of "a dozen primitive log cabins, a bug-infested hovel proudly referred to as the 'hotel' and during the summer months many Indian tepees and tents. Everywhere, starving [sled] dogs roamed at will." Ells, the young engineer with the Dominion Department of Mines who penned those words, was glad to leave. After gathering his hundreds of bitumen samples into sacks, Ells loaded them onto a forty-foot barge and hauled them 240 miles up the Athabasca River to a railhead using a primitive method called "scow tracking." Travel and transport have never been easy in the north, but scow tracking (a verb not found on the internet nor in any dictionary) may be a uniquely Canadian torture. It involved harnessing a dozen men—in this case,

Dene and Métis rivermen in harness, scow tracking up the Athabasca River, October 1913. Note the scow downstream with men aboard.

local Métis and Dene rivermen—to a barge or scow with five hundred feet of heavy manila line, and then driving them upriver, like sled dogs. In an account written for the Department of Mines, Ells described it this way:

> Scow tracking south of McMurray was anything but child's play. Harnessed to the heavy tracking line, men fought their way grimly along the rough boulder-strewn beaches or through a tangle of overhanging brush often ankle deep in mud or waist deep in water. The ceaseless torture of myriads of flies from daylight until dark and the heavy work, which only the strongest could long endure, made tracking one of the most brutal forms of labour.

Their journey took more than three weeks. When they came to impassable rapids—and there were miles of them—they had to portage the tons of bitumen samples and gear around them. Northern rivers are always cold, but it was October, the beginning of winter in the north, and the weather during that last fall before the Great War was particularly poor. The men, including Ells, who had destroyed his feet in ill-fitting boots, were shod in leather moccasins, which wore through almost daily and were replaced from a gunnysack filled with spares. This and the accompanying sackful of mildewed tobacco plugs offered cold comfort. To those in his employ, Ells must have seemed obsessed, if not insane. After all, the scow they were tracking was not filled with furs, fish, or salt—all proven moneymakers from the Athabasca region—but with tons of tar-clotted sand, a substance that, despite two hundred years of trade involving many players and products across two oceans, no one had ever been interested in buying. This obstinate fact has dogged the bitumen industry ever since.

In North America, crude oil has always been accessible, it just took us a while to figure how to ignite it and domesticate it. Historically, it occurred much like bitumen—at random, in natural seeps, and as a noxious contaminant in springs and wells. Prior to the 1860s, the potential seen in crude oil was not so different from that seen in olive oil, seal oil, or whale oil before it: a dubious liniment, a messy lubricant, and, perhaps, a source of light. It was the latter, coupled with crude oil's unbeatable combination of abundance and accessibility, that scuttled the whale oil industry.*

Following the drilling of the first productive New World oil wells in Enniskillen, Ontario, in 1858 and, more famously, in Titusville, Pennsylvania, in 1859,[†] it was optimistically assumed that petroleum (a neologism meaning "rock oil") must be valuable. But, like bitumen during the same period (and internet companies more recently), no one was quite sure how to manage or monetize it. So poorly understood was this smelly, viscous, irridescing substance that, prior to 1860, people drank it, rubbed it on their bodies, recoiled at the stink, and dumped the excess in the river. That didn't stop it from being bottled and sold. "Seneca Oil" was recommended for burns, bruises, sprains, and wounds. "It penetrates, purifies, soothes and heals," claimed one enthusiast. "Try it on barking children."

A handful of chemists, however, saw petroleum's combustive potential, and speculators sensed it, too. Following "Colonel" Edwin Drake's legendary strike in August 1859, farmland around Titusville began selling for staggering sums. So many wells were dug, so quickly and in such close proximity, it seemed as if the local fields had suddenly produced bumper crops of wooden oil derricks. Containing these wells once they'd been tapped was not an exact science, and, in Titusville, the aptly renamed Oil Creek flowed—and sometimes

---

* Whale oil, specifically that from the sperm whale, was light, pure, and clean burning. In both candle and oil forms, it had been—for those who could afford it—the preferred illumination for more than a century. So ubiquitous was it in the parlors of Europe and North America that spermaceti candlelight became the baseline for the lumen, a standardized unit for measuring a light's brightness still in use today. But by the 1860s, with whale stocks dwindling and ships being forced to make longer journeys to more remote whaling grounds, it was also becoming expensive.

† The first mechanical oil rig was erected by a Russian engineer in Baku, Azerbaijan, in 1847. However, a hand-dug oil well dating to 1594 has been identified in the region, which was already famous for its flammable oils, "eternal flames," and fire cults.

Oil derricks near Oil Creek—Venango County, Pennsylvania, 1866

burned—with a surface layer of shimmering petroleum. Meanwhile, wagonloads of leaking barrels turned the surrounding roads into reeking quagmires through which draft horses slogged and staggered, blackened to the withers by a slurry of urine, feces, and crude oil. The combination was toxic, and a notable feature of these sorry animals was that, below the neck, they had no hair. But horses were cheap and, with money flowing as freely as oil, this was considered a small price to pay. Even if they couldn't explain exactly how, those in its presence sensed petroleum's volatile potential with a nerve-jangling intensity. If the future had a smell, it was this: sweet tar laced faintly with rotten eggs.

As abundant as it was, the crude oil being discovered in Ontario and Pennsylvania was not a lamp-ready alternative to the coal- and whale-based illuminating oils already on the market. Crude is not particularly useful as energy or illumination in its raw form, but its constituents are. These constituents, known as "fractions," are rendered from oil much the way whiskey is rendered from mash in a distillery: when heated in a sealed vessel, the volatile fractions will evaporate off at different temperatures. "Lighter" fractions (gases like naphtha, propane, benzene, etc.) will come off first, at temperatures below 500°F. Among these lighter fractions are the main ingredients of gasoline, but in the 1860s, with the automobile still decades away and no obvious use for such a dangerously explosive substance, they were discarded, usually into the nearest creek. Meanwhile, heavier fractions such as heating and lubricating oils need higher temperatures—up to 1,000°F—in order to separate. Left behind, virtually un-distillable, are heavy fuel oils such as bunker fuel, and also bitumen. In order to glean any remaining hydrocarbons from these ultraheavy residues, they must be put through a second, more rigorous heating process in a pressurized tank called a coker. There are a number of these huge vessels at the plants north of Fort McMurray, and their purpose is to "crack" the remaining longchain hydrocarbon molecules into smaller, more useful fractions. Left over at the end of this hellacious and energy-intensive process is a fused mass of ash and carbon called petroleum coke, or petcoke. Petcoke burns like coal, only hotter, but because it is also significantly dirtier, it has fallen out of favor as a fuel for domestic power plants and blast furnaces.*

Occupying a sweet spot on the crude oil fraction spectrum, close to diesel, is kerosene. First produced from coal oil in the 1840s by a Canadian physician and geologist named Abraham Gesner, it was kerosene's utility as a lamp oil that launched the petroleum industry. Gesner really did have lightning in a bottle, and he expanded

---

* Today, petcoke is a pariah of the energy industry that no one wants to be associated with, and mountains of it can be found in railyards and refinery complexes across the border states of the Midwest. The Koch family has seen an opportunity here, and Oxbow Energy Solutions and Koch Carbon (both owned by members of the family) are two of the world's largest exporters. Since 2010, petcoke, along with the terrific pollution it generates, has become one of North America's biggest exports. In 2016 alone, 8 million metric tons were shipped to India, where environmental regulations are poorly enforced, a fact borne out by the lethally toxic state of the country's air. In 2018, China imported more than 7 million tons. Japan and Mexico import even more. In total, the United States exports roughly 90 million tons of petcoke a year, and much of this is derived from Alberta bitumen.

into the U.S. at his first opportunity. But once there, he and his pioneering refinery were soon bought out by an American investor who would join forces with John D. Rockefeller, the Jeff Bezos of his day, making Gesner's Kerosene Gaslight Company one more acquisition en route to Standard Oil's hegemonic control of the oil industry.*

By 1870, it was clear that the same energy and ambition that had so recently driven fire-powered steamships across the Atlantic, and bound the continent with a fire-powered railroad, could accomplish virtually anything a motivated mind might seize upon. The following decade was as pivotal for energy and fire as the 2000s were for information technology and digital communication. It was then that the heyday of petroleum—what could be described as the Petrocene Age—began in earnest: the period of history in which our Promethean pursuit of fire's energy, most notably crude oil, in conjunction with the internal combustion engine, took a quantum leap to transform all aspects of our civilization and, with it, our atmosphere. This period covers, roughly, the past 150 years, and it is peaking now.

Standard Oil, which would beget Esso ("SO"), among many other oil majors, was founded in 1870, the same year as the hamlet of Fort McMurray (where Esso would one day build one of its biggest operations). By then, it was only a matter of time before the coal-fired steam engine was replaced by petroleum power because the advantages were too great to ignore: a ton of coal oil could generate as much steam in a boiler as four tons of solid coal. Engineers had been paying close attention, and, by 1860, they were already designing the future. That year, on page one of the September 22 issue of *Scientific American*, directly above a recipe for Canadian rhubarb wine, a notice appeared concerning a new "explosion engine."

> A Parisian, by the name of Lenoir, is creating a great sensation among his countrymen by the exhibition of a caloric engine . . . [His] little shop, in a bye street, is every day besieged by a crowd

---

* Today, kerosene goes by the updated name of "Jet A," and air travel would be very different without it.

of curious people from all classes . . . According to Cosmos, and
other French papers, the age of steam is ended—Watt and Ful-
ton will soon be forgotten.

The idea that such an engine might one day move people individually—the way a smartphone now moves ideas and
commerce—was just a short step away, and Étienne Lenoir was
already taking it: later that same year, he mounted his engine on a
three-wheeled wooden cart. It was noisy and slow, but it marked
the first time in history that liquid fuel had been combined with
an internal combustion engine and a wheeled vehicle for the purpose of individual mobility, thus achieving a kind of self-powered,
"automotive"—not flight, exactly, but arguably transcendence, especially when you consider how most people got around in those days.
This era-defining, world-changing invention took a more recognizable shape in 1863 (twenty years before Daimler and Benz), when
Lenoir unveiled his Hippomobile, a four-wheeled, multi-passenger
vehicle equipped with a steering wheel up front and an engine in the
rear. Not much has changed: the only difference between Lenoir's
Hippomobile and a twenty-first-century Ferrari is one of refinement.
   Scientific American's praise for Lenoir's explosion engine was
wholehearted save for this: "The practical objections to such motors
are the jerks of its action and the accumulation of heat." The problem of jerking action was quickly solved, but the accumulation of
heat—in the broader sense—was not, and it haunts us to this day.

As the almost medieval fur trade gave way to the industrial age,
interest in petroleum and its combustive properties reached the ears
of commercial prospectors. It wasn't long before the Athabasca fur
traders' offhand references to the methane bubbles they paddled over
and the bitumen seeps they stumbled across were followed up in earnest. For more than a century now, the province of Alberta has seen
its future writ in oil and, to this end, it has had almost as many holes
poked in it as Texas. Most of those holes were dry, but every now and
then, some combination of saltwater, oil, and/or natural gas would

come burping and boiling to the surface. Once in a great while there would be a direct hit, and one of the first of these occurred on the banks of the Athabasca River at a place called Pelican Portage, eighty miles upstream from Fort McMurray. There, in the summer of 1897, drillers with the Geological Survey of Canada hit a void at eight hundred feet, releasing a spectacular jet of methane (natural gas), peppered with walnut-sized chunks of iron pyrite that blasted skyward at the speed of bullets. The sound—something like a fighter jet—was audible three miles away. The well was estimated to be producing more than 8 million cubic feet of gas per day—enough to heat every house and building in modern Fort McMurray. But in 1897, with no place to sell it, no way to capture it, and no way to put this explosive genie back in its bottle, the drillers did the next best thing: they lit it on fire. Before being finally capped, it burned, intermittently, for twenty-one years—a blast furnace in the wilderness, roaring like a rocket booster day and night, winter and summer, while the great river froze and thawed, the low sun wheeled, the northern lights shimmered, and passing hunters marveled and warmed their hands.

It was just a glimpse of what was to come.

One fantasy nurtured by early prospectors in the Athabasca country was that under that layer of worthless bitumen lay a massive reservoir of oil and gas, but even if there had been, it would have made little difference to their fortunes at the time. In 1913, as Sidney Ells slogged southward in heavy harness and rotting moccasins, crude oil was already abundant in Ontario, Pennsylvania, Texas, and most recently, Oklahoma. Alberta was about to make a major strike of its own far to the south of Fort McMurray. Furthermore, the automobile was still a decade away from being the number one consumer of crude oil products in the form of gasoline, motor oil, axle grease, and rubber tires.

Taking all this into account, Ells understood that bitumen was at best a long shot, and, even as the motorcar rose to prominence in the 1920s, the most plausible future he could see for Alberta's bitumen was in road paving. As he explored ways of "upgrading" bitumen into more viable petroleum products, he came to the prescient conclusion that the only way this could be achieved on an industrial scale was with substantial government assistance. Since then, both the Alberta and federal governments have sought partners wherever they could find them. Often, those partners have been foreign corporations, but, one way or another, taxpayers have always picked up the slack. In this way, Canadian citizens have been in harness alongside Sidney Ells and those rivermen, scow tracking Alberta's bitumen industry upstream for more than a century now. And the burden remains a heavy one: according to a 2019 International Monetary Fund report, Canadian taxpayers contributed more than $40 billion (U.S.) in subsidies to the fossil fuel industry in 2015 alone (approximately $1,200 for every man, woman, and child). In the same year, Americans contributed more than $2,000 per person, while China (a nation of 1.4 billion citizens) contributed the equivalent of $1,025 per person to support the fossil fuel industry.

The dramatic transformation of Fort McMurray from trading post to petroleum hub has been as rapid and radical as any on the continent, and it has followed the manic-depressive pattern of all petroleum boomtowns before it. Boomtown stories are part and parcel of American lore, but Canada has its share, and some of the most exuberant examples come from Alberta. This report from 1914, following a bonanza of natural gas that erupted from the Dingman well in the Turner Valley, paints a vivid picture:

> Calgary never saw such a Saturday night (May 16th). It was the wildest, most delirious, uproarious, exciting time that it ever entered into human imagination to conceive. If the city was oil crazy on Friday, on Saturday it was fairly demented. Money fairly spouted from the ground. All day and all night the crowds fought and struggled for precedence in the offices of the most prominent oil companies and clamoured for shares and yet more shares. Relays of policemen barely kept a clear passageway, and there was never a moment when the would-be purchasers were not lined up three deep in front of the counters, buying, buying, buying. No one knows how much money was taken in. The officials of the companies have not had time to count it. They scarcely know themselves how many shares they have sold. Wastebaskets were filled to overflowing with bank notes and cheques, emptied and filled again and again. Clerks and

salesmen went blind and dizzy writing out receipts and filling in application blanks . . . Business men, clerks, car conductors, women, shop girls, anybody, everybody, almost begged the oil men to take their money.

In that single day of incendiary trading, investors drove up the share price of the Calgary Petroleum Products Company by more than 1,600 percent. In such frenzied moments, the utility of the product becomes secondary, and baser motives are laid bare.

Explosive growth of this kind mimics that of fire, and humans seem particularly prone to it. Fire is known to physicists as an exothermic reaction—that is, a reaction that unleashes more energy than was required to initiate it. Just as a single match or cigarette can burn down a house or ignite a landscape, a handful of men with novel trade goods, a drilling rig, or an IPO can set off a commercial explosion of buying and selling. Each of these events has the power to change the fortunes of a region and its inhabitants overnight.

4

And now nothing will be restrained from them,
which they have imagined to do.

—Genesis 11:6, "The Tower of Babel"

The Dingman strike put Alberta on the map as a place with major petroleum potential—enough to attract the attention of American investors. But compared to the instant gratification of gushers like Dingman (or Titusville), bitumen was a hard sell—lima beans to oil's ice cream.

J. Howard Pew sought to change that.

When Suncor completed its state-of-the-art bitumen mining, separating, and upgrading plant in 1967, it signaled a new era for Fort McMurray and the petroleum industry. Suncor was founded by the Sun Oil Company (Sunoco) and opened to great fanfare by its long-time chairman, the American oil magnate and evangelical Christian John Howard Pew. Pew called the quarter-billion-dollar venture Great Canadian Oil Sands, and it represented the largest private investment in Canada's one-hundred-year history. While Sunoco put up the bulk of the capital, 100,000 Albertans were also persuaded to buy $150 million in bonds to back the project. The gleaming plant stood where it still stands today, at a place called Tar Island on the left bank of the Athabasca River, twenty minutes north of Fort McMurray.

Pew, the son of a pioneering Pennsylvania oilman raised in the church. "was a very, very religious individual," recalled Bob McClements, a fellow Pennsylvanian who worked closely with Pew, made a fortune, and retired to Florida. "His conversations," McClements recalled,

> often included two words, faith and freedom, and they were welded together. He believed that . . . after the Declaration of Independence it was a whole new era for the Lord to work His way through individuals. And the . . . new government system allowed him to do it. If I heard him say it once, I heard him say a thousand times: "If the government goes up, freedom goes down."

Pew's notion of "freedom" was similar to that enjoyed by the Hudson's Bay Company and, more recently, by the petroleum industry at large. It is the kind where "free rein" and "free reign" blend together in a glorious, God-given, wealth-generating synergy.* Archival film footage from that opening day in Fort McMurray shows a man undiminished by his eighty-five years. Still engaged in the Lord's Work, Pew was making history, accompanied and applauded by hundreds of government ministers, bankers, corporate executives, journalists, and assorted dignitaries who had been flown in to this remote outpost by thirty chartered aircraft, many from thousands of miles away.

In order to shelter attendees from a north wind bearing rain, a gigantic inflatable membrane called "the Bubble" was supplied by Bechtel, the huge American engineering firm overseeing the project. Inside this man-made atmosphere was seating for six hundred with room left over for a brass band. At one end, on a stage-sized podium, stood a long table covered in white linen and lined with twelve white men, including Pew and senior executives from Sunoco and Bechtel, as well as the premier of Alberta, his chosen ministers, and esteemed clergy.

The circular device on the banner hanging behind them was not the sun, but a bucketwheel, the enormous, infinitely turning shovel used to mine the bituminous sand. GCOS had two of these behe-

---

* Pew also put up the money to launch *Christianity Today*, the televangelist Billy Graham's flagship magazine.

moths: looking like Ferris wheels with teeth, they were among the
largest machines on Earth at the time. More than ten stories tall and
riding on massive caterpillar treads, each bucketwheel was capable of
excavating 100,000 tons of earth per day. They worked around the
clock, and operators recall the buckets' monstrous teeth glowing red
hot from the friction as sparks arced through the winter dark. Around
this image of the bucketwheel, adding one more concentric ring to
the great banner's industrial mandala, was the company name along
with four words that rang, in this setting, like a pronouncement from
God: "MAN DEVELOPS HIS WORLD."

The old scow tracker Sidney Ells was present that day, and he
must have found the transformation astonishing. The whole scene—
the crusading banner; the proud, important, overcoated men with
their fleet of waiting airplanes; the legion of hard-hatted workers and
technicians ready to do their bidding; the daunting assemblage of
buildings and machinery sprawled across acre upon acre of freshly
cleared wilderness, veined with piping and studded with towers and
flare stacks that dwarfed the tallest trees—suggested, on the one hand,
an industrial fantasy come to fabulous, Olympian fruition and, on
the other, a certain kind of political rally popular in the 1930s. This

was a man's dream, a conqueror's dream and, above all, a twentieth-century dream: heavy industry at its most titanic and world altering, with fire at its core.

Many men shared this dream, and one of them was Ernest Manning, a graduate of Calgary's Prophetic Bible Institute and the host of an evangelical radio show called *Back to the Bible Hour*. Manning also happened to be the premier of Alberta, a position he held "at Her Majesty's pleasure" for twenty-four years—a record that still stands. He had been in office since World War II, and he was there in 1947 when Alberta entered the oil game in earnest with the huge Leduc strike just south of Edmonton. He saw Alberta's future then, and it glistened with oil. Standing center stage beneath the bucketwheel banner, Manning addressed the congregation of government and industry faithful: "This is a historic day for the province of Alberta," he said in a tone and cadence honed on Bible radio.

> We're gathered here . . . to officially open this gigantic complex which, for the first time, will tap commercially the vast supply of oil that until now has remained locked in the silent depths of these Athabasca tar sands. It is fitting that we gather here today to dedicate this plant—not merely to the production of oil, but to the continual progress and *enrichment* of Mankind.

It was Alberta's equivalent to a lunar landing, and the premier cast his net wide, declaring it "a red-letter day, not only for Canada, but for all North America. No other event in Canada's centennial year is more important or significant."

But Canada is a big place, a boreal subcontinent larger than India, larger than Australia, and this was an audacious claim to be making in a forest clearing more than two thousand miles from the urban hubs of Ottawa, Toronto, and Montreal, where the young country's centennial festivities were in full swing, and where Expo '67 was drawing record crowds from around the globe. Manning was unfazed by his remoteness from the centers of power, in part because Albertans were used to being ignored by eastern Canada, but also because he was comfortable with prophecy—raised on it, versed in it, and, on the podium that day, he was uttering it. Certainly familiar to Man-

ning, Pew, and many others present that day would have been these
words from Isaiah:

> *Prepare ye the way of the Lord,*
> *make straight in the desert a highway for our God*
>
> *Every valley shall be exalted,*
> *and every mountain and hill shall be made low:*
> *and the crooked shall be made straight,*
> *and the rough places plain:*
>
> *And the glory of the Lord shall be revealed,*
> *and all flesh shall see it together:*
> *for the mouth of the Lord hath spoken it.*

The Great Canadian Oil Sands project lost money from opening
day. By 1973, six years into its erratic operation, half a billion dollars
had been spent, and nearly $100 million was still owing. Between con-
tinual technical problems exacerbated by brutal working conditions,
market volatility, labor trouble, transportation issues, and perennial
funding shortfalls, the plant would not become even remotely profitable for decades. As late as the 1990s, it was referred to by some as
"Pew's Folly." But bitumen is a long game and, despite the mixed
reports coming out of GCOS, Syncrude planned an even larger plant.
The $2 billion megaproject, also overseen by Bechtel, was financed
by a shaky consortium of American oil companies and a last-minute
bailout from the Canadian government. When it came on line in
1978, Syncrude's Mildred Lake complex boasted the largest coking
capacity in the world.

Half a century on, Pew, Manning, and their fellow believers have
made good on Isaiah's prophetic words. Not only are their works visible from the heavens, they are accessible by what amounts to a private
highway. In the intervening years, some of which were lean indeed,
Fort McMurray has become synonymous with bitumen and, for a
while, with lucrative overtime. For those who have settled there, it is
home, a place where ordinary working people can enjoy a kind of ide-
alized middle-class lifestyle that has become all but impossible down

south. The town itself is now unrecognizable; in the fifty years since Pew launched GCOS (which was amalgamated into Suncor in 1979), Fort McMurray's population has increased by fifty times. Not even J. Howard Pew could have anticipated sports cars, casinos, strip clubs, and a Walmart within sight of McLeod House, at one time the most remote inland trading post on the continent.

Syncrude's slogan is "Let's Build Something," and they have— something so extraordinarily ambitious that normal reference points for scale are hopelessly inadequate. It is simply impossible to grasp the magnitude of what is happening in Fort McMurray from ground level. When I asked one young tradesman what impressed him most about working up there, I expected a story about huge machines, crazy overtime, or -40° temperatures. But it was none of those things. What blew his mind, he said, was the view from the airplane window: "the scar on the side of the face of the Earth."

A lot of things are said to be "visible from space," but not from six thousand miles above Earth. At that distance, one is on the far side of the exosphere, the invisible threshold beyond which only remote satellites and heavenly bodies are found. From this celestial height, Earth appears to be uninhabited—no Pyramids, no Great Wall, no Shanghai or Los Angeles; not even the Mississippi River is visible. Canada's great boreal forest shows up as a green smear across the forehead of the globe, and only continental features like the Great Lakes and the Rocky Mountains are easily found. But if you know what you're looking for, you can pick out Fort McMurray's bitumen mining and upgrading complex. From 6,000 miles up, its visibility is comparable to that of San Francisco Bay or Cape Cod, both of which cover roughly one thousand square miles of land and water. Meanwhile, the world's largest open pit mines—abyssal holes like Utah's Bingham Canyon, or the Fimiston Super Pit in Western Australia— are nowhere to be seen.

When you descend to 3,000 miles above Earth, you are still deep in the exosphere, still 2,750 miles beyond the orbit of the International Space Station, and still unable to pick out Bingham Canyon

or the Fimiston Super Pit. But even at this altitude, it is possible to
identify specific bitumen mining projects. Harder to spot are the vast
networks of shafts and piping deployed beneath the region's SAG-D
sites. These barren, pipe-studded rectangles pock the forest like oversized drilling pads. While they have much smaller footprints than the
open pit mines, they are even more energy intensive.

— — — —

The last time bitumen built a city the size of Fort McMurray was
about three thousand years ago, in Babylon. Despite the twin chasms
of time and geography separating that ancient Mesopotamian capital
from this postmodern North American boomtown, their street plans
are surprisingly similar: attenuated gridworks flanking the left banks
of their local rivers. Over the course of two millennia, several sackings,
and a handful of dynasties, bitumen literally bound Babylon together
from the ground up. It mortared the ancient urban center's great,
concentric walls—dozens of feet high and miles long, it underpinned
an extensive road network, and it sealed elaborate canal systems, some
of which survive to this day. Like its more refined successors, this
rudimentary petroleum facilitated ambition on unprecedented scales.
A brief sampling includes the Gates of Ishtar; Emperor Nebuchadnezzar's Hanging Gardens; an artificial riverbed redirecting the entire
Euphrates River; a four-hundred-foot-long bridge to cross it; and the
thirty-story Tower of Babel. "I . . . laid its foundation on the bosom
of the underworld," Nebuchadnezzar declared on a tablet found in
the tower's masonry. "With bitumen and brick I made it high as a
mountain."

Without bitumen, these quasi-mythical megaprojects might
never have been built, and nor would Fort McMurray's modern
equivalent—the enormous and gleaming quarter-billion-dollar pleasure palace that is the Suncor Community Leisure Centre. Alleged to
be the largest rec center in all of Canada, it is in its way a Babel, too:
in 2016, the Canadian census identified eighty first languages spoken
in Fort McMurray and its surrounding worksites, and all of them can
be heard in the pools, weight rooms and saunas, or on the courts,
rinks, and curling sheets of the Suncor Leisure Centre. In this way,

Fort McMurray represents what is arguably the most remote melting pot on Earth—for humans and fossil fuels alike. But if Babylon was a center surrounded by other powerful centers, Fort Mac is the end of the road. The sole purpose of this road—in effect, a dead-end superhighway—is its access to bitumen.

————

On January 31, 2006, nearly forty years after the grand opening of J. Howard Pew's Great Canadian Oil Sands, oil prices were breaking records, Fort McMurray was in the midst of a historic boom, and President George W. Bush was saying something no president had ever said on national television. "Here we have a serious problem," he announced during his sixth State of the Union address. "America is addicted to oil." The president of the world's largest oil-consuming nation then went on to enumerate the ways he would "break this addiction," most significantly by developing alternative energy sources.

Five months later, speaking before the Canada-U.K. Chamber of Commerce in London, Canada's freshly elected Conservative prime minister, Stephen Harper, was celebrating Canada's fourth straight year as America's chief oil supplier, having displaced Saudi Arabia in 2002.* There is a downside that Harper didn't mention: no other Group of 7 country is as vulnerable to the whims of petroleum prices. When the price of oil drops, so does the Canadian dollar, a fact that has drawn stinging comparisons to petro-states like Nigeria and Venezuela.

Like many oilmen before them, George Bush and Stephen Harper are both avowed evangelical Christians and scions of oil industry parentage. Harper was in London that July—the hottest in more than four hundred years of royal record keeping—to tell his audience that British investors were taking notice of "Canada's emergence as a global energy powerhouse—the emerging energy superpower our government intends to build." Warming to his theme, Harper went on to describe "an enterprise of epic proportions, akin to the building

* In spite of this, the U.S. remains the world's largest oil producer.

of the Pyramids or China's Great Wall. Only bigger." As grandiose as Harper's claims may have sounded, he wasn't exaggerating. The prime minister was simply describing what J. Howard Pew, Premier Ernest Manning, and Syncrude's Frank Spragins had long ago foreseen.

But they glossed over some details. "This is not a normal world," explained Louis Rondeau, a veteran scaffolder who has worked on many of Suncor's towering stacks and vessels, inside and out, in all weathers. "Not in any way, shape, or form."

Along with a mosque and a Hindu temple, there are thirty churches in Fort McMurray; most of them are evangelical, and most of them are competing for new arrivals. A young pastor and firefighter named Lucas Welsh explained his church's approach this way: "We've kind of focused on this guy we're calling 'McMurray Mike,'" he said, "a twenty-five- to thirty-five-year-old who has a wife or common-law or girlfriend and a couple of kids. That's our guy we're looking at because Fort McMurray has the youngest demographic of almost any city in Canada and we have seven times less deaths because nobody retires here, nobody dies here; they all leave. In ten years we've done like three funerals."

Between its youth, its ambition, and its superabundance of natural resources, Fort McMurray is a dynamo. Since the early 2000s, Fort McMurray has generated its own gold rush mythology, but unlike so many boomtown legends, most of these stories are true: a hauler driver really could gross a quarter of a million dollars a year; an apprentice pipefitter from Newfoundland really could pull down $150,000 with free flights home twice a month and pocket everything but the taxes; a refugee from Somalia really could get paid $20 an hour to work at a fast food restaurant, and might even get an iPod as a "signing" bonus, so desperate were the local shops for help.

The overtime is irresistible, but the toll those hours take can be crushing, and camp life can be soul-killing. Inhabited almost exclusively by men, the camps are assembled from heavily insulated trailers stacked like LEGO blocks; from a distance they look a lot like polar research stations, or penitentiaries. Up close, they combine elements of a budget cruise ship—salad bar, steaks cooked to order, weight rooms, and a TV in every suite—with a Siberian gulag: a frozen, isolated setting surrounded by high fencing and patrolled by guards where weary, pale-faced men work twelve-hour shifts and take their last cigarette breaks in open jackets at -30° before collapsing into bed and counting the days until they get out—if they get out. Between January 2021 and July 2022, five men were killed at Suncor alone.

The local rock station, 97.9 FM, is tailored to these workers, and a half hour's listening will reveal back-to-back PSAs cautioning against speeding and drunk driving followed by ads for a heavy metal concert promising maximum aggression, and gambling lessons at the Boomtown Casino, "located downtown in the Peter Pond Mall." These are interspersed with ads for six-hundred-horsepower Shelby GTs and extended-cab pickups with V-8 Hemi engines. Rounding out the half hour are an oil news update and some jokes about drinking. Women can be heard announcing news updates. (They could also be seen on an amateur soft-porn webpage called "Babe of the Day," before the station took it down and replaced it with a picture of a kitten.) When the music finally returns it might be Alberta's own Nickelback bellowing their knucklehead anthem, "Burn It to the Ground," an exhortation to alcoholic mayhem that captures perfectly the state of mind of a young pipefitter and his fly-in comrades as they proceed to drink their Airbus A320's bar cart clean after pulling twenty-one twelves in a row. During the boom, the jetways off those outbound flights could feel like mosh pit bowling alleys with horny, swearing, farting men caroming off the walls on their way toward the arrivals lounge and the peeler bars beyond.

Due to a combination of rugged conditions, high wages, and the fact that burning petroleum represents a kind of local virtue, truck and ATV sales are noticeably higher in Fort McMurray than in other comparably sized Canadian cities, but so are rates of assault, spousal abuse, drug abuse, STDs, COVID-19, alcoholism, and suicide. The same steroidal earning power that astounds the folks back home—and gladdens the hearts of truck dealers and tax collectors—burns holes in workers' pockets, and also in their noses. "Fort Crack" is another nickname Fort Mac has earned, and it is a real place, too. While there is an established market for crack and methamphetamine, cocaine is preferable, in part because it clears from the bloodstream more quickly than other blood-tested substances, and blood testing has become much more rigorous since the heady and freewheeling days of the last boom, when it seemed that just about anyone with a pulse could get a job there. Because of this, there is also a market for clean urine. In 2018, a Fort McMurray resident was caught entering Alberta with thirty-one kilograms of cocaine; equivalent to roughly 300,000 lines, it had a street value of between $3 million and $4 million.

Many of those who make the journey to Fort McMurray find the money they've made surprisingly hard to hold on to, and a lot of that intercepted cocaine was destined for people like Jake McManus. I met Jake back in 2010, before he went north at twenty-five with a five-year plan, and I remember thinking, If anyone can pull this off, he can. Jake was a hardworking journeyman, startlingly competent with large, dexterous hands that could fix anything; he had a supportive family, and his father had advised him on a business plan. Up at Site, all expenses paid with as much overtime as he could stand, he was looking at $150,000 a year—to start. He would keep his head down and save it all, he said, and when he looked up again five years later, or even three years later, he would be a player. With a nest egg like that and skills to match, no southern city could say no to him.

When I ran into Jake in Vancouver a year or so later, he was drunk, and there were scars on his handsome face. The money had come and gone—down his throat and up his nose; for a while he'd had a lot of new friends. His hands looked different, too: it was the knuckles; Jake was big, and he wasn't a back-down kind of guy. He was in a program, but it had been hit and miss; his longest drug-free stretch had only been a month. He was still strong and able—he was powerlifting now—and when he fell he would catch himself, but he kept falling. I asked him what happened. It was the hours, the schedule, he said, the ten days off after twenty-eight on, and when he came home, all his friends were at work, nine-to-fiveing like ordinary people. He was often lonely, and not sure how to come down after pulling all those tens and twelves with overtime on top. And then there was the money—so much of it, and so easily spent.

In addition to local and federal police (the Royal Canadian Mounted Police, or RCMP), there is an abundance of private security working in Fort McMurray, particularly around the camps and worksites, and there are many places where you simply cannot go. Most workers are tracked through an RFID system. Photography is forbidden at Site, and is a firing offense. When *Global News* reported on an explosion at a Syncrude upgrader in 2017, in which a worker was injured, not a single witness they interviewed would give his name. Most of these men are Canadian so they have always known cold and darkness, but still, these places can feel like prison and you want to ask them, "So, what did you do?"

But there is no crime in cramming yourself and three friends into a salt-rotted Pontiac Firefly and trading a future on the dole in your moribund fishing village for the Hail Mary dream of the tar sands. A third of Fort McMurray's permanent population are transplants from Newfoundland—to the point that Fort Mac is referred to waggishly as "Newfoundland's biggest city." The resource economy of the island, with its forbidding, iceberg-plagued coast, has never recovered from the collapse of the cod fishery in 1992. For decades now, Newfoundlanders have played the same role in the bitumen industry that Scotland's Orkney Islanders played in the fur trade: hardworking, weather-beaten Atlantic islanders who, with no better offers forthcoming, make willing laborers in frozen hinterlands.

The boreal is a difficult place to work—in any field and in any season. The desire to escape—to a bar in Edmonton, to a drug dealer in Red Deer, to a wavering girlfriend in Kelowna—can be overpowering, and accident statistics reflect this. By the time Highway 63 was twinned from two to four lanes between 2010 and 2016, it had earned a reputation as one of the most dangerous roads in North America. In a single month in 2007, at the height of Fort McMurray's latest and greatest boom, twenty-eight people were killed on it—almost one a day. People started calling it the "Highway of Death" and "Suicide 63." Bumper stickers were made up: "Pray for me, I drive 63." By 2013, 63 and its tributary, 881, had seen so many fatalities that a death map began circulating. It is pocked with red dots—so many, so close together, that it resembles maps of migrant deaths along the Mexican border. Many of these accidents coincide with shift changes, and locals adjust their travel plans accordingly, advising visitors to do the same.

The semis are fast, but the pickups are faster, and there is something lethal about ice and oil and money and men. When a southbound Ford 150, four beers in and seventy-five payments to go, collides with a northbound Kenworth T800 hauling a B-train of empty sulfur tankers, the combined impact speed may be more than 160 miles an hour. A witness to such a collision described it this way: "The front end of the pickup truck was destroyed in a manner I had never seen before: a thousand-petaled peony blossom—tiny strips all peeling away from an epicentre of densely fused aluminum and steel." An impact like that can throw a body a hundred feet or more. Sometimes, the only way to find them in the cold and dark is with infrared sensors, and these are standard equipment on Fort McMurray's emergency vehicles. One veteran paramedic I met has personally attended more than a hundred fatalities—all of them within a fast hour of Fort McMurray. "Most of them are head-ons," he said. When I asked him how he handled it, I immediately wished I hadn't. His eyes, an arresting Slavic blue, fixed on mine as he said, "Do I look normal to you?"

After midnight, when the traffic is lighter, you might encounter Scheuerle ("Shirley") trailers—oversized flatbeds with a hundred wheels and multiple tractors pulling ultraheavy loads, bits and pieces of cokers and fractionating towers the size of Saturn V rockets. These road trains move at a crawl, but nothing else does and, late at night, 63 is generally left to the semis. Most of them are tandem trailers—B-trains, but not the kind you usually see carrying gas or mail. These are shapes you don't encounter elsewhere: tubes of liquid nitrogen; heavily reinforced heated cylinders loaded with raw sulfur; others with long, funneled bellies laden with petcoke, the dirty secret of the bitumen industry. Some are emblazoned with hazardous cargo badges, but it's impossible to read them or any other labeling, not because of the dark, but because every vehicle is coated from end to end with road grit. Thus camouflaged, these massive trucks hurtle through space well over the posted speed limit—all hundred feet and hundred tons of them. This despite the fact that, though the road may be clear, it is frozen solid, every grain and pebble glazed with ice. Road salt is only effective to 30°C; any colder, and repeated pressure from passing tires compresses frost and snow into black ice. On 63, in winter, black ice becomes a kind of unofficial road surface.

The ice is one reason every semi working western Canada's northern routes is equipped with a reinforced grill guard known as a "moose bar" (or "buffalo basher"), because whatever you encounter up there, at those speeds—moose, elk, bear, bison—you are going through it like an asteroid. In daylight, you can see the aftermath: blood smears so long and wide it looks like a paint truck exploded. Huge blue-black ravens, unfazed by the speeding traffic, perch on bones that look like leftovers from the Pleistocene as they feast on the obscene and frozen wreckage.

Even with the traffic, nighttime on 63 can feel extraterrestrially lonely: the all-consuming black hole presence of forest broken only by the firefly eyeshine of wild animals while, farther away, SAG-D sites, bitumen mines, and man camps appear as domes of white light on the dark horizon. High above them, suspended ice crystals cause the residual glow to coalesce into arrays of floating colored beams known as "light pillars" whose otherworldly presence suggests some kind of alien arrival. This Arctic phenomenon, as strange and compelling as it is, was likely never witnessed before the advent of industrial lighting. And yet, somehow, John Milton anticipated it in *Paradise Lost*:

> *. . . from the arched roof,*
> *Pendent by subtle magic, many a row*
> *Of starry lamps and blazing torches, fed*
> *With naphtha and bitumen, yielded light*
> *As from a sky.*

5

Our intent entered the world as combustion.

—Wendell Berry, *Horses*

A fire's beginnings are always humble, and any future beyond the uncertain present is dependent on a tripod of factors over which—at first, anyway—a fire has no control: heat, fuel, and oxygen. These are the ingredients of fire, but a fourth ingredient—a catalyst—is needed to unite these disparate elements into a dynamic, unified whole. Frankenstein's monster needed a jolt of lightning, and so in its way does fire. Absent lightning, a cigarette will do, and so will the hot muffler of an ATV. Under the right conditions, such as a domestic gas leak, even static electricity from a house cat can set it off. Any number of things can ignite a fire, as long as they are, at least for a moment, over 451°F. But ignition—that singular poesis when fire springs into being from nothing, from nonbeing—isn't the true moment of origin. This falls to the chemical reaction that precedes it.

The notion of fire as chemistry does not come naturally because fire is so *itself*—almost a "thing," but not quite. Our experience of fire occurs in the realm of the visible, but it is made possible by the invisible, and there is a world of energy in those vaporous, unseen realms. It is not the tree or house that burns, but the gases those things emit. This is what the heat is for: to liberate the flammable gases from their solid or liquid prisons by transforming them into vapor. In fire's world, everything relevant is breathing, emitting, vaporizing, volatile—not just the air, but the trees, the neighborhood, the house, the Formica countertop, the bag of cat food sitting on it, and, if conditions are favorable enough—hot enough—the cat itself. The higher the temperature, the broader fire's menu; under the right conditions, even a bulldozer will burn.

It is the same process whether it's a pine tree, a plastic chair, a pool of kerosene, or a living room curtain: just as we cook our food to more easily release its nutrients, fire "cooks" its fuel to do the same and, just as a compelling scent draws a bloodhound through the woods, so volatile gases draw fire. In this sense, fire is *pulled* through space. It is here—in the invisible gaseous interface between the thing and its ignition—that fire becomes mobile. This molecule-thin frontier, known as the "reaction zone," enables or denies, from moment to ignescent moment, the promise of fire's limitless, unimpeded growth. This phenomenon—pyrolysis—is the key to understanding the motives and behavior of fire. Simply put, the goal of fire—its three-fold objective—is to string these pyrolytic moments together for as long as possible, as broadly as possible, as intensely as possible.

Fire occurs frequently in Nature. Volcanic eruptions start fires, and falling rocks can generate igniting sparks; under certain circumstances, coal seams are capable of spontaneous combustion. Animals can start fires, too: there are species of hawk and kite that will pluck burning twigs from the margins of bushfires to drop them elsewhere, starting new ones. In 2004, a red-tailed hawk burst into flames after getting electrocuted on power lines near the Southern California town of Santa Clarita. The resulting wildfire burned six thousand acres, injured several firefighters, and forced a mass evacuation. Lightning, by far the most common cause of natural fire, strikes Earth millions of times per day. While only a fraction of these bolts actually ignite the landscape, they still result in thousands of new wildfires. That sounds like a lot, but it's nothing compared to the number of fires humans make.

If we have a superpower—besides our brains and thumbs and speech—it is fire. Without its light and its explosive, *directable* energy, we would not be who or what we are today. For as long as there have been hearths to gather around, cook over, and see by, fire has been, literally, central to our lives. The gap between a million-year-old fireplace in a South African cave and a gas lamp, an engine, a gun, a rocket, or a laser beam is merely one of focus and fuel; all find their beginnings—their world-changing power—in combustive energy that originates in Nature and is guided and refined by hominids—by us. Virtually everything we have accomplished on our increasingly rapid journey of differentiation from other apes, and from our own ancestors, can be traced in direct relation to our ability to focus and concentrate this fierce prosthetic *energy*—"heat" seems far too soft a term for what it has wrought in us and in our world.

Especially now.

The degree to which we have mastered fire borders on the magical and, in this way, with minimal training, we have all become casual wizards. That we are, each of us, firemasters, has become a kind of birthright, and our ability to summon fire at will is ancient. There are Indigenous people across the globe from the High Arctic to sub-Saharan Africa who, using two sticks, or chips of quartz and pyrite, can produce a flame as quickly as the reader can with a Bic lighter. There is no reason to suppose this hasn't been the case for dozens, if not hundreds, of millennia. Today, it is easy to overlook the fire-powered miracles that we daily perform, but they are real, unprecedented, and too many to list. Whether we're boiling tea or crossing an ocean, fire is right there with us. Without it, we really are dead in the water.

It is fire's nature to strive upward—in other words, to aspire, which means, literally, "to breathe desire into," and also "to rise." Fire, it can be said, is aspiration in its purest form: desire burning, and burning desire—to exist, to consume, to grow, to flourish—all as fervently as we do. Fire does not have consciousness, but it does have character, and there are internal and external factors that motivate, guide, deter, and defeat it. One of those factors is us: there is no other natural force or element over which we have such a compelling illusion of control. Fire is the only "natural" disaster that can be initiated by a gesture as casual as dropping a match, and this complicates our relationship to it in some profound ways. Since the dawn of the Petrocene Age just a century and a half ago, we have given fire exponentially more opportunities to engage in all those behaviors we seem to emulate and excel at. But like Ariel to our Prospero, fire is a begrudging servant. Selfish and willful, it yearns, above all, for freedom, which it will take at any opportunity and at any cost.

And it has enabled the same impulses in us. Never in Earth's history, or in ours, have so many fires been ignited in so many places on such a continuous basis. At the most basic level, consider every candle, lantern, and cook fire across the globe: approximately 3 billion people around the world still cook and/or heat their homes with open fires. Then, consider every gas stove, furnace, and water heater; every coal-fired and biomass-burning power station; every generator; every human-caused brush and field and forest fire. Already, we are into many billions of fires per day, worldwide, and that is not even counting matches, lighters, or pilot lights, or more rarified sources like oil refineries, incinerators, and war. Nor does this tally take into account automobiles.

A single four-cylinder car engine turning at an average of 2,400 rpm—driving-to-work speed—will generate around 10,000 combustions per minute—more than half a million per hour. There are well over a billion cars on the road today, along with a quarter billion trucks, buses, and vans, and 200 million motorcycles. Add to this a conservative estimate of 25,000 passenger and cargo jets, more than 50,000 cargo ships, and the unknown but certainly huge number of construction and earth-moving machines, tractors, lawnmowers, motorboats, private planes, helicopters, and weapons. If you ran all of these engines for just one minute, the number of combustions occurring inside them would number in the tens of trillions—each one an individual fire generating heat, energy, and exhaust. Were you to run the world's engines for just one day, the number of individual combustions would elicit an error message from your calculator. Were you to transpose this impossible number to stars, these man-made bursts of heat and light would comprise hundreds, perhaps thousands, of galaxies—every day. In just the past hundred years (a blink in geologic time), our planet has become a flickering universe of fires large and small.

Imagine being able to see them all.

Were these fires and their makers and tenders observed by visitors from another planet, humans could easily be mistaken for a global fire cult—the dutiful keepers of a trillion flames. Were these extra-terrestrial visitors to observe us over time, they would marvel at our pyrolatric devotion, how with every additional building, house, and vehicle, we add thousands of new fire shrines and temples every day. And if, through the flicker and glare of these innumerable flames, they could see Fort McMurray, with its sixty-story flare stacks flying hundred-foot banners of burning gas, they might conclude that, as remote as they are, these night-blinding beacons must indicate one of the fire cult's most sacred sites. Between full-time residents and temporary workers, well over 100,000 devotees toil in that place, their sole purpose to keep those fires burning, or support those who do.

~~~~

Even though we use the terms "oil" and "gas" in casual conversation, as if they were familiar to us, few of us ever actually see them. For most of us, they are abstractions, code words for what we're really talking about, which is fire and money. Whether it is a teaspoon of butane in a pocket lighter, fifteen gallons of unleaded in a car's gas tank, two thousand tons of heavy oil in a freighter's fuel oil bunkers, or five thousand gallons of Jet A in a 737's wings, its ultimate purpose is to burn—to be transformed into fire and the energy that combustion represents. Until then, it waits patiently for one of us—a smoker, a driver, a captain, a pilot, a cook, an arsonist—to summon it forth. Most of us never even see the flames we've conjured up: the car just drives; the plane just flies; the shower just gets hot. The stove burner is not "on fire" but a blue flower blooming—day and night, all year round—so simple and safe and, above all, *contained*, even a child can use it.

These everyday feats become still more impressive when one considers what we're actually burning: even the youngest viable fossil fuels are millions of years old. The French call gasoline "essence"— and that is truly what it is. According to the energy historian Vaclav Smil, every gallon of gasoline represents roughly one hundred tons of marine biomass, principally algae or phytoplankton, that has gone through an inconceivably long crushing, cooking, and curing process deep underground. One way to visualize a tank of gas is to imagine a mass of ancient plant matter weighing as much as fifteen blue whales crammed into a tank next to your spare tire, just behind your child's car seat. A typical driver can burn that in a week, often for the most trivial of reasons. All those distilled plants were grown by the same sun that grows our food today, so what we're burning is, in essence (that word again), ancient, super-concentrated solar energy.

Today, crude oil—liquid sunshine—is the world's most widely traded commodity (though coffee is right up there). It has been found on every continent and under every ocean where it has been sought, often in unfathomable quantities. According to another energy historian, Daniel Yergin, the world economy was worth around $90 trillion in 2019, and nearly all its energy (84 percent) was derived from fossil fuels. We are burning through this energetic trust fund like there's no tomorrow: on any given day, the human race consumes about 100 million barrels of crude oil, while another 40 million barrels are in transit around the globe via tanker, pipeline, truck, and train. More than a third of global shipping is devoted to transporting petroleum products. From this ocean of petroleum are derived the fuels used to power most of the world's cars, trucks, planes, trains, and ships, as well as key ingredients in plastics, fabrics, and fertilizers. Crude oil touches every aspect of our lives, and most parts of our bodies—inside and out.

While bitumen, crude, and all their derivatives are known collectively as petroleum, they fall under the broader umbrella of hydrocarbons. Hydrocarbons are not just oil and gas, they are truly the stuff of life: without hydrogen and carbon, Earth would be a lithic sphere and nothing more—uninhabitable, unrecognizable. Simpler, perhaps, than describing what hydrocarbons are, is describing what they aren't: water, air, rocks, and metals—in other words, things that are not, and never were, alive in the biological sense. But even with 99 percent of Earth's constituents off the table, an awful lot of things remain unaccounted for. "Hydrocarbons" is an awkward concept to wrap one's head around because they take such wildly different forms. As dissimilar as they may seem on the surface, Kentucky coal and Kentucky bourbon are both full of hydrocarbons; so are Irish peat and Irish whiskey. Hay bales, cow farts, library books, and extra-virgin olive oil are all hydrocarbons, too. Likewise, Big Wheels, Barbie dolls, LEGOs, Lululemon, Trojans, Vaseline, WD-40, and every tree on

Earth are also hydrocarbon based. So is human fat. Fats—lipids—
are the organic precursors to all the petroleum we burn. They occur
in plants, too, including algae and phytoplankton. Long after other
components of these creatures have decomposed or been reduced to
inert sediment, the lipids endure, thus confirming a suspicion held by
many people attempting weight loss: fat lasts forever.*

Whatever form they take, hydrocarbons represent potential
energy—specifically, fire. We are, in every sense, and by every metric,
inseparable from hydrocarbons and the phenomena that make them
possible. We owe them our lives because they *are* our lives. It is thanks
to the presence of oxygen in our atmosphere that all hydrocarbons
will burn. Fire, it could be said, is a hydrocarbon's ultimate expression: young or old; alive or dead; solid, liquid, or gas—fire unites
them (and us) in an all-consuming ecstasy of chemical reaction.

For our purposes, oil and gas represent both a commodity and a
Promethean superpower that could be called "fire in waiting." Fire in
the wild is a beast, dangerous and unpredictable: it is lightning and
lava and forests laid to waste. Oil, on the other hand, is fire brought
to heel: cool and contained, it awaits our signal and does our bidding. Or at least that was the original plan. There is a strong case to
be made that there has never been a better time to be a human than
right now. Some might argue this point, but one thing is for sure: in
all of human history, there has never been a better time to be a fire.

The fact that our relationship to fire has become so intimately
symbiotic has profound implications—for humans, and also for fire.
At this point in our shared history, we are almost totally dependent
on fire, and the products it facilitates, for food, shelter, heat, trans-

---

* It is probably not a coincidence that the citizenry of countries burning the most
hydrocarbons tend also to carry the most lipids about their person. In North
America, where more hydrocarbons are burned per capita than anywhere else, a
quarter of all Canadians, a third of all Mexicans, and more than a hundred million
Americans carry better than 25 percent of their body weight in fat. The governments of all three countries boast of their nation's abundant energy, but there is
a surplus they've overlooked: homegrown hydrocarbons account for about one-
fifth of the average North American's body mass. Roughly speaking, that's thirty
pounds times half a billion—and it's renewable. Were he alive today, Jonathan
Swift might have another modest proposal to make.

portation, medicines, and every object we use. Given how mutually enabling our shared existence has become, and that we are both such avid burners, it should come as no surprise that we share an elemental ancestor.

Fire, as far as we know, is unique to our planet. A relative newcomer, it has occurred on Earth for about half a billion years. That might seem like a long time, but if the planet's history were expressed as a century, fire's presence would occupy only the past decade or so. Fire, to be clear, is not the same as *heat*, which is abundant in the universe. The sun radiates heat, but it is not "on fire" in the earthly sense. Earthly fire—our fire—is different; it is not a thing, or a gas, or an element as Aristotle defined it. It is a chemical reaction. But before there was life on Earth this reaction was not possible here. Life—specifically, plant life and the oxygen it generates—is the sine qua non for fire as we know it, and most of Earth's history has played out in an atmosphere hostile to both. When life first appeared, shortly after the formation of the oceans around 4 billion years ago, the young planet's atmosphere contained only trace amounts of oxygen. Earth's first colonists were anaerobic microbes: not only did they not require oxygen to live, it was lethally toxic to them, just as it is for their descendants who inhabit Earth's airless realms to this day. Feeding, yeast-like, on organic compounds unimaginable as food, these anaerobes produced methane and carbon dioxide as waste products (just as we do).

What unites us and fire is oxygen. Oxygen is a chemical element born in the forges of Creation, the godlike heat and pressure found only in the cores of stars. Abundant and mercurial, oxygen could be described as the most promiscuous element in the universe, reacting to, or bonding with, just about everything it comes in contact with.[*] Under the right conditions, it is extremely combustible.[†] Chemists

---

[*] The element fluorine is more reactive, but far, far rarer.
[†] "Combustion," a technical term for fire, is a chemical reaction that begins with the introduction of extreme heat into a source of fuel (anything that will burn) combined with a source of oxygen (usually air). This extreme heat—whether from a lighter, a lightning bolt, or two sticks being rubbed together—initiates a chain reaction causing the hydrocarbon molecules in the fuel and the oxygen molecules in the air to collide and fly apart. It is the process of these excited, suddenly unaffiliated atoms rapidly coalescing into new molecules of CO_2 (carbon dioxide) and

call fire a "rapid oxidation event," but here on Earth, almost everything is oxidizing all the time, just in different ways and at different
rates: a steel I-beam rusts almost imperceptibly for decades; a hiber-
nating frog absorbs oxygen through its skin all winter; a Molotov
cocktail explodes instantly. As disparate as these things are, they are
all reacting with oxygen. Of course, we are oxidizing, too, thanks
to photosynthesis—the alchemical transubstantiation of sunlight,
water (H2O), and carbon dioxide (CO2) into living things that strip
oxygen from water molecules and release it in quantities sufficient
to fuel our cells, and also fire. Because of this, it is fair to say that
both of us—fire and humans—owe our existence to plants, specifically, cyanobacteria. This spectacularly successful phylum of single-
celled creatures was the first to perform photosynthesis on a planetary
scale, starting around 2.7 billion years ago. For these pioneering "solar
cells," sunlight was more than just a heat source, it was a catalyst for
life, and their success had a catastrophic effect on Earth's atmosphere.

All photosynthesizing organisms generate oxygen as a waste product (just as we generate methane and carbon dioxide), but we associate oxygen so strongly with its life-giving properties that it is hard to
imagine it as pollution.* Cyanobacteria's prodigious oxygen production contaminated Earth's atmosphere so thoroughly that most anaerobic creatures—the founding colonists of life on our planet—were
gassed to death. Into these voids evolved new species able to survive
in a lethally oxygen-rich environment. Enabling these early aerobes to
flourish and evolve into more complex creatures was yet another gift
of photosynthesis: the ozone layer, a by-product of the massive oxygen build-up in our atmosphere. Without it, Earth—its life, oceans,
atmosphere, and all—would have gone the way of Mars.

Fire is possible because oxygen is so reactive, but the secret to its
continued success is that there is the "right" amount of it relative to
other atmospheric gases. Fire has a hard time sustaining itself when

---

H_2O (water vapor) that generates a fire's light, heat, and sound. Our eyes experience this process as a blue flame, though it is often obscured by the far larger
orange and yellow glow of burning soot. This process of molecular breakdown and
reassembly we call "fire" will continue in a feedback loop of "burning" for as long
as fuel and air supplies last.
* Most plants, algaes, and fungi also require oxygen to function.

oxygen levels fall below an atmospheric concentration of 15 percent. Meanwhile, in concentrations above 35 percent, dinner by candlelight would be ill-advised. Currently, atmospheric oxygen sits in the Goldilocks zone, just shy of 21 percent: exactly what we need to live and prosper, and exactly what fire needs to burn in ways that have proven extraordinarily beneficial to us—most of the time. We evolved to accommodate these conditions, but fire didn't; fire's behavior is determined by immutable laws of chemistry and physics. So, it is a fortuitous coincidence that we find ourselves overlapping in space and time under such mutually rewarding circumstances.

Long before we climbed down from the trees, fire was climbing up into them. Not only would photosynthesis provide the gas required for the first ignitions, its agents—plant life—would provide storehouses of additional fuel (i.e., hydrocarbons) in the form of wood, stems, leaves, and grasses to keep these fires going. "Going" is the operative word here: oxygen is the gas that powers the engine of motility. Any complex creature that moves for a living does so thanks to oxygen, and the reason animals move is, first and foremost, to find more food to consume and, ultimately, to oxidize—to "burn" in the form of energy transmitted to and through their blood. Fire is driven by the same need. In this sense, fire is not an "element" or a "reaction," it is a hunter. Just like us.

The roster of fellow travelers who share this motivating appetite is a diverse one: bedbug or barn owl, octopus or oil baron, white shark or wildfire, oxygen is our trail boss, driving us through the world on an endless quest for action and reaction, for energy to burn. As much as we may prize individual agency and autonomy, we are, in fact, oxygen's terminal hostages. If in doubt, don't inhale for thirty seconds and notice how you feel. Suffocation is a grave risk faced by all burners and breathers, above and below the surface. Every step we take may break a fall, but every breath we take staves off imminent death. With this barely conscious rhythm of self-resuscitation we save our lives about twenty thousand times a day. Consider the effects of an unplanned blockage to your airway: in the first forty-five seconds of stifled breath you will lose your manners, your language, your loyalties, your love, and, with them, 10 million years of evolution, winding up wide-eyed and flailing, somewhere lower than an ape. In the next forty-five seconds, your evolutionary clock will spin back even faster—perhaps 200 million years—until you resemble a lizard: a savage, soulless vehicle for raw impulse focused with all your being on one immediate and crushing need. And in your final seconds, with your lungs ablaze and not a thought left in your head, you might be reduced to something even more basic—something a lot like fire: an appetite for oxygen, the animating gas, with nothing left of your self, or even your life, but a chemical reaction dying to happen.

Breathing, a biochemical analog to hope, which is just a human expression for "oxidizing," is our single most critical interaction with the outside world—the first and final thing we do. Fish or fowl, our last mortal act is not to move, or eat, or procreate, but to strip oxygen molecules from our environment. Then, like a smothered fire, we flicker and die.

This begs the question: Is fire alive?

After all, it meets so many of the criteria:

- Fire is animate.*
- It can be propagated by others of its kind.
- It can generate offspring.
- It grows.
- It breathes.
- It travels in search of nourishment.
- It persists with the opportunism and single-mindedness of all successful life forms.
- It can lie dormant for long periods awaiting more favorable conditions.
- It can adapt to changing conditions.
- It can die of starvation and/or suffocation.
- It can be killed.
- It can be revived.

And it can turn on its master—who might not be its master at all. Lately, it seems as if fire, our constant, if fickle, companion, our most

---

* The words "animate" and "animal" are derived from the Latin *anima*—the breath of life.

potent and ready enabler, has been biting the hand with a frequency and viciousness for which we are ill-prepared.

Fire may not be alive or conscious in the sense that we are, and yet its behavior manifests a vitality, flexibility, and ambition often associated with intelligent animals. Likewise, we may not appear as flickering sprites composed of light and smoke, and yet, distilled to our essence, we are fire's kin—gas-driven, fuel-burning, heat-generating appetites who will burn as bright and hot as we can, stopping at nothing until we're fully extinguished.

To aspire—to breathe desire into ourselves and into our world—is what oxygen empowers (and condemns) us to do. If you look at how humans—undisciplined and unregulated by education or culture— use resources, they tend to consume whatever is available until it's gone. Of course, fire does this, too. A key difference between us is that fire has no control over its appetite or rate of consumption, but we do, even if it's hard to tell sometimes. Because of this capacity for self-awareness and self-control, science has generously named our species *Homo sapiens*, or "wise man." In Latin, *sapiens* means not only "wise" but also "rational" and "sane." However, given the degree to which our character and culture are now determined by our relationship to fire, and its avatar, the petroleum industry, there is a case to be made for a revised nomenclature. The energy historian Vaclav Smil suggested "hydrocarbon man." I propose *Homo flagrans*. *Flagrans* is Latin for "ardent, fiery, passionate, outrageous." In other words, "burning man."

PART TWO

FIRE WEATHER

6

I balanced between destiny and dread
And saw it coming, clouds bloodshot with the red
Of victory fires, the raw wound of that dawn
Igniting and erupting, bearing down
Like lava on a fleeing population.

—Seamus Heaney, "Mycenae Lookout"

Since the last days of April, fires had been a throbbing presence on the horizon around the city, their smoke plumes bending with the wind as they rose and fell with air temperature and barometric pressure, passing to and fro through a grayscale spectrum: white to charcoal and back again, often many shades together. In the morning and evening when the sun was low, shades of brown and orange would emerge and then recede. For those living in the boreal forest, these shifting, earthborne clouds represented a familiar seasonal awareness, occupying the same mental space as the possibility of a thunderstorm or a blizzard—one among many manageable threats long since factored into the calculus of daily concerns.

After its discovery, and after it proved impossible to subdue, the fire southwest of town was given a code—MWF-009 (McMurray Wild Fire 009), because it was the ninth substantial fire discovered in the Fort McMurray Fire District that year. Fire 009 was not yet a day old, but already it was clear that this one was different. Since first being spotted at 4:00 p.m. on Sunday, May 1, it had increased in size by five hundred times. Much of that growth was toward the city, and preemptive evacuations had been ordered for several neighborhoods closest to the fire. By 10:00 a.m. on Monday the 2nd, the fire had expanded to two thousand acres; an hour later it was over three thousand. Then, it doubled again. Despite the efforts of eighty firefighters, two bulldozer groups, and several water bombers, Fire 009 was zero percent contained.

At 5:30 p.m. on the 2nd, city officials held a press conference, their second fire-related media briefing of the day. Standing at a lectern before a handful of reporters, Mayor Melissa Blake, forty-six years old and halfway into her fourth term, wore her long, auburn hair tied back over a short-sleeve sheath dress with purple accents and matching earrings. Referring to notes on her phone as she spoke, her tone was measured, her pace crisp, and, in light of what was unfolding nearby, her demeanor eerily calm. But Blake was used to explosions; since 2004, she had been the disarming face of this ultimate boomtown, whose convulsive growth had outpaced its American counterparts in North Dakota, Colorado, and Texas. During her tenure, she had ridden a wave of money and manpower that doubled Fort McMurray's population in a decade and turned her hometown into a city as oil surged past $100 a barrel and majors from around the world rushed in to capitalize on the seemingly limitless—and, finally, profitable—potential of Alberta bitumen.

After touching on the evacuation of some eighty horses due to the weekend fires, Blake turned her attention to Fire 009. Bulldozers, she said, had been working day and night to clear a firebreak between that fire and the highway. Firebreaks, created by scraping broad strips of forest down to unburnable mineral soil, are also known as "dozerguards"—or "catguards," because many of the bulldozers used to clear them are manufactured by Caterpillar. Some of the Cats they were using were D10s, hundred-ton machines that can clear a residential street of all its cars, pavement, and boulevard trees in a couple of passes. The dozers' objective was to create what amounted to an unburnable dirt moat between the fire and the city. "Trust the experts in the field," Blake told her audience. "They're doing the best that they can with the circumstances they've got, the equipment that's there, and the resources that are coming in.

"What really has a bigger impact, of course," she added, "is what happens with our weather."

In Fort McMurray that week, nothing mattered more, and Blake's forecast for Tuesday, May 3, bode ill: "Wind is predicted to be coming out of the southeast [away from town] in the morning," she said, "and then it will be southwest in the afternoon."

Afternoon is when temperatures are highest, winds are strongest, and fire spreads most rapidly; "southwest" put Fort McMurray squarely in the fire's sights. Today, most mayors presented with such a forecast would order evacuations immediately, especially after saying, as Blake did, that "extreme fire conditions are once again predicted for tomorrow." But 2016—recent as it is—was a more innocent time. Instead, the mayor emphasized common sense: no open fires, fireworks, or off-roading in the backcountry, no flying drones near firefighting operations, and don't throw your cigarette butts out the window. Because the fire was also threatening the landfill, recycling would not be picked up that week—"a ripple effect," Blake said, "as we continue to manage this crisis." In closing, Blake advised residents who had been preemptively evacuated from the south end of town that they could now return home, where they should continue to "shelter in place"—that is, be ready to leave again at a moment's notice. A notable exception was Centennial Park, a trailer community west of Highway 63 and closest to Fire 009, the sole neighborhood to remain under mandatory evacuation.

Standing with Mayor Blake that afternoon were two men, Bernie Schmitte and Darby Allen. Schmitte was the fifty-one-year-old regional manager of the Wildfire Division of Alberta's Ministry of Forestry and Agriculture; a solidly built Albertan whose bespectacled eyes peered out from beneath a high, clean-shaven dome, he had twenty-four years in with the province. Allen was the municipal fire chief, a frank and sober man of fifty-nine who had carried his rugged yeoman's face and traces of a brummie accent with him all the way from Britain's Midlands. The men were in uniform. Schmitte wore green cargo pants and a yellow Nomex firefighter's shirt that bore shoulder patches embroidered with the word "WILDFIRE" in bright red over an image of flames racing through prairie grass and spruce trees—Alberta's two most flammable landscapes. Allen wore a dark blue short-sleeve shirt and pants held together by a garrison belt and buckled with the firefighter's Maltese Cross.

The current fire situation, alarming at any time of year, had led city officials, in conjunction with the Ministry of Forests and Alberta Emergency Management, to declare a local state of emergency. In order to manage it, they took the additional step of setting up a Regional Emergency Operations Centre, or REOC—a centralized communications hub that allowed members of all relevant agencies to coordinate their response. The situation was considered serious enough that Blake had formally ceded leadership of the fire response, and of the city itself, to Allen. "Thank you very much, Your Worship," the chief said when Blake invited him to the mic. After admonishing people for driving off-road vehicles dangerously close to firefighting operations, Allen went on to say, "We had significant fire today, but the reality is that fire hasn't gotten any closer to town, and that is mainly due to the incredibly hard work of Alberta Forestry. If it wasn't for them, we'd be in a much worse situation."

With that, Allen invited Schmitte to the mic. After giving an update on firefighting equipment and personnel, and saying that more of both were on the way, Schmitte closed on an upbeat note: "We're in for another challenging day tomorrow," he said, "but we have made some progress today, and our guys are feeling pretty good."

"How far away is the fire from the closest residence?" asked a reporter.

Schmitte's calm answer was one and a half kilometers—about a mile. The subject of wind came up again when another reporter asked Allen if he was surprised that there had been no structure loss yet, given how rapidly the fire had spread the previous day. "I don't know if 'surprised' is the right word," the fire chief answered. "We're certainly thankful. We're blessed that we didn't have thirty-five-kilometer winds in the wrong direction."

But thirty-five kilometer winds—twenty-two miles an hour—blowing "in the wrong direction" was precisely the forecast for May 3. At that speed, a fire like 009 could travel a mile in a matter of minutes, especially under the extraordinarily flammable conditions in evidence that spring. Reporters did not bring it up again, but it recalled a comment Allen had made at the morning briefing: "It's way drier this year than it's been for a long, long time," he'd said. "I heard fifty years.

"We've had four significant fires in the last five days or so," he had added, "so that's pretty intense. Hopefully, Nature's done its thing and it'll leave us alone for a little bit."

But hope is a human construct, a coping mechanism in the face of uncertainty that holds no sway in the natural world. And yet, hope, like fear, is contagious, communicable; when expressed by a respected leader, like Darby Allen, it has the power to create an imaginary zone of protection around a group. Hope—the willpower of positive thinking—is clearly adaptive to human survival. To remain cohesive under pressure, communities need trustworthy authority figures capable of leading by example and exhorting others to manage their thoughts and feelings, especially doubt and fear. But there is a fine line between hope and denial and delusion.

There was something else the chief had said that morning, but somehow, it, too, got lost in the noise: "The slightest sign that we see people are in jeopardy, we will evacuate again."

Even as the afternoon press conference was wrapping up, the fire was taking off. Anyone who had the Alberta Wildfire Info app would have seen the alert at 6:17 p.m. declaring Fire 009 to be "OC"—out of control. This meant that no containment had been achieved on any front. Once a wildfire reaches this level of intensity, there will be no further attempts to confront it directly. Instead, firefighting resources—ground crews, bulldozers, and water bombers—focus on the flanks, and on removing potential fuel in hopes of slowing the fire's forward progress. Witnesses that evening described a raging crown fire that sent jet-like whorls of flame spiraling into the smoky, twilit sky. The plume was now gigantic. A local journalist named Vince McDermott posted a photo taken from Lac Laloche, in the neighboring province of Saskatchewan: Fire 009 was plainly visible eighty miles away.

When firefighter Ryan Coutts saw the evening update on the Wildfire Info app, he fully understood the threat this fire posed, but he was one of the few. Coutts, a 220-pound, hockey-playing twenty-year-old with wavy blond hair and sea-green eyes, had been fighting fire since he was in high school. He was based out of Slave Lake, a

small town of 7,500 four hours southwest of Fort McMurray, through
the forest. The Slave Lake Fire Department is largely volunteer, and
its chief, Jamie Coutts, built like his son, only thicker, was an outspo-
ken, gap-toothed, slightly piratical forty-three-year-old with a gray-
ing buzzcut and twenty-five years of firefighting experience. After five
minutes in conversation with either man on the subject of boreal fire,
it is clear that they are people you want nearby when one of these
monsters comes out of the woods.

The Slave Lake Fire Department—comprising just thirty volunteers and nine paid positions, including support staff—was self-reliant
because it had to be. Isolated, and hours from outside help, the town
lies at the east end of Lesser Slave Lake in what could be described
as a fire corridor. As one longtime analyst with Alberta's Ministry of
Forestry and Agriculture put it, "A lot of ash has fallen on that town."
Its small, scrappy fire department has managed to fend off most of
these blazes, but not all of them. In 2011, tiny Slave Lake, virtually
unknown outside Alberta, became nationally famous overnight when
a wildfire, supercharged by eighty-mile-an-hour winds, burned down
more than a third of the town in a matter of hours. Five hundred
homes and other buildings were lost, including the town hall, the
library, and the radio station; fifteen thousand people were evacuated
across the region. In the ensuing days and weeks, what came to be
known as the Flat Top Complex Fire burned nearly three thousand
square miles of forest. The losses to the timber industry were tallied in
the millions of dollars; the losses to the region's burgeoning petroleum
industry approached half a billion and impacted the province's GDP.
It was, at the time, the largest evacuation and most expensive natural
disaster in Alberta history. Amazingly, there were no fatalities save for
a single Quebecois helicopter pilot.

In an effort to convey the intensity of the fire that savaged his
hometown, Chief Coutts rattled off a list of effects: "Metal melted,
concrete spalled, a granite statue was reduced to pebbles—basically,
all moisture was released from everything. I kept hearing 1,600
degrees Fahrenheit. Too hot—that's all I remember."

"It actually incinerated the concrete," explained a volunteer
named Ronnie Lukan. "It pulled the moisture out so you could touch
the concrete and it would peel off."

This is what spalling is; it's a verb you don't encounter much below a thousand degrees.

There is destructive fire, which burns down houses and forests, and then there's transformative fire, which makes familiar objects—like houses—disappear altogether, and leaves whatever's left—the cement foundation, the steel reinforcement rod holding it together—altered at the molecular level. This is what happened in Slave Lake in 2011: large, expensive things like riding lawnmowers couldn't be found because they had, more or less, vaporized. Little remained besides cast iron bathtubs and the warped husks of furnaces and cars. In the aftermath, a formal review was conducted, faults were found, and recommendations were made. Among the many findings was the disturbing determination that the fire was not started by lightning or by accident, but by arson.* A surprising number of people saw the Slave Lake fire as a one-off and said it couldn't happen again. But none of those people serves in the Slave Lake Fire Department.

As far away as it was, the Couttses had been keeping an eye on Fire 009. "We're all sitting in Slave Lake," Ryan said, "itching to go, right? We see a big fire right outside, and we'd been there before—we were there in 2011—we didn't call in extra help and it bit us in the ass, to say the least."

Both Ryan and his father knew what Fort McMurray might be up against, what their strategy would probably be, and how it would likely fail—because they had already tried it. "We had a plan and it was good for forty-five years," Jamie told me, "and then the first time that we actually, physically, had to do our plan, it fell to shit. So from that day, we went forward going, 'There can't be *a* plan, there has to be a bunch of plans.'"

In fairness, Slave Lake's plan had been a sound one: "We would go to the edges of town, we would hook up to all the fire hydrants, and we would spray water out towards the forest, and onto the houses in the first ring," the chief explained, "and that would hold the fire. But in 2011, there was 50,000 burning embers raining down for every acre—50,000 spot fires. So, that's what happened. People think of

---

* According to the National Fire Protection Association, nearly 300,000 fires are intentionally set each year in the United States. There is no data for Canada.

a fire like their campfire, like it's contained, like it's just going to be a wall of fire. But when it gets windy enough, and hot enough, and crazy enough, it's going to come at you—not as a wall of fire, but as a wall of fire with hundreds of thousands of hot embers spewing out the top of it, going half a mile, or a mile in front of it."

This means you can be doing a terrific job fighting the fire, but behind you—a block, a neighborhood, a valley away—your town can still be burning down. Fortified cities have fallen this way for millennia: while soldiers held their own at the wall or on the field, somewhere in the trees, behind the front line, hidden archers were loosing flaming arrows into the heart of the town. In the end, defeat arrives from behind. It is tempting to give credit for this fiendish strategy to great military minds of the ancient world—the Chinese, the Assyrians, or the Greeks—but any shepherd or hunter watching a wind-driven forest fire from a hilltop could have figured it out, especially when those embers started landing on *them*.

Wildfires are commonly described as a single entity moving across a landscape, but big ones can be divided into three distinct parts, and the ways in which they travel and impact human settlements have an analog in medieval warfare, which they may have inspired. Without wind to drive them, most ground fires tend to progress slowly, creeping outward from their source like a roadside cigarette fire. These low flames and smolders represent the foot soldiers—slow-moving and most easily defeated. Wind and dry fuel can get a fire moving, but in order to rise into the treetops, a fire requires "ladder fuels"—brush, saplings, and low branches, which allow the fire to climb upward into a tree's crown. This marks a transformative moment for a fire: up in the crown there is more wind and much more mobility. A "crown fire" corresponds to mounted troops: much faster moving than a ground fire, and far more charismatic, it will charge ahead with flags and pennants waving. Once in the treetops, the fire's heat can generate its own wind as it sucks up oxygen from ahead and below. The wind, in turn, liberates and empowers the fire still further: no longer bound to its fuel source, the fire can now fly ahead of itself in the form of sparks and embers: these are the archers. The analogy can be extended further to include recon units: ember-generated spot fires, and even "sleeper cells" in the form of smoldering roots, which can act on delay, months or even entire seasons later.

Jamie Coutts had been in touch with his counterparts in Fort McMurray on Sunday, May 1. "We knew it was going to be hot, we knew it was going to be dry, we knew it was going to be windy: that's May," Coutts said. "It was close, and we knew they were struggling with the fire, and there was lots of people around [and at risk]. Forestry was hammering it with what they had, but you could see on the news that it wasn't enough."

On May 2, Coutts received a call from Fort McMurray inquiring about the availability of his sprinkler rig, which included 120 garden sprinklers, four gas-powered pumps, and a mile of hose in various lengths and diameters. "Of course, we were already packed up," he said, "so we were like, 'Well, we can leave in ten minutes.' Then they hassled me quite a bit about the money."

The Fort McMurray Fire Department employs 185 professional firefighters, citywide, operating out of four fire halls plus a 911 dispatch center. Its members are among the highest-paid firefighters in North America, a four-way by-product of a strong union, Fort McMurray's formidable tax base (you cannot buy a trailer home for less than $200,000), the town's close association with the multibillion-dollar petroleum industry, and the fact that people living in northern communities get living allowance subsidies from the Canadian government. Furthermore, firefighters' work schedules are such that many members operate private businesses on their off time. Of those 185 firefighters, almost all of whom are white, and roughly 80 percent of whom are men, about 30 are on duty at any given time, with an additional 20 or 30 on call.

With its much bigger, better-funded, full-time department, Fort McMurray was interested in Slave Lake's sprinkler system, which Jamie Coutts had designed and assembled from the ground up, but they didn't want to pay for more men, even men with Ryan and Jamie's experience. On Sunday, May 1, Fire 009 was still a wildland fire and so, technically, Forestry's problem, but it was growing rapidly, and officials recognized the need for additional sprinklers on the town's southern perimeter. They were also aware that the fire weather—the dynamic relationship between temperature, relative humidity, the fuel load in the forest, and the moisture content of that fuel load— was growing more favorable to fire by the day. Even so, many in Fort McMurray's fire department simply couldn't imagine a fire capable of breaching their considerable defenses, which included new firebreaks, transmission tower and pipeline corridors, seismic lines, greenways, a major highway, and the Athabasca River. And there was another reason: this wasn't just any town; this was Fort Mac. Because of its young, skilled, and ambitious population, and the enormous wealth infused by the petroleum industry, Fort McMurray exuded some of the same bullish swagger that major oil hubs like Calgary and Houston (both cities with whom Fort McMurray maintains close ties) are famous for. Because of its industry connections, many of Fort McMurray's facilities are not only named after oil companies, but also built to specifications far beyond what you would expect to find in such a remote and isolated place. The water treatment plant, rec center, hospital, municipal building, and fire department, to name a few, are all top of the line. Furthermore, when you know that the bitumen plants, thirty minutes away, will back you up with their industry-grade equipment and highly trained crews, offers from small-town fire departments—no matter how earnest or urgent—are low priority.

But what Slave Lake was really offering was not so much equipment as experience—a way of thinking about boreal fire that could save lives and infrastructure. "We want them to know about sprinklers, and we want them to know about forest fires," Jamie Coutts explained. "We have some really good trailers that we put together, that have been battle-tested, had lots of deployments. So it just seemed like we should get our stuff up there."

Linked sprinkler systems, in which small sprinklers are rapidly mounted on the rooftops or fascia boards of individual houses, and connected in series to a gas-powered water pump fed from a lake, river, or water bladder, is not an urban firefighting technique. It's a technique used in rural areas where resources are limited and the threat of fire may be widespread across space and time. With linked sprinklers, an isolated community or neighborhood can protect itself, as long as the water supply holds out and the pumps keep running. This is not an approach many municipal fire departments are familiar with because most urban fires originate inside the community—in a house or a building, a specific location where a hydrant, trucks, and firefighters can be brought to bear. The notion that a fire might enter the community from somewhere else, across a broad front—like a tsunami, or a hurricane—is beyond the scope of most structural firefighters' experience, even in Fort McMurray. But in Slave Lake, it's a scenario with which firefighters are now bitterly familiar.

Since their traumatizing experience in 2011, the Slave Lake Fire Department had become evangelists for fire safety and education regarding the hazards of living in the woods. Slave Lake's lean, agile, quasi-guerilla style of post-2011 firefighting differs in crucial ways from its unionized, hierarchal, textbook-following urban counterpart. "For us [now] there is no box," Ryan Coutts told me, "so you don't have to think outside of it, you just roll in there *thinking*. We always surround ourselves with people that can be put into any situation."

These methods, while unorthodox, would prove invaluable in the coming days.

But on May 2, despite the fact that Slave Lake had burned only five years prior, at the same time of year, under conditions nearly identical to those facing Fort McMurray, and despite the fact that fire departments in the region have mutual aid agreements with an established protocol for equipment and manpower loans, there was pushback on the Fort McMurray side. In the end, Jamie Coutts said he agreed to let Fort McMurray borrow his $200,000 sprinkler rig for a pittance—without the Slave Lake crew who knew best how to deploy it. But Coutts had designed and built this system, and he insisted on giving a demo before being sent home. The back-and-forth had delayed the trailer's arrival by a day, and it was during this twenty-four-hour period that the fire south of town, now numbered and named like a tropical storm, was declared by Forestry to be out of control. But as dire as it sounds, "OC" is not the most extreme designation for a boreal fire.

Later that evening, the fire did something no fire ought to be doing at that hour: energized by its own ferocious heat and by winds that should have settled down at dusk, it drove embers all the way across the Athabasca River, where they started a new fire. The Athabasca is one of the north's great rivers, on a par with the Mackenzie and the Yukon; its width, outside Fort McMurray, is a third of a mile. Fire 009, already within striking distance of the highway and the south end of the city, now had a beachhead on the north side of the Athabasca River. While still several miles off, this new spot fire had ready access to the entire west side, including the city's water treatment plant, biggest park, and most densely settled neighborhoods. More important, it demonstrated that, even in its nascent stages, Fire 009 was capable of surmounting any obstacle that Nature or humans could put in its way. Viewed in terms of military tactics, the fire had now arranged itself in what amounted to a double-flanked pincer movement. Fires don't have the capacity for "strategy," per se, but if those flames had been divisions deployed by a seasoned general, his peers would have found little to improve on. After nightfall on May 2, the fire settled down to wait out the dark, nursing itself on the exceptionally dry ground fuel, and solidifying its gains. Sunrise on May 3 would be at 5:33; the rest was up to the wind.

7

Fire synthesizes its surroundings. Those surroundings
are cultural as much as natural, and choices about
fire practices and regimes will inevitably be made on
the basis of social values and philosophies,
as integrated by political institutions. Science can
enlighten that process but will not determine it.

—Stephen J. Pyne, "Pyromancy"

By the time Shandra Linder was pulling out of her garage in Timberlea, en route to Syncrude's headquarters north of town, Jamie and Ryan Coutts had already hooked up their sprinkler trailer in preparation for the four-hour drive from Slave Lake to Fort McMurray. The fire chief and his son were traveling with Lieutenant Patrick McConnell, and their plan was to arrive at Fort Mac's Fire Hall 5 by lunchtime to hand off their sprinkler rig after a quick demo. Knowing that firefighters were in for a busy day, they kept one radio tuned to Forestry's internal channel and another tuned to local news. Shandra Linder, meanwhile, had spring on her mind; her evacuated friends had gone home the day before and there was nothing in her windshield but Alberta's limitless blue sky.

Turning east onto Confederation Boulevard, the six-lane main drag and shopping strip that accesses the rapidly growing neighborhoods of Timberlea, Linder accelerated down the hill, toward the Athabasca River and Highway 63, where she blended into the flow of trucks and buses bound for the major refinery complexes and the sprawling open pit mines that surround them. At that hour, the six-lane highway was crowded with commuters, many of whom were traveling by bus. Diversified Transportation, the company responsible for moving the bulk of the bitumen industry's workers to Site and back again, is the largest privately owned bus service in North America. Its Fort McMurray fleet of six hundred passenger buses was twice the size of Greyhound Canada's.* Several times a day, hundreds of Diversified buses, along with thousands of trucks and cars, plug 63 with traffic. According to Jude Groves, the company's fleet manager, Diversified's cumulative mileage for these journeys—some of which are two or three hours round trip—is equivalent to three circumnavigations of the earth per day. Most of those miles are driven on 63, the only way in or out of Fort McMurray.

Twenty minutes past the Syncrude offices, the highway ends at the First Nations reserve and former Hudson's Bay Company trading post of Fort McKay. Once a commercial oasis in the wilderness, Fort McKay is now completely surrounded by mines, tailings ponds, and upgrading facilities. North of McKay, the public road ends altogether. If you are not affiliated with a petroleum company and want to continue, you must do so off-road, by air, or by boat. In winter, a snow road, sections of which utilize the frozen Athabasca River, continues for another hundred miles up to Fort Chipewyan, the legendary Hudson's Bay trading post and First Nations community on the west end of Lake Athabasca.

At 11:00 a.m., while Linder was in her office up at Site, and the Couttses were passing though the hamlet of Mariana Lake, a press conference began downtown at city hall, in a pair of conjoined brick cubes that squat, bunker-like, on the west side of Franklin Avenue. Standing at a lectern before a dozen reporters and half a dozen cameras, Mayor Melissa Blake opened by saying, "Here we are on another day that's filled with heat, sunshine, and a little smoke in the air—today, of course, being World Asthma Day." After once again advising evacuated residents that it was safe to return home, where they should shelter in place, Blake got down to business:

---

* Greyhound Canada went out of business in 2021.

Fire 009, she said, had more than doubled in size since the previous day, to 6,500 acres, or ten square miles. She then circled back to the weather and Asthma Day, and closed by saying, "It's never a bad idea for everyone in the community to have their plan, should they have it required because of whatever circumstances in the region are."

The circumstances in the region—fires—were implicit, but what that plan could or should be beyond sheltering in place, or signing in to the evacuation center at the Suncor Leisure Centre just north of downtown, remained unclear. What that or any other plan might look like when multiplied by ten thousand—or ninety thousand— was beyond the bounds of imagination. Even as she spoke, the fire was intensifying a half mile west of the only road out of town.

Darby Allen and Bernie Schmitte were present that morning, dressed as they had been the previous day, only Allen was now wearing a green vest fitted with a badge that read "Director Emergency Management." Color-coded vests like this are worn in the Regional Emergency Operations Centre to enable quick identification among people who might not ordinarily work together. It was the only visible indicator that the city was in crisis mode—this and the fact that these three local leaders were gathered together in front of reporters at 11:00 a.m. on a cloudless Tuesday morning. "We know the public want information," Allen said. "We're not hiding anything from anybody . . . We wake up this morning and we don't see anything, and people think it's fine and it's all gone away. And it's nice to have that thought, but I just want people to bear in mind . . . we're in for a rough day . . . It's good to see people taking their kids to the park . . . that's great, but just keep in the back of your mind that we're still in a serious situation here."

The municipal fire chief, known to locals as Darby, had worked structure fires in Calgary and Fort McMurray for more than twenty years, and in Portsmouth, England, for a decade before that; he had seen some big ones, and he had seen a lot of things burn. But the scale of this fire—not what it was that morning, but what all signs pointed to it becoming—was something he was unable to convey and, perhaps, to comprehend. And there was another obstacle: How do you talk about a fearsome thing without instilling fear? How do you prepare the public for a dreadful possibility in one breath, and encourage them to get on with their day in the next? Is it possible to prepare people to flee for their lives without inciting panic? This quandary was the elephant in the room, and Schmitte would describe another aspect of it: efforts to expand the firebreaks between the fire and the city had been seriously hampered by the convoluted terrain and the presence of a southbound pipeline.

In understated tones, the wildfire manager then noted that the forecast called for temperatures in the high eighties. This wasn't just a little bit warmer than normal—it was almost thirty degrees hotter than the average high for that time of year. Meanwhile, the forecast for relative humidity—15 percent—was also record-breaking for that date. Fifteen percent humidity is not typical of the boreal forest in May; it is typical of Death Valley in July. Combined, these conditions were as conducive to fire as is possible anywhere on Earth— comparable to peak fire season in Southern California or Australia. But as radical as these heat and humidity forecasts were, weather stations in Fort McMurray would soon confirm that they had been conservative. The forest that day was a bomb, but it would take wind to detonate it. Six hours earlier, at 5:00 a.m. Mountain Time, Environment Canada had issued the following forecast: "WIND BECOMING WEST 20 KM/H (12 MPH) GUSTING TO 40 (25 MPH) IN THE AFTERNOON." While Schmitte acknowledged that such a wind shift would pose a "challenge," he did not say what would have been crystal clear to anyone familiar with boreal fire: winds of twenty-five miles per hour can cast embers hundreds of yards, all but guaranteeing that Fire 009 would enter the city.

There were other more esoteric scales that Schmitte did not refer to, but was also well acquainted with, and one of these was the Fine Fuels Moisture Code, which measures the dryness of leaves, needles, and other forest litter. On this open-ended scale, anything above 92.5 is considered "Extreme"; the rating for May 3 was 95. There was also the Duff Moisture Code, which rates the flammability of compacted organic matter on the forest floor; the Drought Code, which calculates the dryness of deeper soil, trees, and logs; and the Buildup Index, which measures the density of flammable biomass—all of them were in the 100th percentile for their categories. Meanwhile, the Fire

Weather Index, an aggregate derived by combining more specific fire risk data, was effectively off the charts: where 21 is considered "Very High," and 33 is considered "Extreme," the index that day was pushing 40.

Bernie Schmitte hadn't experienced this before, but he had experienced something a lot like it. Even on its first day of ignition, he had been concerned enough by Fire 009's behavior to take the unusual step of inspecting it by helicopter, in the middle of the night, with infrared goggles. What struck him then, when the fire was still only hours old, was the heat—and the fact that even in the dead of night, usually a quiet time for wildfire, this one was still actively spreading. Schmitte had seen this behavior only once before, in 2001. That was the Chisholm Fire, which had set the record, worldwide, for head fire intensity.

In addition to the Chinchaga Fire of 1950, the continent's largest known wildfire, the forests of Alberta fueled the most intense fire ever measured—not just in North America, but anywhere on Earth. The Chisholm Fire has attained a legendary status, and the story of its discovery is relayed between fire chiefs, fire scientists, and incident commanders with the individual flourishes of a folk tale or an urban myth. Sometime in late May 2001, a senior wildfire expert in Alberta received a phone call from Washington, D.C. Depending on who's telling the story, it could have been from NASA, or NOAA, or NORAD—or maybe the Navy. In any case, while monitoring global weather data from a satellite feed, the American caller detected a mysterious aerosol signature in the form of a smoke plume that penetrated the tropopause—the atmospheric barrier containing most earthly emissions—and kept rising, far into the stratosphere, topping out at forty-five thousand feet. The signal was emanating from a sparsely inhabited area in central Alberta, about two hundred miles south of Fort McMurray. Volcanoes are capable of generating this type of high-altitude aerosol signature, but there are no volcanoes in Alberta.

The American's question, so the story goes, was, "Has Canada just detonated a nuclear device?"

Canada has a nuclear program, but not, as far as the U.S. government knew, a weapons program. There is admirable transparency between the two countries, but what else besides a nuclear bomb, or a volcano, could produce so much explosive energy, followed by such a high and massive injection of smoke and ash? And yet the American had read the satellite data correctly: there had been an "explosion"—many of them, in fact, and the energy released was comparable to multiple nuclear bombs.

This was the Chisholm Fire.

The actual story—who called whom—runs somewhat differently, but the dates are right, and the outcome is the same: on May 23, 2001, an Alberta wildfire, born and raised in the boreal forest, set a terrifying new standard for fire on Earth. Volcanoes can be devastating, but they are fixed in geographical space and usually offer some warning; nuclear weapons are devastating, too, but they are governed by international treaties and detonated with intention by human beings. Fire is different: it has its own agency that manifests as something akin to will, and it is aided, often unintentionally, by human beings. The Chisholm Fire was ignited by a spark from a passing freight train. On that particular irregularity of track, a thousand sparks just like it had probably flashed and sputtered in the gravel ballast below. On a different day, that spark could have landed on a similar tangle of dead grass, or patch of forest litter, and fizzled on the spot. Had someone been there and seen the first arabesques of rising smoke, she could have simply walked over and stepped on it; an animal could have crushed it in passing.

But in late May 2001, during a fire season that had arrived a month early for the second year in a row, conditions were optimal and the moment for early intervention passed. That initial smolder, unseen and unheeded, worked its way down the steadily widening fuse of grass, needles, leaves, and duff spread out before it. Who knows how long or far it crept like that, one diminutive, flameless ignition to the next, until that transformative moment when, quickened by a sudden gust of wind, a drop of volatile sap, or a particularly warm, combustible piece of tinder, it burst into flame: a chimera, engendered by humans and enlivened by Nature. A warm southeasterly wind then lifted that flame from its forest bed and, like a generous host with his hand on your back, urged that young fire to dine at will, with infinite appetite, upon the most abundant and explosive carbon buffet on Earth.

Ignited innocently enough by sparks from a locomotive's wheels, the Chisholm Fire blew up with an explosiveness that is difficult to express in ordinary thermal terms. Under normal circumstances, head fire intensity—a fire's raw energy output—is measured in kilowatts per meter (kW/m) along the leading edge of the fire. (One kilowatt is equivalent to the energy produced by ten 100-watt bulbs, or a 1,000-watt space heater.) Barring excessive wind, a fire of 1,000 kilowatts (a million watts) per meter can be managed effectively by a ground crew, but once it jumps above 2,000 kW/m, even heavy machinery and water bombers may have trouble containing it. By the time it intensifies to 10,000 kW/m—10,000 space heaters' worth of energy per meter of fire—you have an out-of-control wildfire on your hands.

Keeping in mind that a head fire intensity of 10,000 kW/m represents an uncontainable fire, consider this: at its height, the Chisholm Fire generated 225,000 kilowatts of energy—per *meter*—across a front that was described as miles wide. If you're having trouble imagining a quarter of a million space heaters compressed into the length of a yardstick and then multiplied by several miles, you're not alone; six scientists from four countries who studied this fire had the same problem. We are beyond the normal scope of fire here; familiar formulas no longer apply. This is the kind of energy that does not burn but vaporizes, an energy more often associated with lasers, atom bombs, and suns. A professional firefighter and crew leader named Troy O'Connor was on the ground at Chisholm, and he described flames running thousands of feet up into the smoke column: "I told the guys, 'Stop looking at your boots and look up because this may be the only time you see this in your life.'"

Ken Yakimec, a pilot who flew the fire, searching through heavy smoke for its leading edge with infrared sensing equipment, told a colleague over the radio, "I have to be careful because the plane is going to crumple up like a butterfly." According to a final report published by the Alberta government, "The fire behaviour directly east of Chisholm was described as being 'a crowning wall of flame (head and flanks) for several miles.'" At its peak intensity, the fire incinerated

200 square miles of forest (128,000 acres) in seven hours, a burn rate fully three times that of the "Camp Fire" in Paradise, California, in 2018, which was notorious not only for its lethality but also for the shocking speed with which it spread.

In the midst of the Chisholm Fire's most explosive phase, a funnel cloud was observed, and later investigation revealed mass blowdowns of mature trees within the burn area. For fires of this magnitude, we need a different scale of measurement and, in the end, the six authors of a peer-reviewed paper entitled "The Chisholm Firestorm" resorted to megatons, the units of energy used to measure the explosive power of hydrogen bombs. The energy released during the fire's peak, seven-hour run was calculated to be that of seventeen one-megaton hydrogen bombs, or about four Hiroshima bombs per minute.

This, now, is what fire is capable of on Earth. The chemistry and physics of fire remain unchanged, the trees themselves are no different than they were fifty years ago, but the air is warmer and the soil is drier—enough to make the latent energy living and dying in these forests that much easier to release. Historians speak of Britain's Imperial Century, the American Century, and the Chinese Century, but those who study the symbiotic relationship between humans and combustion make a good argument for this one going down as the Century of Fire. Two decades in, the case only strengthens. Across North America, and around the world, fires are burning over longer seasons and with greater intensity than at any other time in human history. As conditions change, and the envelope of our experience is pushed, new language and reference points must be found, and so it is with the Chisholm Fire that subsequent fires are being compared. "Chisholm was the forerunner of worse things to come," Dennis Quintilio, a senior wildfire consultant with Alberta Forestry, explained to me in 2016. "The curves are all going one way."

Quintilio's sober observation has been borne out, not just in Alberta but across the globe. There have been more costly, deadly, and destructive fires, but for sheer wattage, the Chisholm Fire set a precedent for the obliterating energy brewing outside Fort McMurray.

About ten minutes into the presser that morning, right after Bernie Schmitte listed the fuel types present in the forest around Fort McMurray—mainly aspen, poplar, and black spruce—and after

Darby Allen had reassured people that the weekend fire burning inside city limits was "basically out," there came a fateful question from Laurent Pirot, a French reporter with Radio-Canada: "You said it's going to be a rough day. What's the worst scenario?"

The fire chief fielded this one. "Um—well, the worst—I don't really want to talk about worst-case scenario," Allen said, "but what we know is the fire conditions are extreme. The humidity levels are gonna be decreased quicker because the ambient temperature's hotter so that means the fire will be enabled to go more ferociously and quicker than in days previously . . . So, you know, the scenario is that the fire is able to get into areas where we can't stop it."

This was not presented as a hypothetical, but as a fact. For the first time, all present, and anyone listening on the radio, or watching over the Facebook live feeds, were granted a flickering glimpse into the immediate threat Fire 009 posed. But like the plan the mayor had recommended everyone have, which "areas" the fire would get into and what the worst-case scenario might be was left unsaid—not because it went without saying, but because it was unsayable.

And then the moment passed.

From different angles, the reporters continued to probe this fault line between the fire and the city, between what was known and what was being said. A question was asked about the spot fire that had appeared on the north side of the Athabasca River during the night. Allen described this growing fire as being about fifteen kilometers—nine miles—west of Thickwood. "We're confident," he said. "We'll hold that there." No one on either side of the mic mentioned the width of the river at that location, far wider than any man-made firebreak. John Knox, the program director for 93.3 FM, a local country music station, was present at this press conference, transmitting a live feed to his listeners via Facebook. As a longtime resident of the city deeply invested in its welfare, he sympathized with Allen's dilemma. Reflecting on the moment later, he said, "It was almost as if, by talking about the fire, we were giving it oxygen."

Words possess spell-casting, shock-inducing power, even in this jaded age, and the English language has accounted for this: something that is "infandous" is a thing too horrible to be named or uttered. For a mayor or a fire chief, a fire running rampant through the city they are charged with protecting is infandous. For a wildfire manager, it is a once-small fire not only escaping on his watch, but doing so into the largest population center for hundreds of miles around. This possibility wasn't discussed—not only because it was too awful to contemplate, but also because the notion of such a catastrophe was, for many, simply inconceivable. Chris Vandenbreekel's response was typical, even for those with up-to-the-minute information. The boyish, bespectacled twenty-five-year-old news director for Mix 103 FM, a local Top 40 station, had been following the weekend fires closely and was present at every press conference. Even though he had better information than most, he interpreted Darby Allen's comment about the fire getting into "areas where we can't stop it" as a reference to areas in the backcountry. "I don't think it was 'We didn't want to contemplate the worst,'" he told me. "We honestly didn't believe that it could happen because there had never been a full city that had been evacuated before—even national news reporters had never dealt with anything like that."

Actually, no one had. The last time a city had burned in Canada was Toronto, in 1904, but that fire, destructive as it was, had been limited to six blocks. In the United States, the last major urban fire was two years later, when much of San Francisco burned following the devastating earthquake of 1906. But these events were beyond the reach of living memory. The nearest modern comparable was the Oakland Firestorm, a wind-driven blaze that burned three thousand suburban homes and killed twenty-five people in central California in 1991. There were others, but in 2016 such events were still extreme anomalies. For most of the reporters present that morning, their only reference point was the fire in Slave Lake, a town whose entire population could fit into one of Fort McMurray's west side neighborhoods. Vandenbreekel tried one more time to nail down some specifics: "Should residents of Beacon Hill and Abasand [the two neighborhoods closest to Fire 009] be prepared for a possible evacuation order?"

"In general, people—wherever they are living within this wonderful town—should bear in mind that we're in a serious situation," Allen responded. "People need to go to work: Mum needs to take the kids to school, Dad takes the guys to the ballgame afterwards—but just be cognizant that this is a serious situation . . . be mindful that this situation can change in short notice."

It was evident by now that Fort McMurray's fire chief was operating under some considerable constraints, but what those constraints were, or who was imposing them, was not immediately clear. Reid Fiest, a seasoned reporter and news anchor with Global TV, was having a similar experience with the wildfire manager: "How much of the town do you feel is protected right now?"

"I don't want to get into that at this time because of the current conditions," Bernie Schmitte said, doing his best to stay on message. "Our number one focus is the protection of Fort McMurray. Our values are human life and communities."

Fiest pressed him: "How far have you come in getting those firebreaks in place . . . Are you going to be able to contain it?"

It was Schmitte's turn to offer a glimpse behind the veil: "We cannot say that there is any part or portion of the line that is 100 percent secure or contained."

For anyone listening that morning, the situation was now clear: McMurray District Wild Fire 009, which city officials referred to as the "West Fire," which Alberta Forestry named the "Horse River Fire," which was on its way to becoming the "Fort McMurray Fire"— was uncontained, out of control, and perilously close to town. Jamie and Ryan Coutts understood this. Now, just an hour away from Fort McMurray and their sprinkler demo at Fire Hall 5, they had been monitoring Forestry's radio channel all morning. Overnight, a thermal inversion had trapped a layer of cooler, smoke-suppressing nighttime air beneath a "lid" of warmer air, creating the illusion that the fire had died down or even gone out. Inversions like this are a common feature of cooler nights during high-pressure systems; they usually dissipate as the land and lower airs heat up in morning sun. Noon was when the inversion was predicted to lift, and as the cooler, low-level air rose, it would manifest as wind. But on May 3, the weather was in a hurry, and the inversion lifted more than an hour earlier. That morning, winds were light, less than ten miles per hour, and blowing out of the southeast, away from Fort McMurray, but the forecast was accurate: simultaneous with the lifting of the inversion, winds would be swinging southwest, "which," Schmitte said, "is when we can expect to see some full crown fire."

It would be the wind—that ignescent rush of volatile oxygen—that determined the focus and fury of what was to come, and, even as Blake, Allen, and Schmitte were concluding their press conference, it was veering. Over the lunch hour, it would swing to the opposite side of the compass. Meanwhile, the temperature was rising steadily, and the relative humidity was not dropping but plummeting.

When you consider the vast quantities of water flowing and filtering through the boreal forest, and the fact that Fort McMurray is located at the confluence of not two but four rivers, in close proximity to numerous lakes, creeks, and muskeg bogs, it seems counterintuitive that the air could be as dry as Nevada's, but this is the reason Alberta's sky is so famously clear: because the air has so little moisture in it. It has so little moisture, not because of where that air is, but because of where it comes from. This parched air, a feature of the region's high-pressure systems, flows in from the north. "North," in Alberta's case, is a polar desert known as the Northern Arctic Ecozone. Comprising most of the Canadian Arctic, all the way to the North Pole, it is one of the driest places on Earth. The injection of desert-dry Arctic air, combined with the day's unprecedented temperatures, created ideal conditions for the rapid and sustained evaporation of all surface moisture. That morning, the relative humidity had dropped from 30 percent to less than 20 percent in under an hour, and it had yet to bottom out.

As the three officials relayed their information in alternately grave, earnest, and upbeat tones, the forest they spoke of the way islanders speak of the sea, that provided the backdrop for their extraordinary simulacrum of suburbia, and that gave way so meekly to every new subdivision, mining project, and drilling pad, was becoming another thing entirely. That spring, many of the local bogs were bone dry, and those damp and docile trees were firebombs. The wind, a cooling breeze on any other day, was now a flamethrower.

8

The greatest shortcoming of the human race is our
inability to understand the exponential function.

—Albert Allen Bartlett, physicist

Even though its purpose was to alert and inform, the 11:00 a.m. press
conference seemed to have the opposite effect; a crucial sense of
urgency was missing. Reid Fiest, the journalist who had pressed Bernie Schmitte on the fire's containment, was an early casualty of this
muted sense of alarm. Fiest (pronounced *feist*) had flown up from
Calgary with his cameraman on May 2 in response to the partial
evacuations announced on May 1. "We saw stories like Slave Lake,"
he said, "and I remember thinking how quickly that evolved, so I
thought we should be up there in case something bad happened."

Fiest's instincts were good, but as soon as he heard evacuation
orders being rescinded, he second-guessed his decision to come, and
his doubts persisted. "When I left that news conference, I wasn't clear
what could happen that afternoon. I didn't feel the threat." When
Fiest discussed it with colleagues at the head office in Vancouver,
they agreed it was a "non-story." "At that point [around noon]," he
said, "my desk didn't feel it would meet the threshold for a national
story."

While the Fire Weather Index for May 3 may have been a first, the possibility of a high-intensity wildfire confronting Fort McMurray was not. It is, after all, in the nature of large boreal fires to sweep across landscapes in response to wind, heat, and low humidity. Like the direct hits by Hurricanes Katrina, Sandy, Maria, and Ian, it was not a matter of *if* a major fire would arrive at the gates of Fort McMurray, but *when*. In 1995, the Mariana Lake Fire burned 500 square miles of timber south of town, closing Highway 63 and coming close enough to prompt the construction of extensive firebreaks. In 2009, the highway was closed by fire again. In 2011, the Richardson Fire, an epic blaze, burned 2,600 square miles of forest north of Fort McMurray, forcing the shutdown of two major bitumen plants, and resulting in mass evacuations, damaged infrastructure, and claimed losses of more than half a billion dollars. Sandwiched between these swaths of charred timber and muskeg lay the rapidly growing city of Fort McMurray and its satellite communities, surrounded by hundreds of square miles of mature forest that hadn't seen a major fire in eighty years.

Natural disasters may be hard to schedule, but a fire in the boreal is as certain as death: any given tree in the boreal forest can expect to burn once in a century, give or take fifty years. Fire is the principal mechanism by which the boreal forest purges and regenerates itself, to the point that the cones of several keystone conifer species, including black spruce, will not drop their seeds unless they are heated to temperatures unachievable by sunlight alone. Not only do these blasts of intense heat open the cones, releasing the seeds inside, they also indicate that fire has cleared the ground below and opened the canopy above, thus improving the odds of those seeds' successful germination. Without fire and its seemingly random but ultimately regular patterns of return, the boreal forest would collapse. There is in this cycle a kind of codependency that, when viewed from the point of view of fire, upends the notion of what a forest is, and whom it serves. As the naturalist and author David Pitt-Brooke wrote of a similar forest farther south, "Fire-spawned stands of lodgepole pine are, in a sense, locked into a fire cycle. They are creations of fire and, in turn, they create conditions hospitable to future fire. You could almost think of it as a symbiosis . . . a form of farming: fire creates these stands of lodgepole pine so it can eat them later."

But humans and their settlements live and grow by different means and rhythms than forests and fires. The decadal and centennial cycles shaping the boreal forest are rarely considered by city planners or elected officials, not only because their terms (not to mention their memories and lives) are too short, but also because their knowledge of this colonized landscape is incomplete. Virtually everyone who came here to settle this country arrived from far away, often from very different environments. To this day, most newcomers are focused less on the landscape than on what they can take from it. In northern Alberta, the majority of these colonists don't stay long, remaining for only a fraction of a spruce tree's life-span. Most are laid off, if they don't burn out first. There are exceptions, but even longtime residents have an exit plan, and being buried in Fort McMurray is not one of them. This explains why the mortality rate is significantly lower here than the rest of Alberta (though cancer rates are higher). Retirees who don't return to their hometowns take their winnings to warmer places like Florida, Arizona, Southern California, British Columbia's Okanagan Valley, or the coastal islands. Some older Suncor executives followed J. Howard Pew back to Pennsylvania.

In Fort McMurray, as in so many resource towns that live and die by global market prices, ten years can be a career. Most booms and busts play out in half that time, and Fort Mac has weathered a few. In part due to diligent fire suppression and prevention, and in part due to chance, Fort McMurray had managed to subvert the laws of Nature for decades. Whatever fires it hadn't been lucky enough to dodge it had managed to deflect. But 2016 represented a kind of synchronous convergence. In 2015, itself a record-breaking year for fire, an unseasonably warm and early spring had been amplified by an exceptionally strong El Niño, which lingered on through the year, raising winter temperatures and reducing the snowpack in northern Alberta by more than half. So dry was the forest that in December and January, when the boreal forest is normally bound up tight in snow and sub-zero temperatures, five human-caused wildfires were reported.

Even without the added liability of a South Pacific weather system, May is still the cruelest month in Alberta. Over the past twenty years, nearly half of the total area burned by wildfires here has ignited during the month of May, and for a specific reason. In most of the circumboreal, this is when winter snow cover finally gives way to spring's lengthening days and the resurgent flush of leaves and grasses. During this transition, however, there is a critical moment when, with no snow left to cover it, and no foliage yet to shade it, the normally damp, dark forest floor is exposed to the novelty of direct sunlight. This period of a week or two, which occurs before the trees' roots have fully thawed and their branches have budded, has a name—the "spring dip." During this brief window of time, between the river's break-up and the forest's green-up, trees—coniferous and deciduous alike—are exceptionally vulnerable to fire. Under these conditions, leaves, cones, and deadfall take on the characteristics of kindling, and last summer's grass will burn like newspaper. In 2016, the spring dip lined up perfectly with Fire 009.

There are formulas and computer programs that can accurately process all these variables, and thereby predict a particular fire's intensity and rate of spread. One of these programs is called Prometheus, and it was being applied to this fire. But as accurate as it was, it would have no bearing on the outcome in human terms. All the choices made that day were based on sophisticated weather data and live observations filtered through the minds and actions of human beings. Significant here is the fact that programs like Prometheus are designed for use in forests—with trees. No one truly knew what formidable synergy all of these accumulated influences might unleash in a residential setting. It raised a dreadful question: What happens when all the numbers come up at once? What might be released? That morning, it was as if Fort McMurray, with its eighty-year jackpot rollover of unburned trees, and its half-century bonanza of vinyl-sided, tar-shingled plywood houses, had won the lotto for flammability—all six numbers in exact order, plus the El Niño bonus and the spring dip mega ball. The payout for a long shot like that would be in the billions.

The manner in which Mix 103's Chris Vandenbreekel was awakened to the risk this fire posed was probably unique among the citizenry of Fort McMurray. In fact, the next hour in the working life of this young journalist would unfold much like Orson Welles's radio play *War of the Worlds*. Immediately following the press conference at city hall, Vandenbreekel persuaded the wildfire manager, Bernie Schmitte, to join him on his noon-hour radio show, *Fort McMurray Matters*. It was a short walk across the street to Mix 103's offices, which were located in a low storefront on the east side of Franklin Avenue at the corner of Hardin Street. Franklin runs on a north-south axis through the heart of downtown, following the floodplain of the Clearwater River, which joins the Athabasca less than a mile downstream (north), just beyond the sprawling rec center at MacDonald Island.

The downtown core, known as Lower Townsite, was originally laid out to serve barge traffic, fish and fur traders, and a few hundred hardy residents along this still remote stretch of river. Today, Lower Townsite is about a mile long and five blocks wide. There is more to it than meets the eye: together with its government buildings, churches, pocket malls, restaurants, big-box stores, strip club, and mosque, it supports a halal butcher and a continental market that carries everything from fezzes and Korans to fufu flour, manioc, and three-legged cast iron potjies from South Africa. Just up Franklin from Mix 103 is Chow's Varieties, a general store with a stock ranging from candy and hockey memorabilia to cigar humidors and fishing gear. It also carries hundreds of magazines, making it one of the most comprehensive sources of print journalism and comics in the country.* Flanking Franklin on both sides are residential streets occupied by some of the bitumen industry's first permanent employee housing. Just as many cities have Irish, Polish, or Chinese neighborhoods, Fort McMurray's first communities were affiliated with particular companies, principally Suncor and Syncrude. Throughout all the city's changes, Franklin Avenue, named for the doomed explorer Sir John Franklin, has remained the spine of downtown. To this day, both ends are dead ends.

Vandenbreekel's studio window at Mix 103 was plainly visible to passersby, and it offered anyone inside a clear view to the west, across Franklin and down Hardin Street, past the Boomtown Casino, the

---

* Until 2012, when they scaled back, Chow's carried three thousand titles.

Peter Pond Mall, the post office, and Tim Hortons—all the way to the mud-spattered chrome and muted roar of truck traffic that hurtles up and down Highway 63 twenty-four hours a day. On the far side of the highway, an easy potato gunshot from the radio station, is Abasand Hill, a steep, forested ridge that runs parallel to the highway for about a mile before its north end drops away precipitously into the Athabasca River as it loops in from the southwest. There, along Abasand's steep western face, bands of bituminous sand are visible; in the unseasonable heat of the past several days, they had been weeping raw bitumen that glistened like liquid obsidian.

This tapering convergence of Abasand Hill, the Athabasca River, and the low floodplain of downtown appears, from above, to be stapled together by three parallel highway bridges comprising ten lanes and built to bear the million-pound loads of mining and upgrading equipment required by the bitumen industry. This is the only way to reach the strip mines, tailings ponds, and upgrading plants north of town. It also offers the only road access to the densely packed neighborhoods of the west side, where two-thirds of Fort McMurray's population now lives. The south end of Abasand Hill is marked by another steep descent, this time into the Hangingstone River, a narrow tributary of the Clearwater. Rising up from the far bank of the Hangingstone, due south of Abasand, is Beacon Hill, another ridgetop neighborhood. Up on these breezy, park-like plateaus, thousands of bitumen workers have raised their children, and even been raised themselves, in a subarctic approximation of the suburban ideal. From these tree-lined heights, residents had, through a screen of leafless spring forest, a disconcertingly good view of the fire.

Downtown on Franklin, through Mix 103's storefront window, Schmitte and Vandenbreekel had an equally good view of Beacon Hill and Abasand, and the smoke now rising up directly behind them. After a brief introduction, Vandenbreekel jumped right in: "It's been a tough slog for you guys over the past few days with all these wildfires around," he began. "Of course, this massive one"—he glanced out the window—"we can see the smoke starting to pop up again here."

By now, the smoke-suppressing inversion had lifted, and the inverse gap between temperature and relative humidity was widening by the minute. Thus liberated and empowered, the fire had begun to burn in earnest, preparing to pick up where it had left off late the previous evening. As the smoke plume billowed higher in the shifting airs over Abasand and Beacon Hill, and Schmitte elaborated on information he had covered in the press conference, there was, in his lucid, thorough, and media-friendly delivery, a sense of events being experienced in the third person. At seven minutes past noon, nestled inside Schmitte's aura of dispassionate calm, the worst scenario Vandenbreekel could envision was "Maybe it'll get close enough to Centennial Park that it'll burn a trailer or two."

This was about the time Jamie and Ryan Coutts rolled into Fire Hall 5 with the sprinkler trailer. Hall 5 is located off the highway on Airport Road, five miles south of downtown. It is positioned on a low hilltop that offers an excellent view in all directions. When Jamie looked to the west, his worst fears were confirmed. At the 11:00 a.m. conference, Schmitte had described the fire's new forward operating base on the north side of the Athabasca River as being about twelve acres in size, and he had explained that ten firefighters were en route. But that was old news. While it might have been true at ten or eleven, it was now past noon, a long time in the life of this kind of fire, one that was currently in the midst of a pivotal transition plainly visible through the radio station window. At about 12:15, Vandenbreekel said, "Looking out here in front of the Mix office at Franklin and Hardin, we can see that column really getting up today, and this is what you predicted at the press conference—that, just past noon, this fire was going to explode up again and crown. When the smoke goes up like that, is it a signal to the city that there is more danger?"

Schmitte demurred—he was a wildfire manager, not a safety officer. "As the intensity starts to increase," he said, "then you start to see a change in the column. A white column means that the fire intensity is not real high; a black column means that there is extreme fire behavior."

"So, what we're seeing right now," Vandenbreekel said, "is a little bit of a mix—kind of a brown with a little bit of white on the edges. So, what would you say the intensity of this is right now?"

"We're getting up into what we would call a Rank 4 fire—intense surface fire," Schmitte said. "You're also going to see some full-tree candling and some intermittent torching. Once the column turns black, then it's a full crown fire—the entire tree is being consumed by flames."

"Within a few minutes of him saying that," Vandenbreekel told me, "the smoke is turning blacker and blacker. I can see Bernie glancing at it, and he's starting to sweat."

~~~~

What Schmitte and Vandenbreekel were witnessing out the window—live, on the radio—was a moment similar to the one in Stanley Kubrick's film *2001: A Space Odyssey* when Hal, the spaceship's AI computer, becomes self-aware and takes over the ship. In firefighting parlance, this moment is known as "crossover." For a wildfire, and everything in its path, it represents a point of no return.

Fire, like virtually every living thing, is solar-powered: like us, it wakes and sleeps on a diurnal cycle, and, like the plants it feeds on, it gains and loses energy in direct relation to the sun's presence and strength. Because of this, there is, in the daily life of a forest fire, an arc it traces—a kind of circadian rhythm that takes its cues from the sun, the mother of all fires. Wildfires tend to lie low at night and wake up in the morning, lifting their heads as the sun rises and opening up, flower-like, as the dew evaporates, the humidity drops, and the temperature rises. The warming air and earth in turn generate breezes, particularly in hilly terrain like that along the Athabasca River, where the warmer, sunlit hilltops draw cooler air upward out of the shaded river valleys. As the air heats "up," something must fill the void below, and this process of atmospheric in-filling is what we call wind. The temptation to think of wind as being blown or pushed across a landscape is a holdover from ancient times: Aristotle thought this, too. But wind is *drawn* across a landscape, and it is inhaled by fire.

When a wildfire is burning in optimal fire weather, there is a moment when what once took effort becomes almost effortless—when, like Hal the AI computer, the fire says, in effect, "Thanks, I'll take it from here." Instead of burning in spite of its environment—wet wood, cool temperatures, the dead, smothering air of an inversion—the environment becomes an ally attending to the fire's every need: high temperatures, low humidity, dry fuel, and wind. Crossover is truly a crux in the day for fire and firefighter alike, and it is predictable almost to the minute. Technically speaking, crossover occurs when the ambient temperature in degrees Celsius exceeds the relative humidity as a percentage. A common example would be when the air temperature rises above 26°C—roughly 80°F—and the relative humidity drops below 26 percent, which is already kindling dry. Typically, the winds will pick up under these conditions, at first due to the uneven heating of the landscape and later due to the increased energy generated by the fire itself. The effect on a fire is dramatic—roughly analogous to a motorboat attaining sufficient speed to transition from pushing through the water to skimming over it.

Once crossover is achieved, the fire is set free, able to move and grow exponentially faster and with far more agility. As the gap between temperature and humidity widens and the attendant winds build, it's like throttling up on a motorboat, only on a geographic scale. Crossover is what enables a fire to escape its natal valley, prairie, or forest block and take over a landscape, along with everything on it, in one hot, dry, windy afternoon. It can happen with startling speed, and once it does, the fire, which just a few hours earlier may have been smoldering in place, or plodding through leaves and underbrush at a slow walk, becomes a juggernaut, sometimes hundreds of feet tall, and moving like the wind. With so much less work to do in the form of drying and preheating fuel, and aided by the oxygen rush of a rising wind, the fire can now devote its full combustive energy to consumption and growth. If unregulated free market capitalism were a chemical reaction, it would be a wildfire in crossover conditions. Alberta's bitumen industry follows a similar growth pattern, with market forces standing in for weather.

The moment of crossover is easily identified by a fire's ascent from the forest floor via so-called ladder fuels of bushes and low branches into the trees themselves. Once in the treetops, closer to the wind, the flames appear to leap improvisationally, almost lemur-like, from fuel to fuel, higher and higher into a forest's architecture, until they are bounding through the treetops. This is a crown fire and, just as a troop of lemurs or gibbons line up their next jumps on the fly, the crown fire's heat runs ahead of it in an anticipatory way, drawn forward by the wind, preheating unburned trees and releasing clouds of combustible gases in preparation for the next explosive ignition. In the case of a big boreal fire, thousands of these preheated ignitions will be occurring simultaneously across a front that may be several miles wide. As with the Chisholm Fire in 2001, the energy releasing at lunchtime on May 3 was equivalent to a nuclear explosion.

The effects are spectacular. Under these circumstances, what Bernie Schmitte described as "full-tree candling" will turn trees, almost instantly, into hundred-foot pillars of flame. With sufficient wind and heat this can happen to an entire line of trees simultaneously. Meanwhile, high above, wind, heat, and combustible gases combine in a terrible synergy that enables the fire to become airborne, not merely in the form of flying needles and embers, but as actual fireballs and spontaneous explosions that those in the business of wildfire call "dragons." These Godzilla-sized and -shaped eruptions of combusting gas bursting from the crowns of superheated conifer trees can be three hundred feet high and are hot enough to reignite the smoke, soot, and embers above them, driving flames hundreds, even thousands, of feet higher into the smoke column.

The more extreme the crossover differential between high temperature and low humidity, the more rapidly the preheating and ignition are able to happen, and this can create a feedback loop—a spiral, really—of ever-increasing heat and wind. Once a crown fire like this is fully under way, it is unstoppable. As one civilian witness to the fire's crossover on May 3 said, "There is no way a man is going to put this out." The best one can hope for is to head it off with catguards, by bombing downwind fuels with water and retardant, or by lighting backfires in order to consume the future fuel supply. But, as the firefighters in Slave Lake learned so painfully in 2011, you can do all these things, but you can't stop the flying embers. As light and mobile as a spore—or a virus—each spark has everything it needs to start an entirely new fire hundreds, even thousands, of yards downwind. In crossover conditions like those on May 3, each one of these fires would be fully capable of becoming as large and vigorous as the parent fire that disseminated it.

This kind of high-intensity crown fire, the same kind that had so impressed the residents of Beacon Hill, Abasand, and Thickwood the previous evening, is classified as Rank 6. This is the highest ranking for a wildfire, and, while the fact that Rank 6 crown fires can jump major rivers had been addressed by Schmitte, it was not done in a way that indicated what an ominous development this was. At about twenty minutes past noon, after a brief recap of firefighting personnel, Chris Vandenbreekel asked Schmitte which part of the fire posed the most immediate risk to the community: the eastern flank, which was less than a mile from the highway, the Centennial Campground, and the south end of town; the northern flank, which was pushing up against the Athabasca River, the backside of Beacon Hill, and Abasand; or the spot fire on the north side of the river, which threatened the west side neighborhoods. "It would be the northern flank," Schmitte answered. "It's quite a distance from Fort McMurray, but that portion of the fire hasn't had any tanker activity on there. So, the plan today is to use the air tankers that we have and try to cool down that northern flank, and get the catguard in as well." As for the spot fire threatening the golf course, Schmitte believed that his men had it well in hand. "We're pretty confident," he said. "We're going to put everything we've got into containing that fire today."

Just a few minutes later, at 12:23, Chief Allen issued the following message on Twitter: "Wherever you live in #ymm* bear in mind that we're in a serious situation. Get your kits ready." It is unclear how many people actually saw this; it was retweeted only thirteen times.

That the Rank 4 fire Schmitte and Vandenbreekel were observing out the radio station window would shortly attain Rank 6—analogous to a Category 5 hurricane—was not a possibility but a certainty was a fact well understood by Schmitte and anyone familiar with fire weather in the boreal forest. In fact, it was happening at that very moment, literally, before their eyes. And yet, at twenty-five minutes past noon, Vandenbreekel was still trying to get a fix on the risk this fire posed. "What should Fort McMurray–ites be doing," he asked, "to prepare for eventualities with this forest fire?"

"I'll just reiterate what Chief Allen stated at the press conference," Schmitte said. "People should carry on their normal day, and also be prepared."

This advice, paraphrasing the British wartime slogan "Keep calm

---

* YMM is the Canadian airport code for Fort McMurray.

and carry on," taps deeply into Canadian national virtues, which favor "Peace, Order, and Good Government" over "Life, Liberty, and the Pursuit of Happiness." As reassuring as these words were, by lunchtime on May 3 they had become a hollow prayer, and perhaps something far more dangerous.

With the humidity dropping and the temperature rising, a fire that had already been burning out of control for two days and nights was accelerating toward the fourth-largest city in Alberta, and it was doing so in record-setting conditions whose specifics had been detailed and refined over months, days, and now hours. But unless you were a wildland firefighter, a senior civic leader, or one of several hundred temporary evacuees from smaller weekend fires, you would have had only a passing awareness. That morning, the other 88,000 inhabitants of greater Fort McMurray were living in a parallel reality, as oblivious to their environment as the passengers on the *Titanic* were to theirs. Just as icebergs were clearly visible from the deck of that doomed ship, smoke plumes billowed over the green sea surrounding Fort McMurray. For those who took note of them, they were an abstraction—a feature rather than a factor, worthy of an Instagram post but nothing more.

Scarcely an hour after he had asked it, Chris Vandenbreekel's calm but earnest question about whether or not residents in Beacon Hill and Abasand should prepare for evacuation was already moot. Holding things together at this point, all across town, was the inertia of habit, decorum, and a disciplined workforce. And why not? The skies above were still clear, and most people were indoors anyway—at school, or work, or shopping, or else up at Site, thirty miles away. Even though it was understood that this fire was close by, that today would be hotter and drier than yesterday, that the forecast called for the winds to shift toward town, the citizens of Fort McMurray were going about their day as instructed. Even if Schmitte was sweating— and he had good reason to be—he sounded calm enough on the air, and that was all listeners had to go on, if indeed anyone was listening at all.

Meanwhile, the winds were making their predicted swing to the southwest, a development that would, as Schmitte told Vandenbreekel, "challenge that perimeter" along the northern flank. But there was nothing in his tone or inflection to indicate that the fire was going to challenge that perimeter exactly the same way it had "challenged" the Athabasca River the night before. Even Vandenbreekel, who had been tracking both the behavior of the fire and of Schmitte, took everything he'd heard at face value. His impression after the wildfire manager hustled out of the studio just before 12:30 was that Schmitte "truly believed they were going to be able to battle it back."

Still feeling the symptoms of spring fever, Shandra Linder knocked off work at lunch and organized herself for a two o'clock meeting at Syncrude's downtown office in the Borealis Building. With no traffic, it's a thirty-minute drive back down the highway. Linder was meeting with Syncrude's head of emergency response, not to discuss the fires currently burning around the town, but, rather, the more abstract danger of social unrest. Their longtime competitor, Suncor, was having serious union troubles, and there were ongoing threats posed by First Nations and environmental groups who, for years, have protested the social and environmental impacts of the industry. Should one of these entities decide to block Highway 63, it could threaten the safety and productivity of the mines and refineries to the north, all of which operate twenty-four hours a day, 365 days a year, and require massive movements of personnel at every shift change.

It was around half past one when Linder descended Supertest Hill into the river valley. She had planned to do errands after the meeting; in the back of her car were reusable grocery bags (it was "Discount Tuesday") and her dry cleaning. Still fifteen miles north of downtown, she was rounding a long, gradual bend in the river when her windshield filled with something she did not recognize. "When I got that first view of it," Linder told me, "I went, 'Hang on a second.' It was clear overhead, but to the south it was black—black smoke with streaks of red in it. It took up the whole sky. I'm trying to make sense of this thing; I'm thinking maybe it just looks closer than it is; I'm thinking maybe the red is the sun shining through smoke. But it's not the sun; it's in the wrong place. And then I realize it's flame."

Linder had lived up here with her husband for two decades;

together, they bought a house on a greenway and built a life. They had probably seen a hundred fire plumes, but this didn't look like any of them. "So, I pull over to the side of the highway, and I call my guy, Byron, and I say, 'Are we having this meeting?' And he's like, 'No. We are not having this meeting. It's not happening, don't come in.'

"And I'm like, 'I'm looking at a wall of flames here.'"

It was a short call, in part because no one was quite sure what to say next. What do you say on a beautiful spring day when the city where you live and conduct business appears to be on fire? After hanging up, Linder pulled back onto 63 and continued on to the Confederation exit, where she turned west, back up the hill and out of the river valley. "Okay," she said, "not right, but here's what I do— I'm going to go home anyway, I drive past the dry cleaners—well, I'm gonna drop off my stuff at Sam's place 'cause in our minds the fire's still on the other side of the river. So, I'm dropping my stuff off and while I'm standing there, a guy comes in from the golf course in full golf apparel, and he's like, 'Holy shit! We just got evacuated from the golf course!' Now he's at the dry cleaner and I'm like, 'What are you doing here?' And he says, 'Well, I'm here to pick up my dry cleaning.' I said, 'What golf course were you on?'—because we only have two in town. He says, 'Thickwood,' meaning the fire has just jumped the river.

"So," Linder continued, "Sam the dry cleaner, who's our buddy, is like, 'Whoah!' And he's on the phone to his wife: 'Honey, it just jumped the river—get out, get out, get out!' 'Cause they live in Wood Buffalo [neighborhood], which is *right next* to the river. In all this craziness, Sam still takes my stuff, logs it, and he's like, 'Tuesday good?' I'm like, 'Yeah, next Tuesday's great.'"

At two o'clock in the afternoon on May 3, Linder tucked her laundry ticket into her wallet and aimed her Porsche for home.

Disaster is, almost by definition, a kind of existential dissonance. For the individual, it is cognitive dissonance made manifest: a disruption to one's personal and physical world order so profound that you don't know where to file it, how to measure it, or even how to react—because you have no precedent, because it's simply too big and violating to grasp. Shandra Linder's job is demanding and complex; the stakes are high and she functions well in a crisis, but even though she had just visually and verbally acknowledged the presence of a catastrophic fire, and even confirmed it with others, she didn't fully register what it meant, or how it might impact her. Because the fires she knew—that most people in Fort McMurray knew—were brown or gray, and always in the distance. This fire was none of those things; it was too big, too black, too red, too close.

As the minutes ticked by, the advisories being transmitted over local radio and Facebook Live grew increasingly out of phase—not only with the wildfire they were purporting to describe, but with the experience of those actually trying to fight it. Nearly a century ago, Hermann Hesse anticipated this modern dilemma in his existential novel *Steppenwolf.* At the end of the book, the protagonist encounters his idol, Mozart, who appears to him playing a primitive radio. "When you listen to wireless," explains the long-dead composer, "you are a witness to the everlasting war between idea and appearance, between time and eternity, between the human and the divine." In their attempts to use mass media to manage responses to an unmanageable—one could say "divine"—entity, Fort McMurray's leadership fell victim to this "everlasting war," and so did the quality of their information. Because radio, even in the hands of a live broadcaster, not only lacks the dynamic nuance of the event it is trying to describe, it is always a beat behind. As a mathematician named Aubrey Clayton wrote, "The problem with exponential growth is that it means most of the change is always in the recent past." For this reason, any kind of update—any kind of news at all—is, by its nature, a kind of incomplete history: by the time one has gathered, organized, and relayed it, the world has moved on. But in most cases, the world doesn't move as quickly as it did in Fort McMurray on May 3. On that day, the inertia of the present was overmatched by the impatience of the future.

9

Macbeth shall never vanquished be until
Great Birnam Wood to high Dunsinane Hill
Shall come against him.

—*Macbeth,* act IV, scene 1

With the wind clocking steadily toward the southwest, and cross-over well under way, Fire 009 was in the process of introducing the residents of Fort McMurray to the new reality of twenty-first-century fire. It is one thing for the temperature on a given day to nudge into record territory, it is quite another to break the standing record by nearly 10°F. Meanwhile, the relative humidity was dropping into the low teens. The hottest hours of the day were still to come, and there would be many of them: sunset wasn't until after 9:00 p.m. The disconnect between the reality of this fire—its objective potential based on the physics and chemistry in play that day—and the leadership's estimation of it had less to do with information or attitude than with vision. Following the terrorist attacks of September 11, 2001, the U.S. Senate's *9/11 Commission Report* declared, "The most important failure was one of imagination." This conclusion could be applied just as easily to the financial crash of 2008, the election of Donald Trump, Russia's invasion of Ukraine, or Odysseus's Trojan Horse trick—any number of accidents, catastrophes, victories, and defeats.

Depending on conditions, wildfires may move as slowly as hurricanes, but they can also appear or accelerate with explosive suddenness—arriving in a community at flash-flood speed. The difference is that, while floods generally follow rivers, and hurricanes follow weather patterns in specific parts of the ocean, fire can occur anywhere there's fuel, and fire's menu is astonishingly broad, including virtually everything under the sun save dirt, rock, metal, and water. Furthermore, the paths fire takes are determined almost exclusively by the wind, which can blow in any direction. In this sense, fire is the most versatile and whimsical of disasters. Able to self-generate from a single spark, explode like a bomb, turn on a dime, and fly over obstacles, fire possesses an unparalleled capacity for random movement and rapid growth—qualities that make it particularly difficult to anticipate and respond to at the municipal level. Because of this, and perhaps because of their successful young city's innate sense of invincibility, the citizenry of Fort McMurray discovered that their city was burning mostly by personal observation and word of mouth.

~~~~

While Bernie Schmitte was on the air acting as an impromptu information officer, he could not be in regular contact with his agency's shortwave radio channel. Jamie and Ryan Coutts, however, had been monitoring it all morning on their four-hour drive up from Slave Lake, and the scenario they heard unfolding wasn't jibing with the information coming out of Fort McMurray. "We're streaming the press conference in the truck," Jamie Coutts recalled. "We listened to Darby Allen, and he's like, 'If you're going to school, go to school. If you're going to baseball, go to baseball. Go about your normal business and we'll let you know.' Then, about an hour from the city, we're like, 'That guy's fucked, man.' We had the Forestry channel on and we could hear them call it 'past resources.' So we're driving in there completely in the know, because we have the right radio channel, we have the right knowledge, we know how this works. Those guys are sitting up in the REOC [at Fire Hall 5], asking us if we want a fucking tour."

Jamie had earned the right to say this because, as a fellow fire chief—one who had seen his own town burn—he had paid his dues. He understood, perhaps better than anyone else in the country, the bind Allen was in, and the price he and his city stood to pay. The most extreme designation for a wildfire is "past resources." Also referred to as "beyond resources," the term means that not only is the fire out of control, it is actively unsafe to be in its vicinity. "It's pretty much the worst thing you can hear," Ryan Coutts told me. "It means there's nothing Forestry can do."

Under these circumstances, the only appropriate action is to withdraw and wait for a change in the fire's behavior, which usually means a change in the weather, or nightfall—a time when bombers don't fly and ground crews rarely work. "Beyond resources" is not an uncommon call to make on an aggressive wildfire in a remote area, but it has dire implications when there's a city in the way.

Mark Stephenson, a firefighter and acting captain with the Fort McMurray Fire Department, saw it coming, but like so many on the municipal side, he didn't fully grasp the implications—either for his city or for his own home. Stephenson lives up in Abasand, so he'd known this fire for days. A dedicated father, husband, churchgoer, and former serviceman, he hunts and traps, and back then, when he had the time, he looked after a vintage muscle car waiting out the winter in his garage. That spring, most of his free time had been taken up by home renovations. On Saturday, April 30, the day before Fire 009 was discovered, Stephenson recalled standing on his back porch with his five-year-old son, watching the bombers angling in for their drops on another nearby fire—so close that they could see the pilot's head in the cockpit. A lot of wild things come close in Fort McMurray: Stephenson has seen wolves out his kitchen window. Downtown, in the springtime, you can stand in the Walmart parking lot and watch bear cubs in the poplar trees across the Clearwater River. They look like small, black-furred men as they climb unnervingly high into the treetops while their mothers pace back and forth below.

Stephenson—bearded, sized and sculpted like a gladiator—seems built for life on another scale: too big for his chair, too big for the coffee in his hand, too big for his extra-large T-shirt. Even in quiet conversation, he exudes a kinetic potential that seems ideally suited to the unreasonable demands life can sometimes impose on a person. This, after all, is someone who, finding himself with no keys and needing to get into his locked garage, kicked the door down—not the side door, the metal front door. At 12:30 p.m. on May 3, this was still in the future, but only just.

Stephenson happened to be at the Slave Lake crew's sprinkler demo, which was held at Fire Hall 5, where the REOC had been set up. You could see why they wanted to offer a tour: Hall 5 is a brand-new, state-of-the-art, LEED-certified facility so swank it could pass for a modern art museum. Attending the demo along with Stephenson and a training supervisor named Dave Tovey were about twenty-five firefighters, many of whom were new recruits, or "probies," firefighters still on probation. In truth, most of a municipal firefighter's career is spent training, getting ready, and being ready. There is little actual firefighting. As one frustrated firefighter said of working at the airport, "It's just a hell of a lot of polishing." In the fire hall, you might hear this joke about how firefighters occupy their considerable downtime: "Eat till you're tired; sleep till you're hungry." But municipal firefighters serve the city in many other ways; in Fort McMurray, they attend a disproportionate number of motor vehicle accidents.

While Forestry's wildfire crews can now be fighting fires in March or earlier, and begging for mercy by August when there's still a good two months of fire season to go, the Fort McMurray Fire Department might attend three or four structure fires a *year*—across the entire city. In light of this, one could be forgiven for wondering if, somewhere back in bookkeeping, they got the salaries reversed. Prior to May 3, there were city firefighters with ten years in who had never been on duty for an active house fire. Perhaps it's no wonder that they could behold that black and angry roil churning the air above the treetops—just beyond the highway now—and have no idea what lay in store for them. Perhaps this is why, even when it was clear to Jamie and Ryan Coutts that Centennial Park and Beacon Hill were in imminent danger of being overrun by fire, officials at Hall 5 were trying to send the Slave Lake crew—battle-hardened boreal warriors with proven skills and methods—back home.

The sprinkler demo didn't last long. At 12:30, just as Bernie Schmitte was leaving Mix 103's downtown office, Acting Captain Stephenson was looking westward from the Hall 5 parking lot and seeing the same sight that had shocked Shandra Linder, and that had confirmed the Couttses' worst fears. "We were looking down Airport Road towards the dump and you could see the fire walking towards the city," Stephenson recalled. "It was like [the 1998 film] *Armageddon*—you know, when you see the asteroid come in and, all of the sudden, there's a new fire plume—every couple minutes there's a new plume—getting closer and closer to the city."

Stephenson found this alarming, but not nearly as alarming as what came out of the mouth of the stranger standing next to him. "I just happened to be standing next to one of the Slave Lake guys, and he looked me straight in the face and said, 'You guys are losing part of your city today.'"

Stephenson wanted to be clear on this point: "It was not like, 'Yeah, you guys *could*.' It was, 'No, you *are* losing part of your town today.' Just from experience, he knew."

Whether they wanted it or not, the Slave Lake crew got their tour.

The Regional Emergency Operations Centre had been set up in a high-tech conference room on the second floor of Hall 5. Packed into it, in addition to Darby Allen and the recently arrived Bernie Schmitte, were dozens of people in color-coded vests representing various city and provincial services and agencies, ranging from electricity and gas to law enforcement and Alberta Emergency Management. "So, we went inside," Jamie Coutts recalled, "and a deputy took us down to one of the windows at the end, and he's showing us: 'This is Beacon Hill and that one down there is Waterways, and over here is Abasand,' and we're watching this smoke column right over the top. Finally, I said, 'I gotta go.' So I left those guys standing there and I walked right into the REOC, and I'm like, 'You know what, guys? I'm watching this smoke column; we gotta get going with these sprinklers. We have to do this right now. It can't wait.'"

As the smoke column grew steadily darker and higher, taking up more and more of the western sky, it became clear to just about everyone present that urgent action was needed. Crews and equipment were then dispatched to Beacon Hill, along with Slave Lake's sprinkler trailer.

Before Jamie followed them, he had a moment alone with his fellow fire chief. "I told Darby Allen right in the parking lot, 'You're an idiot no matter what. If you evacuate everybody and nothing happens, you're an idiot. If you don't evacuate everybody and the fire comes, you're an idiot. You're going to be wrong and everybody's going to hate you no matter what, so follow your gut. If your gut says, 'We need to get people out of here,' then get them out of there. If the fire doesn't come, then, whatever—no matter what call you make, and no matter when you make it, everybody else is smarter. Everyone else would have figured it out before you. Everyone else would have done it different than you. Doesn't matter. When you're the guy that's in charge, you just do what you know is the best in protecting people and you roll with that. That's the job. The job is to do what's right for people regardless of what the people think.'"

Then, Jamie jumped in a truck with some probies and raced up the highway to Hall 1, just behind his son, Ryan, and Lieutenant Patrick McConnell.

By 1:00 p.m. there was serious discussion in the REOC of evacuating Beacon Hill, Abasand, and Thickwood, north of the golf course. It was a big decision with many considerations to weigh. Many of those involved didn't know each other—at least not well—because assembling a REOC is a response to crisis, which was, thankfully, an uncommon event. It was a room full of alphas, all leaders in their respective domains. While there were some women present, there was a lot of testosterone in the room, along with mounting fear.

The Gordian network of intertwined systems that make up a city is truly daunting. Initiating an evacuation, even from a portion of Fort McMurray, would disrupt many thousands of people—everyone in that area from newborn infants and senile grandparents to pets and belligerent drunks. Complicating matters was the fact that most people were at work or in school. How do you get the message to them? How do you handle their emotions? What if they insist on going home? What if they refuse to leave? What if there's an accident at a crucial intersection? If you shut down gas and electricity, what are the implications? What about the hospital—should you use vans or medevac? What about looters? And law enforcement? There were men in jail and women in childbirth; there were supermarkets full of perishable food, and perishable customers; there was a pet store, an animal shelter, and tanks of live lobsters.

A large evacuation, mismanaged, could create its own disaster.

The fire wasn't the only elephant in the room that afternoon; the bitumen industry was another. The tension between the two put Darby Allen, the public face of public safety, in an almost impossible position. Leaders in any community are duty-bound to acknowledge possible disruptions, but disruption is a touchy subject in a place that lives or dies by global oil prices and the skittish whims of investors, many of whom are based offshore. After all, the world is big, there's still a lot of flowing oil in it, and the challenges posed by bitumen are many and daunting. It did not help that the market was experiencing a glut in supply that had sent synthetic crude prices tumbling to $16 a barrel that winter before rebounding to a still-anemic $30 in late April. Thirty dollars a barrel was well below the break-even rate for bitumen upgrading; new projects were being shelved, and more layoffs were imminent. So close on the heels of this bearish news, no investor or CEO was going to want to hear about a shutdown due to fire.

Because these operations are so large and complex, and because they involve such volatile and toxic substances, such extraordinary sums of money, and so much manpower, interruptions of any kind—personal or mechanical—have serious, potentially career-ending, consequences. Syncrude, one of the smaller companies in the area, employs more than five thousand people and produces roughly 300,000 barrels of synthetic crude per day. Any one of these factors would make an official think twice before calling for a major evacuation. Considering the financial implications alone, there were, literally, millions of incentives to maintain the status quo, to keep calm and carry on.

This was a lot to bear, and Darby Allen wasn't Winston Churchill, nor was he a charismatic captain of industry; he was a British-Canadian fire chief a year away from retirement whose midsized town had ballooned into a city, now threatened by one of the biggest urban catastrophes in modern times. What the chief had going for him was a frank and open countenance that had been visible in Fort McMurray long enough to become known and trusted by the people around and under him. But even though Allen was surrounded by some experienced crisis managers in the REOC, none of them had dealt with a threat of this magnitude, or immediacy, in any form.

The Global TV reporter Reid Fiest's experience resembles thousands of others over that hour. He was editing a local news story in a borrowed office on the western edge of Beacon Hill when "someone came banging on the door and said, 'You need to come out and see this!' When we had gone in, about noon," Fiest recalled, "there was smoke in the air, but nothing overwhelming. When we walked out [around 1:30], it was huge—a huge plume of black smoke."

Even with this abundant evidence—on a day notable for its unseasonable heat, and with the memory of Slave Lake clear in his mind—Fiest still observed this fire from behind a bulwark of incredulity. It was a common theme in people's responses that day—perhaps because this new possibility was simply too huge and disruptive to assimilate. Fiest, situated as he was, had a front-row seat to a fire that was now approaching as inexorably, and as quickly, as a storm. "That's when I realized this is potentially the worst-case scenario," he told me, "but I didn't quite believe it yet. I had never been in a situation like that." Fiest's uncertainty didn't last long: "Over the next twenty minutes, you could see it growing, you could really hear the wind changing, and we began to see the embers falling on the grass and on the trees. That's what really triggered me to the fact this was going to spark a massive fire. That's when I phoned my desk and said, 'I think you need a story from us . . . because I think the worst case is about to become reality.'"

There are sights and sounds—frequencies of perception—that are instinctively wrong, that tap into our primal, animal sensibility and tell us it's time to go—*now*. It can be the tone in someone's voice, or the pitch in a gust of wind; it can be a screech of tires, the lurch of an airplane, or a sudden movement by a stranger; it can be the size, color, and proximity of a smoke plume. In none of these situations is there an established threshold, or fixed redline. Instead, there is a kind of subjective, sensory equivalent to crossover—a moment beyond which we are moved involuntarily to a state of high alert—and this moment seems to be the same for the majority of people. In Fort McMurray, starting around 12:30, this threshold was being crossed by everyone who looked at the southern sky.

At this point, most of the city's residents were still at work, or at lunch, and their children were still in school. But that was about to change, and it would do so like the moment of arrival in a sci-fi movie, or in H. G. Wells's *War of the Worlds*: people stopped in their tracks, heads turned to the sky, beholding something whose size and import they could neither limn nor scale. It wasn't Martians, or Godzilla, but it was a monster and they knew it.

The firefighter Evan Crawford was off duty that day, at his home in Beacon Hill. "All of a sudden there's a big plume of smoke behind me on my back deck and I thought, 'What the hell is going on here?' I thought it would *maybe* come to the city limits and we'd get spots here and there. I never thought it would hit us with that much force."

"When I got in the shower, the sky was blue," a west side resident named Sandra Hickey recalled. "When I got out, the sky was black."

The city's schools were filled with children, some of whom, no doubt, were staring southward from playgrounds and classrooms at the place where the sky should have been. At the St. Martha Elementary School, a good seven miles north of the fire, firefighter Ryan Pitchers was in the midst of showing his pumper truck to a class of kindergarteners, and, even from that distance, he was alarmed by what he saw. A photo he took at 1:22 shows what appears to be an enormous storm cloud bearing down on the city. No evacuation notice had been issued. "The teachers asked us, 'What should we do?'" Pitchers recalled. "I suggested, 'Get ahold of the parents and be ready to go.'"

The kindergarten visit was over.

With crossover now complete, the fire would be setting the agenda for what was to come. Minutes later, Pitchers's crew and their pumper truck redeployed to the Dickinsfield neighborhood in Thickwood, three miles south, and that much closer to the spot fire that had jumped the Athabasca River. The fire that Bernie Schmitte had described that morning as a minor fire worthy of a ten-man deployment was now unrecognizable. Before joining the fire department, Pitchers had done a tour in Bosnia as an infantryman, and he drew on that vocabulary to describe what he saw coming at him out of the forest: "That's tighter than a duck's ass pucker factor right there."

Looming over the city now was a towering black cloud shot through with streaks of orange and seething with flames. Somewhere behind it, lost to view, was the sun. Directly in front of it were thousands of homes and dozens of schools. Now a fully developed Rank 6 crown fire, its kilotons of energy were generating internal winds approaching gale force. There was no longer any question of whether or not it was coming into town. Very few people have stood downwind of a Rank 6 boreal wildfire. Now, tens of thousands of people were in that position, caught by surprise in the middle of their daily lives. The shock was palpable and traumatizing.

All morning, time had been moving in a peculiar way, but this is the nature of Nature on a deadline: things unfolding gradually across the intersecting horizons of landscape and time until that moment when, with astonishing suddenness, they merge and the event is upon you. You wonder where all the time has gone, when in fact it hasn't gone anywhere, it is the events within it that have appeared to amplify in speed and scale—because they now include you. This is one of the supreme challenges facing humans in how we manage the physical reality of our planet: the deceptively simple tension between time, rate, and distance. A hurricane can be plotted and tracked a week out, where it remains an abstraction on a network weather map, and yet when it is upon you, time and events achieve a kind of singularity and, suddenly, nothing else exists; its immediacy—its *presence*—is overwhelming.

There are literally thousands of cell phone photos and videos from this moment of collective anagnorisis in Fort McMurray, and they all show the same thing, just from slightly different angles—the post office steps on Hardin, the parking lot of the Showgirls nightclub on Franklin, downtown office windows, drive-through lineups, and back porches in Abasand, Beacon Hill, and Thickwood. It's a huge and overbearing presence blocking out what, half an hour earlier, had been a picture-perfect Alberta sky.

The prevailing southwesterly was being further energized, and also confused, not only by the fire's self-generated thermals but also by a third wind source, what meteorologists call a "low-level jet." This is, essentially, a localized jet stream directly influenced by regional weather conditions. Occurring a thousand feet or more above the ground, and blowing anywhere from forty to more than one hundred miles per hour, a low-level jet can generate tremendous turbulence inside a fire's smoke column. If it persists, it can have a dramatic effect, not only on the plume's behavior but also on the fire's. A steady injection of the jet's energy (and oxygen) can hasten and amplify a fire's transition from a largely forest-level event to a meteorological one, a full-blown weather system that, like a hurricane, comes with its own engine (the burning forest). Hurricanes are energized by warm seawater and tend to collapse when they hit land; in the case of a boreal fire, only a weather shift, a great lake, or an ocean can stop it. It gives a whole new meaning to the poet Andrew Marvell's line, "Had we but world enough and time." Boreal fires have both. The wind was now blowing toward the northeast; in that direction lay unlimited fuel from Fort McMurray to Hudson's Bay, seven hundred miles away. With a good five months left of fire season, this fire could, in theory, burn from the heart of the country all the way to the sea.

At the moment, with not one but three different wind sources energizing it, the fire was poised to overrun Beacon Hill and Abasand. Nearly ten thousand people lived up there between schools, shops, churches, and even a gas station. For vehicles, there was only a single access road to each of those neighborhoods, but for the fire, there were no such limitations. In this way, the topography of Fort McMurray was perfectly suited to fire's most basic aspiration: to eagerly follow heat and volatile gases, both of which are compelled upward at every opportunity. In wintertime, the steep banks of the Hangingstone River funnel sound like a bullhorn, but in the heat of this freakishly warm spring day, they funneled wind like a bellows, supercharging the flames with invigorating gusts of preheated oxygen, driving it up the thickly forested slopes and onto the hilltops. With the low-level jet injecting more air and oxygen from above, the effect, directly over those neighborhoods, was like a huge convection oven—a convection oven set to a thousand degrees, and filled with shoeboxes.

10

Which kid you going to pick and choose and run?

—Ali Jomha, evacuee

The dominant forest type on the Boreal Plain is known as mixed-wood forest, and three of its most common species—balsam poplar, aspen, and black spruce—grow in dense groves in and around Fort McMurray. Of these three, black spruce is the most volatile—so volatile, in fact, that boreal firefighters call it "gas on a stick." With branches that grow all the way to the forest floor, these trees, already sticky with flammable sap, have their ladder fuels built in. Black spruce evolved to burn and will do so ferociously, with little provocation. Aspen and poplar, meanwhile, have been called "asbestos trees." With their moisture-laden leaves and relatively pulpy, water-retaining wood, they can function almost like a brake on forest fires—to the point that Indigenous communities encourage aspen groves around popular hunting and fishing camps, a practice some settler communities have imitated.

There was a lot of poplar and aspen growing in and around Fort McMurray, but on May 3, in the middle of the spring dip, it didn't seem to matter. Radiant heat—the kind projected by a forest fire—travels at the speed of light and, that afternoon, it was causing these normally fire-resistant trees to desiccate and ignite instantaneously, almost before the fire was upon them—something local firefighters had never seen before. By the time the fire reached the back side of

Beacon Hill, the heat was so intense that it was causing these trees not to catch on fire, but to explode.

Chris Vandenbreekel, still downtown at the Mix 103 studio, recalled when the seriousness of the situation dawned on him. "It wasn't until around the 12:30 point, as Bernie Schmitte was leaving my office. The wind picked up toward the town and that was kind of an 'Oh crap' moment." (Understatement had been a running theme that morning.) "It wasn't long after that," he said, "that I started getting phone calls from Abasand and Beacon Hill saying, 'The fire looks really close, what should we do?' So I was making calls to [municipal] Communications, and they were saying, 'Yeah, the firefighters are fighting it; it's not a huge danger to the communities just yet.'"

Vandenbreekel dutifully relayed this information to anxious callers. So far, there had been no evacuation order.

Paul Ayearst lived in Beacon Hill, and his neighbors were among those making anxious calls to Mix 103. The forty-six-year-old Ayearst was a rarity in Fort McMurray: he was raised here, arriving as a child in 1975 when, as he put it, "Dad drug us up here looking for work." The elder Ayearst found it at J. Howard Pew's Great Canadian Oil Sands, where he started out as a beltwalker. All day long, he strode the black and broken earth, inspecting the miles of conveyor belts and thousands of rollers carrying freshly dug bitumen from the ten-story, three-hundred-foot-long dragline shovel (which replaced the smaller bucketwheel) to the giant crusher at the plant. It was, literally, a dead-end job, and, with the company's help, he trained up to millwright. Like his father, Paul Ayearst made a home for himself in town, where he got into selling and servicing heavy equipment, a critical and lucrative business up here.

A little after 1:00 p.m., Ayearst was downtown, leaving Moxie's Bar and Grill on his way back to work, when he registered the plume. "You could see that big cloud coming over Beacon Hill, so I figured, you know what, fuck it, I'm going home—be with the family."

When Paul arrived, his wife, Michele, a homemaker, had the radio on. Paul went into the backyard and tinkered with his garden hose. Like everyone else in Beacon Hill, Paul and Michele were in a holding pattern, waiting for an update on the fire. As a local kid, Paul had grown up with fires on the horizon the same way Texans grow up with tornadoes and Newfoundlanders grow up with blizzards: at a certain time of year, they're always around—sometimes closer, sometimes farther, but rarely impinging on your daily life. It was the same for Michele, who had grown up right in Beacon Hill, and they still lived within sight of her childhood home. Their house stood on a loop called Beaveridge Close. Theirs was an established neighborhood of what now would be considered "starter homes," but because this was Fort McMurray, the Ayearsts' forty-year-old, single-story bungalow was appraised at $650,000. Just to the south of them were two elementary schools, some playing fields, a minor league hockey arena, and a Mormon church. Beyond that was the thickly settled south end of Beacon Hill, and the only road off the hilltop. All told, about 2,200 people lived up there, in six hundred houses. An island unto itself, Beacon Hill was one of the smaller neighborhoods in Fort McMurray, but it was more than twice the size of any other community for two hundred miles around.

It wasn't until 2:05 p.m. that the first evacuation orders went live. Issued by the Municipality of Wood Buffalo, which includes Fort McMurray, they called for the mandatory evacuation of Beacon Hill, Abasand, and Grayling Terrace, a small enclave on the left bank of the Hangingstone River, just west of Fire Hall 1. Centennial Trailer Park, due south of Beacon Hill, had remained under mandatory evacuation since the previous day, and now the one hundred trailer homes located there in angled rows were in the process of burning to the ground. The Ayearsts did not hear the first alert, perhaps because they were out of earshot, or because the station they were listening to did not air it. The first inkling they had of an evacuation was over the PA system of the nearby Good Shepherd Elementary School. At around 2:20, they could hear the names of children being called, notifying them that their parents had arrived.

"That's when it started getting really crazy," Chris Vandenbreekel recalled. The Mix 103 studio also houses Cruz FM, a classic-rock station, and, with each new update, Vandenbreekel would announce it on the air and then run across the building to tell his colleague at Cruz. By now, other local stations were relaying the information too. "It was Beacon Hill and then Abasand, and we have people in our station who live in those neighborhoods—'Oh my God, should we go back?' 'No, you can't go back'—the phone is ringing off the hook, I can't handle all the calls and we're trying to get all these evacuation orders out, and there's more and more neighborhoods being called."

It had been less than two hours since Bernie Schmitte had left Vandenbreekel under the impression that "they were going to be able to battle it back." In between fielding listeners' calls, Vandenbreekel called his wife and told her to evacuate. Because there was now real concern about Highway 63 being cut off by the fire, Vandenbreekel advised his wife to go north, toward the mines and the camps. By then, north was the only direction where there wasn't visible fire.

Word was now catching up to events, and, all across the west side of town, schools began evacuating on their own. Emma Elliott, a parade leader and commander with the Air Cadets, and still just a student herself, was sent to pick up her brothers in Thickwood. "The sky was red," Elliott told me. "I just saw a red sky and my car was covered in ash. I turned on the wipers and I felt like I was beside myself—otherworldly. Parents were panicking. There were these twelve-year-old kids talking about their houses burning down. I said to my little brother, 'Don't look up.' "

When Elliott got her brothers home, her mother was packing the car. "There was a look on my mother's face I'll never forget. It was like, 'What if we don't get out of here?' "

Paul and Michele Ayearst had two children, a college-age son in a man camp north of the city, and a grown daughter who called from downtown at around 2:30 with the news that an evacuation order had been issued for Beacon Hill. Even though the alert had been issued a half hour earlier, it had taken a while for it to trickle through participating radio stations, Facebook pages, Twitter accounts, and word of mouth. While Paul could clearly see that the smoke plume was growing larger and darker behind Good Shepherd Elementary, he could not, from his vantage, see exactly how close the fire was. Nonetheless, he took the order seriously and started gathering suitcases. It was then that their phones began ringing steadily: the fire had already made the news, and friends and family were calling them from other cities, checking to see if they were okay. Michele was growing visibly upset and Paul was trying to calm her down, trying to keep the focus on packing. He went to their safe and gathered up jewelry and their passports. By 2:35, five minutes after their daughter phoned, Michele, as Paul recalled, "was almost hysterical." He was hugging her, trying to comfort her, but he was also trying to fill his duffel bag. "I'm like, 'Calm down, we're okay, just take a breath, you need to get ahold of yourself. We're not in any danger, you're fine.'

" 'No,' said Michele. 'The fire!'

"I said, 'I know there's a fire coming, but it's not a big deal.'

" 'No, no,' she said. 'The *fire!*'

"I said, 'Yes, there's a fire. Just focus on what we have to do and give me twenty minutes so we can get out of here.'

" '*No!* You don't understand!' Michele was screaming now. 'The fire's on our *street!*' "

Paul Ayearst, a lifelong resident of the boreal forest, a hard worker, conscientious citizen, and dedicated family man, found himself in the same cognitive dilemma as so many others that day: he had heard the warnings, he had seen the fire growing bigger and closer, and yet, on some crucial, active level, he did not, or could not, acknowledge the immediate and terrible implications of a Rank 6 boreal fire at his doorstep. The danger confronting him, his home, and his beloved family did not fully register until his weeping wife shouted the fact into his face. Even then, his brain, and its fierce loyalty to the status quo, resisted.

But reality does not require human belief in order to be real. The fire was breaking over Beacon Hill like a wave.

"When I looked out the door of the garage," Paul told me, "I see a wall of flame coming at us. By then, our street was empty and there was four Forestry firefighters running around screaming to everybody to get out. A cop was there, screaming. He's yelling and screaming, 'We gotta go! We gotta go!' "

The fire was now at the end of the block, consuming everything in its path. Embers were falling, igniting spot fires in all directions. Michele Ayearst drove away in her car with the dog; their daughter, who had just arrived home from downtown, drove off in her own car with her cats. Both women fully expected Paul to follow in his truck. But, even with the fire bearing down on him, driving before it a blizzard of burning embers and ash, the battle of wills between Paul's worldview and the insistent, intrusive fact of this fire remained undecided. "I'm there," Paul told me, "literally got a cop dragging me by the arm—'Get the fuck out of here now!'—he's forcing me out of my house. So I gave the cop like, 'Get your fuckin' hands off me!'—gave him a shove, and I went running back towards the house.

"He said, 'What the hell do you think you're doing!'

"I said, 'I'm locking the door.'

"I never thought the house was going to burn so I'm locking the door. Then, I jump in the truck, and I'm contemplating, 'Do I leave my Harley? Do I take the truck? Do I take the Harley?'"

It has been observed that people in shock or overwhelmed by traumatic events will focus obsessively on small details. But for a lot of men in Fort McMurray who make their living in and around machines, their motorcycles, vintage cars, and snowmobiles are not small details. Shedding tears over a beloved Harley-Davidson is not unheard of here. But on May 3, this deep allegiance to the internal combustion engine posed a major problem: residents owned so many vehicles, recreational and otherwise, that it was impossible to drive them all out. There was an inescapable irony in Ayearst's dilemma: that same combustive energy that thrilled, empowered, and enriched them was now manifesting itself in the most primal, potent, and destructive way imaginable. It had taken a hundred years of grit and sweat, hundreds of square miles of bulldozed forest, hundreds of billions of dollars of investment, and trillions of cubic feet of natural gas to access, and finally profit from, the stubborn potential of bituminous sand. And now the remaining "overburden" was unleashing more combustive energy in a single afternoon than all the upgraders combined.

Up above Highway 63, directly opposite Hall 1 and the floodplain of downtown, the neighborhoods of Beacon Hill and Abasand Heights had become, with astonishing speed, the frontline of the fire. Residents were now evacuating en masse. Each community was accessible by only a single road, and they were now completely jammed with fleeing cars and trucks, making their way down to the highway with agonizing slowness through showers of wind-driven embers and billowing clouds of smoke. The same trees that, an hour ago, had made these neighborhoods such attractive places to live were now bursting into flame and crowning with a speed and explosiveness more in line with a napalm drop than with a natural fire. As each new tree and grove ignited, it sent waves of heat over the traffic that felt almost three-dimensional in their intensity, and turned the residents' sole escape route into a gauntlet.

In the end, Paul Ayearst chose his truck, escaping his burning street by driving through the neighboring school's playing fields. When he joined his wife and daughter in the line of traffic, both lanes of the access road were filled with cars, every tree in sight was in flames, and boulder-sized fireballs were rolling over the roadway. Ayearst showed me a picture: "There's my daughter, there's my wife, there's a fireball going over top of the truck and my wife's driving into it. That's intense heat." Inside the vehicles, which had now become fire shelters in addition to escape pods, dashboard thermometers were indicating temperatures that had likely never been registered in a vehicle with live occupants. The thermometer in Paul Ayearst's truck hit 151°F. "You could feel the heat radiating through the glass," he said.

Just ahead of Ayearst were his wife and daughter, each in their own car, trying to maintain their composure, their efforts to communicate and reassure each other stymied by the fact that virtually every other person in Fort McMurray was trying to do the same thing. Local cell towers were not only overwhelmed with signal but in danger of burning down themselves. "My daughter's right in front of me," Paul explained, "my wife's in front of her, and I'm trying to talk to them and I couldn't."

Opening the window was out of the question. Barely fifteen minutes had elapsed since his daughter had called from downtown, and they were now trapped in a corridor of fire, inching forward at the pace of a Tim Hortons drive-through. Ayearst had no choice now but to acknowledge his new reality: "I'm stuck over here," he said, "sitting in the fire. I got a wall of flame coming at me and I've got my wife and daughter in front of me and traffic's not moving, and it's like, 'How do I keep them safe? How do I get to my son?'" With these questions, Ayearst was speaking for thousands of parents that afternoon.

At that moment, sealed inside his vehicle was now the only safe place to be, but the line between sanctuary and death trap was narrowing. The window glass was too hot to touch and embers were blistering the paint. Just outside, the same world where they had raised their children, where some had grown up themselves, appeared now to be made of fire. Gone was the sun and sky, replaced by this rogue element, which had commandeered the trees, the houses, the very air. As the Ayearsts and their neighbors made their way slowly down the hill, Paul heard a thud and his truck shuddered. He thought he'd been hit, and he had—not by another vehicle, but by a fleeing deer, its fur smoking and aglow with embers. Running blindly out of the flames, it had collided with Ayearst's passenger door before regrouping and barreling on through the traffic. That impulse, to run away—anywhere—was shared by many others that afternoon, but it was held in check, in part, by the still-deeper fear of being burned alive. But self-discipline, company training, religious faith, and peer pressure were factors, too. These weren't fly-in workers with no stake in the place; Ayearst and his neighbors had put down roots here. If they broke ranks and took off across the median, they would betray a communal trust. If they did it, who wouldn't?

A camp manager and type A marathoner named Dave Dubuc was confronted with this very question while trying to get out of Timberlea. His wife, Joanne, their three children, and two cats were in the truck with him, traffic was at a standstill, the fire was moving fast, and getting trapped and overrun was a real possibility. For Dubuc, the choice was stark and therefore simple: "I have to save my wife and children," he thought to himself as he prepared to turn on his company truck's roof light and escape across a sidewalk and through a park. Dubuc is a problem solver with the mental toughness of a competitive athlete, but his wife is strong in a different way: when Joanne realized what her husband was doing, she admonished him. "Don't be that guy," she said. Dubuc settled himself then and stayed in line—for hours. It wasn't easy. "There were times, driving out," he said, "that the fire was burning six feet from my tires."

With Hell on your heels and small children depending on you, your priorities shift, and so does the math. Ali Jomha is a local imam who runs a halal butcher shop downtown on Franklin Avenue. He is a pious man, deeply involved in the Muslim community, and he bears a striking resemblance—in both looks and manner—to Al Pacino. On May 3, caught in traffic with everyone else, the fire forced Jomha to consider an imponderable dilemma. "If you're driving in the car, flee-ing," he proposed to me, "and you and your wife have in the car four kids, five kids, and for God's sake something happens to the car in front of you—which kid you going to pick and choose and run? My brother has five kids; my other brother has five kids; I have three. Going bumper to bumper, we saw the fire crossing the highway. If one vehicle burns, it will transfer to another. If I have to run away from the fire, how many kids can I take? That's a strange feeling."

What would astonish people later, in addition to the bravery of police and first responders who helped guide the evacuation, was the relative order that drivers maintained as they crept along, bumper to bumper, with children and dogs whimpering in the back seats and many in tears themselves, while sparks and firebrands rained down on them, flaring and rattling off their hoods and windshields. This kind of civic-minded restraint and courage was the norm, not the excep-tion, and the many dashcam videos taken that day bear this out. Paul Ayearst and his family caravan, stuck as they were at the end of their own slow-motion convoy, would be among the last civilians out of Beacon Hill. Somewhere behind them, where home and safety used to be, came the sound of things exploding.

11

You own the fuel, you own the fire.

—Urban fire specialist's adage

Downtown, at Mix 103, "things were getting worse and worse," Chris Vandenbreekel recalled. "We couldn't see the sun anymore, there was ash raining down, and there was this red hue downtown. Hosts and different staff members couldn't handle what was going on; one staff member pulled her knees up to her chest and just started rocking in the fetal position. They were worried that their houses were gone, whether their families were safe. There were a couple of us that were just like, 'No, we gotta keep going, we gotta keep doing this,' and we're getting more and more evacuation orders: Waterways [the oldest community in Fort McMurray, due south of downtown] gets the evacuation order; downtown is close to getting an order. Wood Buffalo [near the golf course] getting its evacuation order— Thickwood . . . It was crazy, trying to do this as a new employee."

Vandenbreekel had been in town for all of seven months; there were still neighborhoods whose names and locations he did not know.

A mile and a half south of the radio station, in a wedge of flood-plain formed by the Hangingstone River and Highway 63, firefighters from all over the city were converging on Fire Hall 1. Centrally located, with ready access to Beacon Hill and Abasand, Hall 1 is the oldest in town; inside, it has the feel of an open-plan clubhouse. In the lounge, a semicircle of La-Z-Boy recliners are aimed toward a wide-screen TV surrounded by shelves and cabinets full of trophies and memorabilia, the walls hung with portraits of past chiefs. Steps away is a well-equipped kitchen that adjoins a sunny dining area with a long communal table. The weight room is through a side door, and bunkrooms and showers are down the hall. It looks like a wonderful place to spend a weekend with good friends. But its comforts are deceptive: you can be settling into a weight session, a meal, a movie, or a nap—at any time of day or night—when the alarm will sound. Depending on the tone, it will be for ambulance or fire, and many members are trained for both types of calls. While everyone has a shift, no one has a schedule, and all live at the whim of the next car accident, heart attack, or barbecue fire.

On the afternoon of May 3, Hall 1 ceased to be merely a fire hall and became instead something closer to the Alamo. Across the city, fire hall intercoms were barking: "All halls! All halls! Bring every apparatus and proceed to Hall 1!" Because only about 30 of Fort McMurray's 180 members are on duty at any given time, it took some time to track people down. A couple dozen firefighters were out of town altogether; one chief was bobbing at a swim-up bar in Mexico when he learned that his city was on fire. Everyone who could answered the call and, by 2:00 p.m., the parking lot at Hall 1 was overflowing with every kind of pickup and fire truck. Chief Troy Palmer, based out of Hall 3, across the river, was off duty and had yet to be alerted.

Palmer's days of wrestling hose and climbing ladders were behind him; as a shift chief in his forties, he was focused on the big picture, on assessing the scene in its entirety. His piercing blue eyes informed an analytical mind, and, when he wasn't on duty, he read a lot, particularly military history. Historic campaigns and strategy provided him with frameworks for making sense of the world, and they would come in handy for making sense of this fire: the chaos, which mimicked the fog of war; the all-too-common mistake of underestimating one's adversary; the experience of being outflanked, outgunned—of being besieged. The terrible realization that you are, in fact, surrounded.

Even though Beacon Hill was now on fire, with smoke and embers crossing the highway, the situation was hard to appreciate from Palmer's home in Timberlea, eight miles away. Out there, in the northwest corner of town, the subdivisions had been going in so fast that maps couldn't keep up with them, but just beyond the surveyor's stakes lay raw wilderness. If Palmer walked out his front door and headed down Chestnut Way, past the rows of one- and two-story starter homes to the edge of his subdivision, he would come up against a kind of terra incognita. To the north, beyond the current frontier of Walnut Crescent, lay a DMZ of scraped, drained, and leveled earth awaiting the next development boom. To the west stretched a virtually limitless expanse of dog-hair forest and muskeg—moose and beaver country that favors the amphibious and well insulated, and discourages casual exploration. Once beyond the tenuous membrane of suburbia, you were bushwhacking—all the way to Buffalo Head Prairie, two hundred miles to the northwest. En route, you wouldn't see another swing set, cul de sac, paved road, or settlement of any kind. Out there, out of sight, the mining companies and wild creatures hold sway, roaming at will across a landscape unmarked save for the ubiquitous hatchwork of seismic lines.

Troy Palmer's four-day shift wasn't scheduled to begin until 6:00 that evening, but he had a feeling he would be called in sooner. On May 2, he had seen Fire 009 from the tower at Hall 5 on Airport Road—the growing plume, the water bombers, a little too close for comfort, angling in for their drops across the steadily shrinking gap between the fire and the community he was hired to protect. Palmer's eight-year-old daughter had been staying with him that week, and, even though May 3 was a school day, he decided to keep her home. She said she wasn't feeling well, but the fires were also on his mind and he wanted his daughter close. Palmer knew that all across the index, fire weather conditions had shifted progressively further into fire's favor, and he knew his brothers and sisters in the department were going to be busy. At lunchtime, he called his ex-wife and asked if she could get off work and pick up their girl as soon as possible. "She was at the door picking her up when I got my call from our computerized dispatch at two in the afternoon," Palmer told me. "Normally, it'll say, 'Overtime opportunity. Call by such-and-such a day, such-and-such a time.' But this time, the computer called and it goes, 'Emergency, Hall One. Emergency, Hall One.' I've never heard it say that before. So, when my ex left with my daughter, I jumped in my truck. Of course, by now they were starting to evacuate. Of course, I get stuck in traffic."

Under normal conditions, Palmer had a twenty-five-minute drive ahead of him—but nothing was normal that day. The delay, and the solitude, gave him time to consider the TeleStaff call he was responding to: there had been no specific information and no accept/decline option. This was not an offer; it was a Mayday. Whatever was happening across the river trumped rank, priority, the union, family— pretty much everything that gave shape and meaning to these men's and women's lives. Alone in his truck, stuck in traffic, Palmer had yet to understand that this wasn't so much a fire alarm as it was the mobilization of a militia in the face of an invading army. It wasn't until he reached the river that he would understand what they were up against.

By then, word was catching up with the fire, and evacuation orders were, for most people, already moot. When Palmer made a left off Millennium Drive onto Confederation Way, he found himself in a slow river of cars, trucks, and SUVs that filled all three eastbound lanes. Everyone was headed toward the highway, and also toward the fire, because that was the only way out of town. Scattered throughout this creeping procession were other firefighters who had just received their TeleStaff alerts.

Palmer had made this cross-town drive just the day before. This was Forestry's third day working Fire 009, but the fire department had been busy, too, marshaling resources and setting up sprinkler lines around key infrastructure lying in the fire's predicted path. Working overtime, Palmer had spent much of May 2 at the brand-new transfer station and recycling facility south of town, *removing* dozens of pumps and sprinklers along with thousands of feet of hose that the previous shift had set up and primed on May 1. Due to a change in the predicted winds, that entire defensive system had to be drained, coiled, and repacked. But that was yesterday. Today, it was as if the narrative of this fire had leaped ahead of itself, carried on the wind into a future that arrived far more quickly than the human beings watching it, and fighting it, had anticipated.

The gravity of the moment hit Palmer as he rounded a gentle

bend in the river by the water treatment plant, just downstream from the evacuated golf course. The plant—new and state of the art like so much of Fort McMurray's infrastructure—lies on the left bank of the Athabasca River. In addition to purifying city water, the plant's huge new pumps initiate the pressure needed to drive water across the city and up onto the hilltops and ridges where so much new development has taken place. Less than five hundred yards farther downstream was the highway and Fort McMurray's single most crucial piece of connective tissue, the bridge, the only road link between the west side neighborhoods and the rest of the world. Directly above the bridge, opposite the treatment plant, loomed the steep face of Abasand Heights. From this vantage, veins of bituminous sand are visible in the ridge's eroded west face while the gentler slopes are covered in forest, but that's not what Palmer saw. "All along the river, the entire hillside was burning," he told me. "That's when I phoned my ex and I said, 'Pack up the trailer and pack up our daughter. You should be leaving town. Don't bother going south, because that's trapped already. Go north.'"

"North," beyond the mines and man camps, was a dead end at the First Nations community of Fort McKay, thirty miles up the road, but that was a problem Palmer's ex, Vandenbreekel's wife, and thousands of other evacuees would need to deal with later. At that moment—about 2:45 p.m.—Centennial Park was in flames, Beacon Hill was igniting, and wind-driven embers were showering Abasand's rooftops in visible gusts, like so much flaming snow. This midday squall of flying fire was now impacting the entire west side of the municipality, from the Gregoire neighborhood, down by Airport Road, all the way to Dickinsfield, one subdivision south of Timberlea, where Palmer lived. Palmer was trying to make sense of something he had never seen before, but Jamie and Ryan Coutts, who had arrived at Hall 1 about an hour ahead of him, were witnessing a repeat of the same scenario that had incinerated a third of their town in a matter of hours. Only this fire was ten times bigger.

By the time Palmer made it over the bridge and through the traffic to Hall 1, hundred-foot flames were cascading down Beacon Hill toward Highway 63. One eyewitness described it as a "wave of flame." Another, who caught it on video as it came down the highway side, described it as flowing "literally, like lava." The video bears this out: the trees ablaze over a hillside that appears molten and in motion, rolling down to the roadway. Meanwhile, wind-driven embers showered the bumper-to-bumper traffic and the businesses downtown as the firefighters massing at Hall 1 deployed their hoses, not to spray down the surrounding buildings, but to protect their fire trucks and the hall itself—a first for any member of that department. "Standards were gone," recalled Palmer, who arrived in the middle of this. "Normally, you get a 911 call, they sort it out very structurally. Now, it was crazy: 'We need help here! We need help there! We need more backup in Abasand!' And the look on people's faces was just unbelievable—fear, uncertainty: What do we do? It was so unconventional."

In order to avoid situations like this, many state, provincial, and municipal fire services follow a protocol called "Blue Card Command." Developed by the Brunacini brothers, a legendary firefighting family out of Phoenix, Arizona, and building off of California's popular Incident Command System (ICS), Blue Card Command is a comprehensive approach to decision making and resource deployment, providing a step-by-step, if-then order of operations for each member responding to a call, whether they are first on scene or backup, whether it is a house fire or a chemical spill. The Blue Card system is widely used across the U.S., and Alberta adopted it in 2011, but as versatile as it is, there was no protocol for a situation like Fire 009. "Blue Card is generally how we manage things," Palmer told me, "but this fire—it was kind of like, 'Okay guys, throw the textbooks in the fire because that's as good as they are right now.'"

Instead, firefighters, police, and other first responders fell back on deeply ingrained hierarchies. "Everyone understands the fundamentals of paramilitary organization," Palmer explained, "so you knew if you showed up and you saw two stripes [captain], that's the guy you got to answer to—even for some comfort and for some reassurance—'Okay, now I know where I got to go and what I got to do.' But beyond that, it was just—whatever."

There was a lot of "whatever" in Fort McMurray that day. In the face of this unprecedented chaos, Palmer resorted to fundamentals: "Okay," he thought to himself, "I need my equipment to fight fires." Soldiers are trained to bond with their weapons; firefighters

never respond to a call without their bunker gear. But Palmer's bunker gear—the forty-plus pounds of protective clothing and breathing
apparatus worn by structural firefighters—was still at Hall 5 on Airport Road, where he had left it the day before. Because many other off-
duty firefighters were in the same situation, Palmer figured he would
drive down there and grab all the gear he could find, and he requested
permission to do this from Hall 1's battalion chief. The chief pointed
toward a rescue truck and Palmer grabbed the first firefighter he could
find and drove off into the whirl of smoke and embers.

As the fire burned its way through dense stands of black spruce
and into the trailer homes of Centennial Park and then the houses of
Beacon Hill, the smoke grew progressively blacker and more acrid.
For the fire, it was a virtually seamless transition from one highly
volatile fuel to another. The dark and increasingly toxic smoke created an artificial twilight suffused with an unearthly orange glow.
Visibility was reduced to the point that headlights were required; even
though sunset was still seven hours away, streetlights were illuminating automatically. There appeared to be several layers of smoke and,
down at street level, gray clouds rolled quickly over the thickening
traffic. The air was filled with drifting ash and large embers clattered
off the hoods and roofs of vehicles, and even cracked windshields.

Driving south toward Hall 5, Palmer didn't know where the
smoke would end. The man next to him, a relatively new recruit,
was a stranger to him. "Initially it was like, 'What are we going to
see?'" Palmer recalled. "'Are we going to see the other side of this?'"
Ascending the long hill out of the river valley, they finally got upwind
of the smoke; Palmer knew where he was, but nothing was the same.
"It was just eerie," he said, "because everything was burning on [the
west] side of the highway." By then, the four-story Super 8, the Den-
ny's, and the Flying J gas station were smoking ruins. Meanwhile,
spot fires were igniting on the east side of the highway. There was not
a fire truck to be seen. Palmer was especially dismayed by the "Welcome to Fort McMurray" sign. "One side had burned completely," he
said, "and it was just hanging there with little pieces burning on it.
The tourist information booth had already burned."

Palmer knew this road intimately, but now he felt disoriented.
Landmarks are how we navigate and scale ourselves to landscape;

their steady, dependable presence projects a sense of permanence and inviolability, which helps to anchor us in space and time. We don't call them "songlines," as they are known in Australia, but they serve a similar psychic role: we know when we see a familiar feature that the next one—a sign, a gas station, a tree, a hotel—will be coming along shortly, and in this way we stitch our homeworld together. Even though we may only name them when giving visitors directions, these are the waypoints that guide us through our daily rounds. Between their disappearance and the smoke, Palmer was becoming a stranger in his own hometown. This was happening now on a city-wide scale: the memory palace was burning down.

When Palmer arrived at Hall 5, he raided the gear locker, filling the back of his rescue truck with all the bunker gear he could find. "Then," he said, "we drove back down to Hall 1 and threw it all in the middle of the floor. That's the last time I looked at bunker gear for the rest of the fire." It's hard to articulate how irregular this is, but it is comparable to a soldier going into combat without a helmet and flak jacket. "Because we never needed it anyway," Palmer said. "That's how unconventional it was."

In structural firefighting, the kind Palmer and his fellow firefighters were trained for, there is almost always more equipment and manpower than there is fire. When you get a call, it is usually to a house or a building, and if you show up with a pumper but you need a ladder, it will be there in minutes. Likewise, if you need a Bronto—an aerial platform—or another pumper, the resources are there to gang up on the fire and knock it down. Yes, you may lose the building—if the fire doesn't destroy it, the water usually will—but, guaranteed, you will beat that fire: one more victory for the team. And this delineates two of the biggest differences between municipal firefighters and their counterparts in Forestry: time and scale. Structural fires are usually dealt with in a matter of hours; forest fires, despite a stated goal of "out by ten,"* may require days or weeks to bring under control. In the boreal forest, it can take months to fully subdue a major wildfire. Likewise, where structural firefighting is comparable to a rugby match—muscular, confrontational, and contained, with the goal

---

* Ten a.m. on the day following the fire's discovery

visible at all times—wildland firefighting is more like lacrosse as it was originally conceived: not on a field per se, but across a landscape where it was not so much "played" as waged—like a running battle whose outcome is neither visible nor certain. These differences are evident in the equipment they deploy, and also in the firefighters' respective builds. Forestry crews may spend days at a time in the bush or the mountains moving quickly through country accessible only by foot or helicopter, while municipal firefighters are seldom more than a hose length from their trucks and a short drive from their kitchens. Form follows function: where wildfire fighters are built for endurance in rugged terrain, munis are built to scrimmage and scrum. This is one reason you won't see many powerlifters like Mark Stephens on a Forestry crew, but you will often see more women. Because of these and many other factors, the techniques, equipment, and even the mind-set required to fight their respective fires are fundamentally different. When these worlds collide, as they did that afternoon in Fort McMurray, lines and methods blur, and massive destruction and confusion can ensue.

This zone of collision has been given a name: the "WUI" (rhymes with *phooey*), an acronym for "wildland-urban interface" (though some call it "wildland-urban *in your* face"). On a map, the WUI represents the fault line between the forest and the built environment, but over the past thirty years it has also come to represent the sweet spot in North American real estate development: hiking trails out the back door and a scooter-friendly cul-de-sac in front. Today, more than a third of American homes and more than half of Canadian homes are located in the WUI. It is a beautiful place to live, until it goes feral. When a wildfire enters a residential community, the result—for the fire—is a smorgasbord of kiln-dried fuel topped with tar shingles, garnished with rubber tires and gas tanks. Meanwhile, the result for homeowners is shock and disbelief as the very structures designed to provide comfort and shelter turn on them in the most frightening way possible. In this situation, a bigger house does not mean more protection, it means a greater concentration of fire-ready fuel. These effects are most dramatic in residential areas built from modern, highly flammable materials in the middle of ecosystems that naturally regulate themselves with wildfire—places like Australia, the American

West, and the Canadian boreal forest. As homeowners in these regions are learning every summer now, when the WUI burns, it does not burn like a forest fire or a house fire, it burns like Hell.

We learned this lesson the hard way in the eighteenth and nineteenth centuries, when forests, houses, and towns burned repeatedly across North America. But we forgot it in the relatively cool and rainy twentieth century as suburban expansion coincided with improved fire suppression. "Watching houses and then communities burn [is] like watching polio or plague return," wrote Stephen Pyne, fire's most eminent living chronicler, in 2016. "This was a problem we had solved, then forgot to—or chose not to—continue the vaccinations and hygiene that had halted its terrors." These forgotten practices were simple ones: Don't build your house in the woods, surround it with open fields; not only does this make for convenient planting and pasturage, but those cleared lands also serve as excellent firebreaks. Roofs made of tin or slate go a long way toward foiling embers.

None of these time-tested methods were used in Fort McMurray. With the exception of downtown, virtually every neighborhood is a perfect example of a WUI community. By 3:00 p.m. on May 3, Beacon Hill was a raging WUI fire; Centennial Park was all but gone; and Waterways, Abasand, and Thickwood were next. When Troy Palmer arrived at Hall 1, sometime after 2:30, Jamie Coutts and

the Slave Lake crew were already in Beacon Hill, and Paul Ayearst's family was trying to escape it. Barely a mile down the highway from Hall 1, the single entrance to Beacon Hill was a serpentine two-lane ascent with breakdown lanes and grassy shoulders flanked by mature forest. By the time Coutts got up there, the moment for setting up sprinklers had clearly passed. Both the neighborhood and the fire-fighting effort were in a state of shock and chaos. Traffic was a confusion of fire trucks and rescue vehicles, evacuating residents, RCMP officers going from house to house looking for stragglers, and other residents racing home from downtown. It seemed as if each person was at a slightly different stage of awareness of what was happening, and of how they might respond.

It is hard to overstate the totality of the disorientation people were experiencing: the roar and crackle of the fire; the wind, searingly hot and alive with sparks and ash; the black and acrid smoke that stole breath and reduced visibility to a car length; the flames a hundred feet high across a front that seemed to have no end or edge. It was as if the world had been remade in fire, and now it was coming for Beacon Hill.

A rule of thumb for structural firefighting is that the gallons per minute (gpm) of water should match the British thermal units (Btus) being generated by the fire. On Beacon Hill, this equation was so hopelessly skewed in the fire's favor that sending a fire department to face it was like sending plumbers to confront a bursting dam. Most of the hose streams deployed were evaporating long before they reached the flames. The situation facing those patchwork crews in Beacon Hill—now a textbook firetrap—was not so different than that facing the firefighters marching resolutely up the stairs of the World Trade Center: it was nothing they had dealt with before, it didn't feel right, and the prospects weren't good, but there were people up there, so they went. Service before self.

In such a lopsided situation, the firefighters' ignorance may have been a kind of blessing. "I remember their faces," Jamie Coutts told me, "and I mostly remember because that's what my face must have looked like when it slammed into our town in 2011—because it was just disbelief, right? It was like, 'Yeah, we're going to lose ten houses here, but we'll get it.' And I'm thinking, 'No, you're going to lose *five hundred* houses here. You guys don't get it.'"

Once the crews were on scene, it was difficult for them to see the houses, or even each other, and it was impossible to hear anything over the growing roar of the fire, which witnesses compared variously to a jet engine and a freight train. The unremitting din was further intensified by the constant snap and crack of sundering timber—trees and houses alike—and by the sporadic explosions of electrical transformers and fuel tanks. The heat between those burning houses, now comparable to the planet Venus, was unbearable, and so was the smoke. It quickly became clear that, not only was there no way to fight this fire, there was no time. There was time only to get the residents out before they were overrun.

This spontaneous transition from firefighting operation to lifesaving operation has become a hallmark of twenty-first-century urban fire—from Portugal and Greece to Australia and California. Some residents had fled already, but who, exactly? In order to find out, firefighters were going door to door. The RCMP were on hand—both to manage traffic and to aid in the searches. The risk of becoming trapped, of being surrounded by the fire on this thickly wooded hilltop, was real and growing more likely by the minute. Steve Sackett, a firefighter in the first wave with the Slave Lake crew, was reconnoitering with Jamie Coutts when the Shell station by the entrance road exploded, producing a shock wave that could be felt hundreds of yards away. "We could feel it in our bodies," Sackett later wrote, "and it followed with a fireball hurling into the sky . . . Without flinching, [Coutts] continued talking to us, saying, 'I hope nobody died.'" Fire trucks, invisible in the smoke, began sounding their airhorns in three-blast intervals—the Mayday signal. The only way off this isolated plateau was that single entrance road at the south end, right by the burning gas station. The only way out was through; the entire hilltop was igniting.

12

We don't have a forest fire problem, we have a home
ignition problem. As soon as you come to that
realization, it changes your view on wildfire.

—Ray Rasker, cofounder of
Community Planning Assistance for Wildfire

Of the four most abundant fuels available to the fire on May 3,
three of them were trees: black spruce, balsam poplar, and aspen.
The fourth was houses. Of those three tree species, the houses of
Fort McMurray most closely resembled black spruce. They were not
infused with highly flammable sap but with something equally combustible. Most house fires originate inside the home, and people who
study these ignitions draw a sharp distinction between modern furnishings and so-called legacy furnishings. The latter are considered
antiques now: wooden tables and chairs; lace curtains; couches with
cotton upholstery, stuffed with cotton batting, or maybe even horse-
hair. These will burn, but nothing like modern furniture, much of
which is made of plastic, or wood "products" bound together with
glues and resins, upholstered with polyester or nylon, and stuffed with
polyurethane. Today, it is common to find oneself sitting or sleep-
ing on furniture composed almost entirely of petroleum products.
As bizarre as it sounds, many people prepare for their day by dressing
themselves from head to toe in highly flammable petroleum-based
materials.

In 2005, Underwriters Laboratories conducted an experiment in which fires were set in a pair of model living rooms, one filled with legacy furnishings, the other with modern equivalents. The fires were started with candles in the sofa cushions, and, for the first minute or so, the only real difference was in the quality of the smoke, which was much darker and more acrid in the modern room. At three minutes, the modern room is clearly in trouble, with the sofa fully engaged. Then, just twenty seconds later, something unexpected happens: the modern living room bursts into flame, engulfing everything from floor to ceiling as thick, oily clouds of black smoke billow from the entryway. What only seconds earlier had been a living room with a burning sofa was now raging like a refinery fire. Anyone inside would have been killed. It is terrifying to watch. Meanwhile, next door in the legacy room, the smoke is minimal and the fire, while clearly growing, could be doused with a bucket of water or a small fire extinguisher. It would take another twenty-five minutes for the legacy room to reach a similar state of involvement.

This phenomenon of sudden and total combustion in an enclosed space is known and feared by municipal firefighters. They call it "flashover," a reference to the flashing, or reflective nature of radiant heat, and it poses a lethal hazard for anyone inside a burning building. Because fire feeds on the gases released by heated fuel (as opposed to the fuel itself), a localized fire—like a burning sofa—will radiate heat that causes off-gassing in the fuels around it. If those fuels are petroleum based, they will volatize more quickly and at lower temperatures than most organic substances, causing a room to fill up, and then blow up, like a vapor-filled gas can, in a surprisingly short period of time. Though far more rapid than wildfire's "crossover," flashover represents a similar moment in the life of a house fire, when it effectively takes its host hostage and starts dictating the terms.

Given the nature of the industry that built Fort McMurray, it seems only natural that the houses of its workers would be built from petroleum products and related chemicals: shingled with tar, clad with vinyl siding, illuminated by vinyl windows, the wood itself impregnated with glues and resins, the floors covered in linoleum, polypropylene carpeting, or highly flammable laminates, lacquers, or varnishes; the appliances, furnishings, clothes, toys, recreational equipment, garden furniture, bedding, and food packaging—virtually all petroleum based. When a boreal fire is projecting thousand-degree heat and blizzards of burning embers into a recently built neighborhood, the houses stop being houses. They become, instead, petroleum vapor chambers. The result resembles a black spruce grove—or two-story, two-thousand-square-foot versions of the modern living room in the Underwriters Laboratories video.

All across the west side, from Abasand to Timberlea, witnesses described phenomena that seemed to defy physics. Lucas Welsh, a handsome and garrulous twenty-nine-year-old firefighter with Suncor, saw entire houses disappear before his eyes, but this was only one of many extraordinary experiences he would have that afternoon. "We got the call from the city around lunchtime," Welsh told me. "It was crazy: 'Send everything'—without any description of what they needed us to do."

Welsh and his three-man crew found themselves leaving the massive Suncor facility, which was still under blue skies, and racing down 63 in a thousand-gallon pumper truck—lights and sirens—with no idea where they were supposed to go. Even when they came in view of the fire, it was still unclear. "We're firefighters," Welsh said, "we're trained to put out fires. And one of the weirder things I remember is passing fires—on both sides of the road—cars on fire, houses on fire. I was driving by them, and you're not supposed to do that. I was ignoring them because, over the phone, we got sent to Beacon Hill, and if we stopped every time we saw a fire we'd never get anywhere. So, we pulled up to Beacon Hill and it was on fire—like, *all* of it. It was superhot. You couldn't see five feet in front of you. The heat, the smoke, not being able to see—it was like sensory deprivation and sensory overload both at the same time."

Firefighters carry charcoal filter masks with them (along with full face masks and breathing apparatus), but Welsh and his men gave all of theirs away to police and traffic cops who were standing at intersections in the midst of the smoke and fire, managing the traffic. "We were there probably twenty minutes helping evacuate people," Welsh said. "You couldn't see anything; I lost one of my guys in the first five minutes." Welsh and his men were equipped with radios, but because they were operating on Suncor's private system and the repeater was back at Site, they couldn't communicate with each other. Eventually, the missing man emerged from the smoke, coughing and hacking. "Then we got a call from command saying, 'Get out.' It was a May-day: 'Get out. It's too dangerous.'"

And so, with the citizens and fire crews evacuated, Beacon Hill was abandoned to the fire.

"After that," said Welsh, "I got sent to Wood Buffalo."

The neighborhood of Wood Buffalo is on the extreme west side of town, just above the golf course where the fire had jumped the Atha-basca earlier that morning; it is about as far west as you can go and still be in Fort McMurray. "I was annoyed," Welsh said, "like, 'Why am I going all the way over there?'"

After jumping the river, burning over the golf course, and level-ing the clubhouse, the fire continued barreling northward, through mixed forest and muskeg, directly into Thickwood, a catchall name for a dozen tightly packed suburban neighborhoods, which include Thickwood Heights, Westview Heights, and Hillcrest. As their names imply, these neighborhoods are built on high ground above the river, and the fire had to run uphill to get there, just as it had in Beacon Hill and Abasand Heights. In the southwest corner of Thickwood, exposed to open forest—and the fire—on two sides, was Wood Buf-falo Estates.

In an astonishingly short period of time, the question had changed from "Where's the fire?" to "Where *isn't* it?" Fire was now burning with unsurpassed fury from one end of town to the other. Houses were burned, burning, or in imminent danger of igniting from Gre-goire Drive in the southeast all the way to Thickwood Boulevard in the northwest, across a front more than five miles wide. As Welsh ran against the gridlocked traffic back up 63 and across the bridge, the fire was burning toward the landfill, it was burning at the doorstep of the water treatment plant, and it was burning downtown on Hospital Street, where the six-story medical center was about to be evacuated. From Beacon Hill, the fire had jumped the highway—four lanes wide plus breakdown lanes, on- and off-ramps, and a broad median—and was burning eastward through Waterways, all the way to the baseball diamonds in J. Howard Pew Memorial Park. Meanwhile, embers and ash were falling on downtown and across the Clearwater River, settling over the two-hundred-year-old ruins of McLeod House, the first trading post in the Athabasca region.

After crossing the bridge, Welsh and his crew took the first cloverleaf exit to Thickwood Boulevard, a six-lane strip running through the heart of the west side's southern neighborhoods. As the pumper made its way up onto the west side plateau, Welsh and his men could see to their left, above the shopping centers, pocket malls, and many hundreds of homes, a towering wall of illuminated smoke that appeared to block out everything to the south. A lot of that smoke was coming from Beacon Hill and Abasand, and a lot more was boiling up out of the forest just to the west, but it was hard to tell how close it was. Speeding down Thickwood now, to the western edge of town, they turned south at the Esso station onto Real Martin Drive toward Wood Buffalo. Welsh knew this part of town like the back of his hand, because he had grown up here. He had a house of his own in Dickinsfield; you had to drive through it to get to Wood Buffalo.

With about eight hundred houses and condos between them, Wood Buffalo and Dickinsfield had significant strategic value, because both communities were surrounded by shallow ravines lined with trees. For residents, these natural greenways represented prime real estate that could accommodate walkers, bikers, and cross-country skiers of all ages. For the fire, they represented fuses to a bomb. That bomb was Birchwood Trails, the city's biggest park. Thickly wooded, with no roads and convoluted terrain covering a square mile, Birchwood Trails was essentially a captive boreal forest ringed by dense suburbs—Thickwood to the south, and Timberlea to the north. On a day like this, it would be suicidal to confront a fire in there. But the price of not confronting it would be catastrophic. Together, Thickwood and Timberlea were home to two-thirds of Fort McMurray's permanent population: sixty thousand people, living in twenty thousand houses and condos. It was still early in the afternoon and the temperature was rising, along with the wind, even as the humidity continued to fall. Should those woods ignite under these conditions, the park would go up like a munitions dump, taking the west side with it. By the time Welsh's Suncor pumper turned onto Real Martin Drive, the fuses were being lit.

But Welsh, still stunned by the holocaust he'd just witnessed in

Beacon Hill, had yet to realize that houses just to the south of him were also on fire. "I didn't know that the fire had crossed the river," he said. "The river's almost a kilometer wide so I didn't think for a second that this side of town was in danger. Nobody had told us. We felt stupid. I called my wife and I'm like, 'You should probably get your stuff ready to go, they just sent me to our neighborhood.' And so she packed a bag and packed up our two boys and they left." The Welshes' home was so close that had his wife turned left out of their driveway, toward Thickwood Boulevard, her escape route would have taken her right past her husband and his crew. "She knew that," Welsh said, "so she turned right. She didn't want the boys to see me in perceived danger. If I had seen them it would have been heartbreaking."

Separations as sudden and poignant as this were now happening all across the west side as the closure of 63 and the jamming of cell phone signals, among many other factors, prevented families from communicating, even when a loved one's car was in sight.

Welsh and his crew pulled up to a hydrant at the south end of J. W. Mann Drive, a side road between Wood Buffalo and Dickinsfield lined with tightly packed houses. The street, which fronted open forest and muskeg to the west, appeared to be evacuated. If there were other fire trucks around, it was unclear where they were. As far as Welsh knew, he and his crew were first on scene. He still couldn't see how close the fire was, but he could hear it through the smoke. "Our first indication that the fire was headed towards us," he said, "was the sound. A forest fire sounds like a freight train, and we heard a freight train increasing in volume."

Welsh and his men were directly in its path. As they tried to get their bearings in the whirl of smoke and ash, the roar of the fire was drowned out by the even louder sound of a large airplane passing way too low. "A water bomber dropped its load right in front of us," Welsh said. "Missed us by twenty feet. It hit the tree line and we all got splashed, and I said, 'What the heck is he doing? Why is he dropping right here?'"

This was Welsh's first indication of how desperate the situation had become on the west side, and how close the fire was to Birchwood Trails. Water bombing is a wildfire-fighting tool; it is seldom used in residential areas (though it has become much more common in recent years). Even a small helicopter bucket of one hundred gallons weighs eight hundred pounds; dropped from ten or twenty stories up, eight hundred pounds of water or retardant can crush a roof or kill a person. A Convair 580 air tanker, like the one that had just buzzed Welsh and his crew, carries twenty times that. The pilot likely had no idea there were people below him, and it was impossible to see, anyway. Welsh soon understood the need for such an unorthodox drop: "That's when I got on a roof, looked over the trees, and there was a wall of fire coming towards us."

Somewhere inside that burning wall, entire blocks of houses were being obliterated. These were the same streets where his friends and neighbors lived, where Welsh and his brother, also a firefighter, used to ride their bikes. Now, fire was looming over the treetops, over the rooftops, over everything. It was the biggest thing in town, the biggest moving thing most residents had ever seen, and it was consuming their city unabated, cutting them off from the rest of the world. If the fire wasn't stopped in its tracks, it would run up the ravines into Birchwood Trails, igniting the inner cores of Thickwood and Timberlea as it went. J. W. Mann Drive formed a barrier between the fire and one of these ravines precisely where Welsh had parked his truck. From his point of view, the flames appeared as a wave bending over him, about to break. It wasn't an illusion; fires—like flowers, like the residents of Fort McMurray, like so many living things—lean toward energy, "because that's where the unburnt oxygen is," Welsh explained. "So, it's being pulled—drawn if you will—forward by the oxygen."

As the fire was drawn into the deep reservoir of unburned fuel and oxygen that was Wood Buffalo, the sound changed: the steadily intensifying freight train roar was complicated by a much more intimate symphony of chaos—the sound of houses detonating. "The forest is this single, solid sound," Welsh said, "but the sound of a house burning is tons of different sounds. The wood that's burning crackles, you hear glass break all over the place, there's pressurized things letting loose that make wisps, and plastic siding is melting and making a different sound, then it hits the insulation and that creates like—" Welsh is a man rarely at a loss for words, but here, they failed him. "It's all different sounds," he said at last. "You're hearing explosions, pop after pop after pop. Propane tanks are going, tires are popping, gas tanks in garages, compressed air tanks."

One of the things that set this fire apart was the sheer number of vehicles and gas grills. Virtually every house and condo had its own grill supplied by a twenty-pound propane tank. Likewise, almost every garage and driveway had vehicles that had not been driven out. As each one was engulfed, its alarm would yelp briefly before melting as the tires and gas tanks blew up in rapid clusters: five tires and a gas tank, over and over again—followed or preceded at random intervals by the grill's propane tank. Because many residents were tradespeople and hunters, garages and basements were also mined with ammunition, fuel cans, and gas cylinders used for welding.

In the language of firefighting, the term for one of these detonations is "BLEVE" (*blevvy*), an acronym for "boiling liquid expanding vapor explosion." Among firefighters, BLEVE is used as both noun and verb. Most fuel tanks are designed with safety valves, allowing for a controlled release of burning gas vapor in the event of a fire. But all fuel tanks have a breaking point, and this fire, due to its intensity, pushed many of these tanks past theirs, resulting in much more random and dangerous explosions in which the tanks were sometimes reduced to flying shrapnel. By midafternoon, the number of BLEVEs—large and small; gas, propane, diesel, and acetylene—was already into the thousands.

The firefighter John "Toppy" Topolinski, who put thirty-seven years in with the Fort McMurray Fire Department, recalled that "the explosions were constant." Pat Duggan, a captain and former soldier who did a tour in Yugoslavia in the 1990s, said the steady thumping reminded him of mortar fire, "only without the incoming warble." BLEVEs were occurring with such regularity that firefighters quickly learned to use them as a way to track the progress of the fire through the impenetrable smoke. Each burst represented another rapid infusion of combustive energy into a fire that was now taking over the land and air so thoroughly that it would soon be creating its own weather. It was so hot inside this storm-sized combustion system that ordinary household objects were becoming, not just fuel, but torch-like accelerants. Foam mattresses, plastic garbage cans, bags of Doritos—all preheated to combustive temperatures—were not so much igniting as exploding into flame, each detonation contributing yet more energy to the cataclysmic whole.

The fire, pushing northward from the river into Thickwood, aided by the southwest wind, was wrapping around the western edge of town, burning into Wood Buffalo on its way north toward Dickinsfield and the much larger neighborhood of Timberlea, where Chief Troy Palmer and Shandra Linder lived. Barely three hours had elapsed since the first trailer home had ignited in Centennial Park, and already the entire city—all sixty square miles of it—was at risk of being engulfed by the fire. This was the beginning of a long period of brutal decisions and heavy sacrifice, not so different from a medieval castle's defenders retreating from the walls to the castle keep. As blocks and neighborhoods were overtaken, firefighters cut their losses, fell back, and devoted their limited resources to critical infrastructure. "We were overwhelmed, right?" said Evan Crawford, the firefighter and Bronto operator. "But we were in our own little world at this point—you're doing what you can in your own little spot."

In this way, isolated groups of firefighters were fighting pitched battles all over the city: for schools, for the hospital, for the airport, and—a true sign of desperation—for their own fire halls.

⸺

In addition to being a firefighter, Lucas Welsh is a pastor and bandleader at Fort City Church on Thickwood Boulevard. Like most of the Christian churches around the city, Fort City is evangelical. The church itself is based out of a wood-sided, 1970s-era showroom located at a busy intersection shared by three gas stations. Inside, there is no pulpit, but there is a formidable sound system and, right next to the bandstand, a deep wading pool with stairway access for full-immersion baptisms. Next door, on the edge of the church's one-acre parking lot, a snow-white doublewide serves as the church office. On a bright and freezing January day, Welsh and I sat together in a back room there, alone save for a pair of enormous and glittering Styrofoam skulls—one black and one white—that loomed over Welsh from opposing walls, as if to better hear what he had to say.

I had gone into that office thinking we were going to talk about what goes on in the mind and soul of a firefighting believer when an end-times-class fire descends on his hometown. And we did talk about that, but it was when the conversation turned to physics—a science as mysterious to Welsh as the presence of God is to me—that we zeroed in on a kind of essential truth about the transformative power of fire. Under certain circumstances, Welsh explained, fire can alter the nature of things so completely that they are transformed from objects in space to moments in time. Few people on Earth have seen what Welsh and his crew saw that day and lived to tell about it, much less articulate it as he has.

"We avoided heat," Welsh explained to me. "It's a different type of firefighting: if one house is on fire, you're going to go in there and try to put that fire out. We didn't go into any houses. There was no time. There was one point where we were looking at a house—an entire house was disappearing—in five minutes."

I asked Welsh what he meant by "disappear." His answer was unambiguous: "Fully there, totally normal, to fully gone was five minutes."

"That sounds physically impossible," I said.

"It does," agreed Welsh. "You throw wood on a campfire and you can sit around that for an hour before it's just coals. These houses were disappearing in five minutes."

A typical house weighs fifty tons or more, not including the foundation. It is composed of thousands of component parts and dozens of different materials, many of which are rated for fire retardancy. In Fort McMurray, houses are built—and certified—not only to resist fire but also to fend off wind, rain, and subarctic blizzards for decades. But on J. W. Mann Drive, and in many other places on the west side, these large and sturdy structures were incinerating like milk cartons in a bonfire, converting almost instantaneously from inert matter to pure energy in the form of combustive heat and its signature by-products of light, sound, soot, and vapor. I asked Welsh if he knew anyone in the department who could explain how this might be possible. "Firefighters aren't really well known for being scientific or intellectual," he said, "so I don't know that there'd be anybody who could weigh in on that."

The people who could were back at the fire hall, in offices with their names on them. The guys on the truck are generally following their instructions, but out there, on the western perimeter of town, directly in the fire's path, there was no fire chief, no fire inspector, and no manual. There were just four men, rows of houses going down like incendiary dominos, and a fire the likes of which they had never seen.

What Welsh and his crew were witnessing firsthand—and would soon be timing on their watches—was a hazard that had not been considered by Underwriters Laboratories: an entire house behaving like an enclosed room; in other words, flashover occurring *outside*— involving not just one house, but rows of them. To manifest on this scale, the environment itself—the atmosphere above and around these homes—had to take on the same thermal characteristics as the man-made, twelve-by-twelve-foot living room in the UL fire video. All across its now miles-wide front, isolated vortices of fire, fuel, and terrain were creating virtual "rooms"—combustion chambers, really—that enabled the immediate and total incineration of everything inside them. What is striking and alarming about this behavior is that there is nothing "containing" a neighborhood or a city, and yet the fire, with its abundant fuel and vigorous convection of superheated air, achieved an equivalent effect of broiler-like enclosure. Fueled now by the houses themselves, the fire "energized" everything in front of it, from cedar hedges to three-story homes, from basements all the way up through the smoke column, which was fast becoming a storm system of heat spiraling into the stratosphere.

More alarming still, these transformations were occurring less than ten degrees from the Arctic Circle, in early spring, despite ice-covered local lakes and riverbanks, and even though freezing temperatures had been recorded less than a week earlier. Fort McMurray was now operating under different laws than the subarctic world around it. The city and the surrounding landscape had become something akin to a fire planet—not a biome but a "pyrome" whose purpose was not to support life but to enable combustion.

Welsh and his crew may have "avoided heat" because the heat was otherworldly and all-consuming, but they couldn't just let these houses burn. As Welsh said himself, "You're not supposed to do that." One of the unspoken obligations of a firefighter—of any first responder—is to not run away. Even if your equipment and manpower are inadequate to the task at hand, even if your adversary is disintegrating entire houses like a Martian death ray, your duty is to somehow stand between it and the citizenry and infrastructure you're charged with protecting. Courage and self-sacrifice are not the exceptions, but the expectations. Often hopelessly overmatched, small bands of professional firefighters from the city and the bitumen plants, combined with volunteer crews from nearby hamlets, were stepping into the breach and holding the line. Resources had been scattered so widely and spread so thin that, in many cases, firefighters did not know who their comrades were; sometimes, there wasn't a ranked officer to be seen. As Crawford, the Bronto operator, put it, "procedure" that day amounted to "If you find something, you're in charge of it."

Due to the overwhelming volume of calls going through cell towers, along with other infrastructure issues (one local tower was lost in the fire), some squads of firefighters found themselves in empty, burning neighborhoods with no outside communication for prolonged periods. Coupled with the heavy smoke, flames, and general chaos, there was a sense of being caught behind enemy lines. The speed of events and the attendant collapse in communications caused a vertical, military-style command structure to swing ninety degrees to the horizontal. In some cases, over-the-radio orders to abandon a given area were simply ignored by the men on the ground. Wood Buffalo–Dickinsfield was one of those areas. This was not someone else's war, but their hometown they were fighting for. "It was chaotic and it was personal," Welsh said. "I love this city. It's my home and that was my neighborhood; my kids and my wife were five hundred yards away from me and evacuating."

Two of the most crucial battles were for these two neighborhoods. Welsh and his crew were not alone in this fight, but they would have to hold Mann Drive, one of the gateways into Birchwood Trails, on their own. Even though the Slave Lake crew had become vocal champions of this spontaneous and egalitarian approach, those in command of the Fort McMurray Fire Department were still operating in a world where freelancing of this kind was frowned upon. But that afternoon, the brass in the REOC might as well have been in Ottawa.

With only each other to rely on, Lucas Welsh and his crew devised a way to respond effectively without getting themselves killed or burn-

ing up their equipment. Together, they performed a kind of spontaneous ontological leap that redefined what a house is, and they did it not
by thinking like firefighters, but by thinking like fire. Fire's world is
a stark and elemental place where all carbon-based things, including
firefighters, are potential fuel. And, just like a match, a candle, or a
stick of firewood, each house is quantifiable—not in dollars, or tons,
or square feet, but in burn time.

"We were losing," said Welsh, "so we started making decisions
based on that number—five minutes per house: You'd say, 'How long
is it going to take? How many houses do I need to get ahead of it, to
give myself twenty minutes to set up and stop it?' So that ended up
being our benchmark for time, and we changed our tactics. When
we'd roll up on a burning house, we wouldn't even try, we'd go four
down. We would sacrifice four houses to stop the fire from progress-
ing." At the fifth house, Welsh and his crew would stop and set up:
"You'd take a hydrant, you'd start wetting down that fifth house,
you'd wet down everything around it and then you'd set up to stop
that fourth house. The only thing left standing of that fourth house
was the wall adjacent to the fifth house. It was a tactic that started
working."

The tactic may have been working, but the houses Welsh and his
crew were sacrificing were those of his neighbors. They represented
the nest eggs, retirement plans, and irreplaceable memories of people
much like himself, and so these losses, witnessed and also incurred by
so many firefighters that day, often felt like personal failures. There
was simply not enough machinery, manpower, or water to apply this
tactic on every block, so these small victories were scattered among
greater defeats.

But Wood Buffalo's susceptibility to fire had been foreordained. I
asked Welsh about the spacing between houses in that part of town.
"Fort McMurray's bad for that," he said. "Land's so expensive here,
you're talking a meter and a half (five feet) between walls of houses
sometimes."

In other words, it is possible to walk down J. W. Mann Drive on
the rooftops. These "zero lots" are a common feature of homes built
within the economic blast radius of Alberta's last petroleum boom.
Compressed, firetrap neighborhoods like these can be found from

Fort McMurray all the way down to Calgary, where a mile-deep ring of cookie-cutter dwellings mushroomed around that city's downtown during just a few years of frenzied building between 2000 and 2010. In the case of a fire, such narrow gaps between structures can have a bellows-like effect, sucking air and oxygen at high velocity through the gaps, further invigorating the fire. When those houses are clad in vinyl siding, as so many in Alberta are, it literally adds fuel to the fire: firefighters refer to this cheap, weatherproof material as "solidified gasoline."

In May 2016, despite their small and crowded lots and often hasty construction, houses on Fort McMurray's west side were commonly priced between $500,000 and $1 million. Starting in the early afternoon of May 3, millions of dollars in real estate were converting to combustive gases every minute, and there was no end in sight. Through the smoke, fire, and chaos, no one had time to consider the possibility that this might go on for days, or weeks, or that, elsewhere in the city, houses could burn even faster than they were burning in Wood Buffalo.

13

I do my best to do my damnedest
and that's about all I guess.

—Tim Hus and Corb Lund, "Hurtin' Albertan"

In rapid succession, Beacon Hill, Waterways, and Wood Buffalo had become raging interface fires, and Abasand Heights was next. Before it became known as a popular neighborhood for workers in the bitumen industry, the name belonged to Abasand Oils, Ltd. (a contraction of "Alberta Sand"). Opened in 1936, Abasand Oils was one of the first commercial bitumen processing plants, upgrading raw bitumen into diesel fuel. The plant was abandoned in 1945 after burning down twice. In 2016, the well-established neighborhood of Abasand Heights followed the tree-lined ridge northward for a mile, but most residents were concentrated around the crescents and cul de sacs at the south end, near an Esso gas station, a mini mall, and the Father Beauregard Elementary School. About five thousand people lived up there in 1,800 homes and condos.

By 2:45 p.m., with Beacon Hill in flames, and evacuation orders expanding by the minute, Abasand residents were fleeing en masse—everyone except Wayne McGrath. McGrath (pronounced *McGraw*) was a welder and millwright with nearly twenty years in at Suncor. Thanks to long hours at doctor's wages, McGrath and his neighbors were able to afford homes of their own in Fort McMurray's superheated housing market. Credit was easy to get when you worked for Suncor, and, in addition to real estate, it paid for some very expensive toys. McGrath's large garage doubled as a hobby shop where he tore down, rebuilt, and modified cars, trucks, boats, snowmobiles, motorcycles, and ATVs. Nearby stood missile-shaped tanks of oxygen and acetylene for welding projects. These, plus various gas and oil cans, and the fuel tanks in his menagerie of machines, made McGrath's garage a dangerous place to be with an open flame. McGrath understood this, and fire was a hazard he was comfortable with—as a welder in the bitumen industry, where the threat of fires and explosions comes with the job, he had to be. This is why he had flame-resistant Nomex coveralls hanging in his garage, and why, on that exceptionally warm Tuesday morning, he happened to be wearing them as he switched out the snow tires on his truck and changed the oil. McGrath was lubricating himself, too: he had the day off, so he mixed himself a couple of screwdrivers to go with the sunshine as he puttered away, getting his fleet ready for summer.

McGrath was tall, wiry, and hard, his head clean-shaven. He grew up, as more than a third of Fort McMurray's residents have, 2,500 miles away on Canada's underemployed, often impoverished east coast. McGrath was from Labrador, which is so close to Greenland that iceberg bulletins are a regular feature of marine weather forecasts. He grew up inland in a town so new that his older brother was only the second baby to be born there. It's called Labrador City, another one-industry town where the overwhelming presence of ice, granite, and iron ore necessitate a rugged approach to life. "There's four boys," McGrath recalled, "none of us were angels. None of us are left there now." Not everyone in Labrador City was sad to see the brothers go. "We left a mark there for sure," he said.

McGrath was forty-five, but you could still feel the hell-raising potential jazzing under the surface like full-body sewing machine leg. It helped explain his affinity for snowmobiles and motorcycles, which was as visceral and intuitive as a hunter's bond with his dogs. He understood these machines intimately, inside and out; in return, they completed him in ways that other things, including people, could not. Very few things made Wayne McGrath cry, but his Harley was one of them. In the course of our conversation, another thing came clear: given an opportunity to back down, he wouldn't.

All weekend, McGrath had been aware of the growing plumes on the near horizon, but he wasn't too concerned, because those fires were in the forest and he was in town. In McGrath's mind, there was an invisible barrier between the two. Every year the forest burned somewhere up here, sometimes spectacularly, but never in the fifty-year history of Abasand Heights had the greenbelt been crossed. The possibility that this barrier might be an imaginary one arose shortly after noon. "Just like that," McGrath said, "the wind changed. I looked up and seen smoke, and my neighbor, who's pretty on the ball, is like, 'This ain't good.'"

Not long after that, a police car rolled through and McGrath flagged it down. He asked the officers what was going on with that fire. "Just smoke," they said, "no cause for alarm." When McGrath relayed this to his neighbor Ralph, they looked at each other and at the smoke billowing up out of the forest, darker now, and leaning in their direction. Maybe those cops didn't understand what they were looking at, but McGrath and his neighbor were beginning to. McGrath's home stood on Athabasca Crescent, at the extreme southern edge of the neighborhood, closest to the fire. Across the greenbelt, on all sides, this long, narrow whaleback drops down steeply into serpentine river valleys two hundred feet below. To the west was the Horse River, for which the fire was named; due south was the Hangingstone and, just beyond it, Beacon Hill. No one who witnessed it, firefighter or civilian, could quite believe how quickly Beacon Hill was overcome by fire.

There was no longer any doubt that Abasand was next, and the two-lane access road was now choked by three lanes of evacuating traffic, forcing emergency vehicles to use the shoulders. While his neighbors, Ralph included, fled for their lives with their children and pets, McGrath, who lived alone at the time, marshaled his fire extinguishers and watered down the surrounding trees and fencing. There, on his choice corner lot, directly in front of a fire that had emptied his neighborhood in less than an hour, McGrath prepared to make a stand. He had a lot to lose, and he was damned if he was going to. "I had about a hundred grand of tools in my garage," he said, "and five sleds"—snowmobiles—"'cause I'm a toy guy." These vied for space with a quad ATV, a vintage car, a late-model Dodge pickup, and McGrath's beloved Harley-Davidson Road Glide. Just outside, on a trailer, was a canary-yellow jet boat, which came in handy in this largely roadless but river-laced country.

"My neighbor gave me a hand to push my Harley out of the shed," McGrath said. "It wouldn't start because it was put away for the winter. We pushed that up into my garage. I put the battery in my antique car, a Cutlass Salon—a '77—fired that up, put that in the garage. So the smoke's getting pretty intense. I seen neighbors just fleeing. Fire trucks came by. My buddy Gavin phoned me, said, 'Wayne, they're evacuating Abasand, time to get out,' and I was like, 'No, buddy, I ain't goin'.' He said, 'No, Wayne, serious, you got to go.' 'Nope. Not close to me yet. I'm not letting my place burn.'"

Unless you saw it firsthand, it was hard to appreciate how extraordinarily combustible the woods had become that day, but when the embers started falling, McGrath understood. "I was looking out the window at my garage and off in the woods, *poof!*—a three-foot flame just appeared out of nowhere and, just like that, trees started going all around. I got the sprinklers going, got in my garage, and, just like *that*, the flame went all around the house. Melted my siding, cracked my glass. I opened the garage door and it filled full of smoke pretty fast. I phoned my son in Ottawa and I said, 'I'm in a predicament here, bud.' He's watching on the news. The garage filled with smoke, I got down on the floor, had a couple of swigs of vodka, and said, 'I don't know if I'm getting out of this, buddy.' Flame came right to my fence. It was insane heat. My son's a lifeguard so he's freaking out, telling me to breathe and breathe and breathe. I stood up anyway, went to the door, and it was like a hurricane. It created its own weather, right? I went out and the flame had burned all the trees, all the grass right to my fence, and then it was gone. It was running. I went out [with the garden hose]. I fuckin' . . . beat it. I fuckin' beat it. I had my son on the phone: 'It's gone, man, it's gone!' Didn't burn my fence, came right to the grass. I have a big tree inside my yard; that caught a little bit of fire that went out because I'd had a hose on there right off the bat. I open the door and it's like—fire's gone!"

But it hadn't gone far.

McGrath drew a picture of his lot, the houses behind it and across the street. "My fence is like this," he said, as his fence, his yard, his house and garage—all he had—was resurrected in his mind. "I got a big greenbelt here, then there's a house here, really close to the trees. So, within minutes, this one's on fire, and then this house—my hose wouldn't reach it. I go over, I can't do it, I can't reach it. I got to save my own place. The next house caught on fire, which is right in the back of my fence, right behind my shed. Not much I could do. So my common fence is here," he said, jabbing at the page, "I put a fire out right here. This is all trees and this one caught on fire. I'm watering down the trees. Lots of trees. This house was on fire. This trailer over here was on fire—Harley-Davidson trailer—I went and put that fire out. There was another small fire a couple of doors down here in the backyard. I put that one out and he had a whole bunch of tires in the backyard so I threw all of them out in the street, out here. The back of my shed was on fire, I put that out. This tree over here was on fire, I put that out. Insanely running around for an hour. Insanely. Like, pure adrenaline."

This was the speed of events—a manic, desperate game of whack-a-mole, only with fire and mortal stakes. Firefighters described the same scenario and the same pace, all across the city, and it never let up.

It is around this time that McGrath, realizing perhaps that he was fighting a losing battle, paused and pulled out his phone. "So, folks, here's the video," he began. "I tried, I tried. Pretty sad. This is what's all around me." The phone, topsy-turvy now, panned around his property. Everything, right up to his white picket fence, was either on fire or burned black. Through his aggressive intervention, McGrath's property had become a historical artifact, a relic of domestic order and green grass surrounded by the future: an entire neighborhood burning to the ground. In the background, blue flashes could be seen amid the orange flames as breaker boxes and transformers exploded. "It's all around," he continued over the ferocious crackle and shatter of burning houses. "Not sure how much longer I can keep it up . . . But . . . those that know me . . . know I'll do my best. So here goes, but I don't like seeing what's down there. Anyway, I'm out. Love you all."

It is the final transmission of a resolute captain fighting to save his doomed ship after everyone else has abandoned it. There, sweating in his Nomex and armed with a garden hose and sprinklers, his blood buzzing with adrenaline and alcohol, McGrath braced himself for what was coming. "By this time," he told me, pointing to another neighbor's house on the map, "this house was on fire. I get up on my shed and I'm watering trees, and then the propane tank blew in my neighbor's barbecue. It almost knocked me off the shed. If I didn't have on them coveralls I'm not sure I would have lived. I got burned pretty good—I was all red in the ears, lost my eyebrows, and knuckles are burned. So I had to douse myself and the flames are coming. Prepped my trees and my fence. Took about five, six minutes per house to collapse—yeah, that's true. The wind is going this way [northward], and the fire kept going down here [northward, on the opposite side of the street], down, down, down about five houses and I'm thinking, 'I'm saving my corner!'"

This is how a running fire behaves: a fury of devastating energy followed almost as quickly by a smoky, simmering calm. With the most volatile fuels exhausted, the fire, like a crazed wolf in a sheep pen, rushes off in search of more victims. It looked like McGrath's risky bet had paid off: his neighbors had fled, but he had stayed and fought for what was his, and he had won.

"Then," he said, "it crossed."

McGrath pointed again to his hasty diagram, and the way his finger traced the fire's progression down the street, its sudden pause, followed by its astonishing shift in direction, was chilling. It recalled the velociraptor in *Jurassic Park* peering at her prey through the window of that kitchen door, suddenly realizing there was another way in. "I'm up on my roof with the sprinkler system," he said, "and I look down and it starts coming *back* up the street."

In order to do this, the fire had to burn against the prevailing wind. Ambient weather no longer appeared to be a factor; the fire was now in charge, empowered by its own rapacious agency. Even though it had all of Abasand at its disposal, the fire doubled back toward McGrath, like it didn't want to miss anything, like it wanted to finish the job. McGrath's property would be a feast. "I had five or six propane tanks behind my garage," he recalled, "along with two bottles of acetylene and a bottle of oxygen, a bottle of argon, and I had a tank of nitrogen for doing shocks. So, it's five, six houses down to the corner here [where the fire turned]. Like I said, five, six minutes per house to collapse. I never seen nothing like it. You could hear glass windows shatter. It was like a windstorm; it created its own vortex of shit. So, one or two houses went, I got down off the roof, I ran inside my house not knowing what to grab. I should have grabbed so much more. Couple shirts, I had a small hockey bag, filled it up full of stuff. I even looked at my photo albums and all my computer shit—think I grabbed it? No. Looked for my passport, couldn't find it. Come back outside, it's about three houses away. I get back up on my roof. I'm there with the water hose just going and going and going. By this time it's all in flames here."

Even in that back corner of Abasand, clearly a lost cause, first responders were still searching for people who might have been left behind. They weren't expecting to find someone like Wayne McGrath. "Three firemen walking down the street wearing the masks," he said. "They see me up there. One guy pulls his mask off, 'Get outta here!'

" 'Fuck you! I'm saving my fuckin' house!' I knew, though—once the shingles were on fire—my little hose just wasn't doing much. So I get down, and me and him—I had words with the guys and they're like, 'You have to leave or you will be forcefully evicted.' Whatever. I wasn't going out without a fight. I'm almost in tears, but I still didn't cry—never cried 'til I looked back and seen my roof on fire. Anyway, I looked and the house next door—that one's on fire. Nothing I could do. I said, 'I want forty-five seconds.' He said, 'You better be getting medication.' I went in my garage, swung open the garage door, fired up my Cutlass, and put it out in the parking lot, and he said, 'Now get!' I said, 'Nu-uh.' Back in, grabbed my Harley. Pushed it out. Then I got in my truck and looked back and my house was on fire. So, when I was driving away, I looked in the mirror and called my brother. That's when it hit me."

Like Paul Ayearst in Beacon Hill, Wayne McGrath was almost certainly the last civilian to leave Abasand. As McGrath made his escape, steering his truck through fire, smoke, and tears, he was leaving behind everything he owned save his truck and his hockey bag. Behind him, fully engaged by the fire now, were close to a million dollars' worth of welding and mechanic's tools, fast machines, real estate, and household goods. The fury of the fire as it found its way into all those tanks of gasoline, propane, acetylene, and oxygen can only be imagined.

Variations on this debacle, which had already played out many hundreds of times in Waterways, Centennial Park, and Beacon Hill, would occur even more dramatically in Abasand, which was twice the size. And this was only the beginning.

As stubborn as he was, not even Paul Ayearst had stayed long enough to see his own house burn. Other than Wayne McGrath and some unlucky firefighters, few people did. But they bore witness in other ways. As safe and small-town feeling as Fort McMurray's neighborhoods are, and even though there is only one road out with a long way to anywhere else, home security systems are surprisingly popular. These devices, linked to smartphones, were programmed to notify their owners when something went amiss, but in the case of the fire, it wasn't always clear what that was. Sometimes, the only clue was a loss of contact with their home. In some cases, homeowners had their phones linked to security cameras, and these revealed the fire's behavior from points of view unseeable, and unsurvivable, by human eyes. They told stories otherwise impossible to tell.

One in particular looks like it could have been shot by the director of *The Blair Witch Project*. Taken from a security camera in an upstairs living room, just a few blocks from McGrath's home, the point of view appears to be from a mantel or bookshelf, looking across the room toward an inviting sofa flanked by end tables with lamps and photos on them. A large framed print hangs above the sofa on a brick-red wall. To the left, next to one of the end tables is a cabinet with a fish tank on it, and just to the left of that is a plate-glass window. Everything appears to be in order: the table lamps are lit, and so is the fish tank, as if whoever lives there just stepped away for a moment. But something is outside the plate-glass window. For the first thirty seconds or so, it looks almost like a movie screen with a defective black-and-white film playing across it. Vague shapes race past in flickering shades of gray, but it is impossible to see what, or who, they might be—almost as if the glass is frosted. Another window, on the right-hand side of the room, shows a gray sky and spruce boughs tossing fitfully beyond the glass. If you didn't know there was a fire nearby, you might guess it was a winter storm.

As the movement by the left-hand window intensifies, so does the contrast—until it is clearly smoke, pressing up against the glass before racing on, light to dark and back again. A faint crackling can now be heard. This sound, combined with those restless clouds so close against the window, gives the impression of a motivated presence just outside. Somewhere can be heard the faintest trace of voices, but their tone sounds unrelated to the events before us. There are no other sounds save an increasingly loud crackling. With your eyes closed, you could mistake it for the splatter of heavy rain. Then there is a bang, and a few sparks flicker past the window. This is followed by another bang, louder this time, and a piece of something—a gutter? vinyl siding?—drops across the window, followed by louder bangs and more strips of siding. And then the fire is there, right outside, bobbing this way and that, like it's trying to see inside the room. The window, about four feet square, is triple-paned for subarctic winters, and we can hear it cracking in the heat. The outermost pane gives way first and falls out with a crash. It is clear now that the fire is trying to get in; the remaining glass is going alternately black with smoke and blindingly bright with flame. That window, with its jarring play of light and shadow, could be submitted to an experimental film festival.

The window does not shake or rattle, but the sound and rapidly moving smoke outside make it seem as if it does. Suddenly, the fire punches through the second layer of glass, making the same sound and hole as a fist. There has been no three-dimensional intervention of any kind, only this vaporous, spectral presence, and yet it is battering its way into the room. This is what horror is—a malevolent entity from another dimension breaking through to this one. More pieces are broken out, and then the last layer gives way. The smoke enters first, creeping up and across the ceiling as flames now probe the hole in the glass, wrapping around the broken edges like gloved hands, darting in and out of the window a foot from the fish tank. Fire is now visible outside the right-hand window as well. The spruce tree is in flames and strips of molten vinyl siding can be seen dangling outside the window, stretching and dripping in a surreal way. Somewhere, a smoke alarm goes off, much more quietly than one would expect.

Flashover is now minutes, or moments, away, and yet the nanny cam continues recording and transmitting as more pieces of window glass fall out and shatter. Smoke is rolling across the ceiling and into every corner. Even so, the fish tank remains visible, holding our attention because it contains living things left behind, and because something—it is hard to tell exactly what—is happening inside it. The smoke, thickening and darkening, continues to crowd in, shrinking the room and blurring the details as the sofa ghosts in and out of view. The tank light is still on and so are the table lamps, their pale beams creating cones of visibility, which feel now like zones of protection over the photos they illuminate. Who are those people? Everything around them is becoming increasingly opaque. The crackling grows sharper and more insistent. The fire is in the room now; flames are brushing against the fish tank and sparks streak by in the turbulent air like bright leaves in a storm. There is a sizzling sound; the fish tank—the entire room—is now far past the boiling point. The nanny cam, pushed well beyond its design specifications, cuts to black.

Scarcely five minutes have elapsed since the room was habitable.

————

Acting Captain Mark Stephenson, the same firefighter who had heard the Couttses' ominous prediction at the Hall 5 sprinkler demo, was dispatched to Abasand with a truck and men. Stephenson had not expected to be sent to his own neighborhood that day, but once on scene, he responded to this chaotic situation for which there was no training the same way so many of his brothers and sisters did—by resorting to hardwired fundamentals. There wasn't much else to go on; instructions from the top had been minimal, along the lines of "You're on your own. Get yourself a water source and defend." Stephenson was doing exactly that—connecting a hose to a hydrant—when he made a sickening discovery. While opening the valve, he encountered a sound he'd never heard coming from a hydrant before: instead of rushing water, there was only the breathy hiss of sucking air. The hydrant was dry.

By now, hundreds of Abasand's houses and condos were on fire. Running to each one was a half-inch waterline connected to the city's water supply. As each house burned down, those lines would break, and water would run freely at a pressure comparable to a powerful garden hose. The same thing had happened in Centennial Park, Beacon Hill, and Waterways. One broken line wouldn't make a difference; a hundred might not even be noticeable; but a thousand or more flowing unchecked, hour after hour, could compromise a pump's ability to maintain pressure and push water up onto those hilltops. They could even drain the reservoir feeding it. With no water, the firefighters were like soldiers deprived of ammunition.

Meanwhile, the relentless showers of embers were igniting anything that would burn. Flowerbeds and garden paths covered in what had recently been damp or frozen cedar chips and mulch were transformed into wicks and fuses—pathways for fire—many of which ran toward houses, sheds, or woodpiles, and adjacent to wooden fences. Effectively disarmed, with the fire spreading in all directions, firefighters had no choice but to shift into evacuation mode. But Stephenson, who seems constitutionally incapable of giving up, took matters into his own hands, something he was used to doing. A captain and dangerous goods specialist named Ryan Pitchers described how, during one of the weekend evacuations, Stephenson had wrestled a "monstrous horse" into a trailer all by himself. Before meeting Stephenson, I had never heard of a person kicking down a garage door, nor what he described doing once he got inside. With no water pressure and his neighborhood in flames, the logical thing to do—the only thing to do—was to find his chainsaw and start cutting down the neighbors' fences as fast as he could. "I think I knocked down about three or four fences," he said, "before Captain Collins came screaming up in his truck. He's like, 'We're abandoning Abasand! I got the pickup here, let's get into your house and grab everything we can.' I'm like, 'No, let's just go to work.'"

When it was suggested that this must have been a hard decision to make, Stephenson was matter-of-fact. "It was a no-brainer at the time," he said, "'cause I'm not paid to protect my house. Maybe it's because I spent so much time in the military, but I have pride in what I do. If it means sacrificing myself or something of my own for somebody else, it's my job, right?"

There was no aw-shucks or foot shuffling about this. It was a frank description of how Stephenson sees his role, both in the brotherhood of firefighters and in the larger community. Character has a lot to do with it, but it helps knowing that the union has your back.

"We're an IAFF [International Association of Fire Fighters] department," explained his union brother Pitchers, "and it is a very, very tight-knit group. Not to say that we're not with the [bitumen] plant guys and the volunteer firefighters, but I mean IAFF is a brotherhood and a sisterhood. We see a disaster, we want to be on the next frickin' truck to go and help."

There are a lot of people like Pitchers and Stephenson in Fort McMurray, and a lot of sacrifices would be made that day. Shortly before abandoning Abasand, Stephenson pulled out his phone and shot a brief video of his house in flames—a kind of farewell to the home he had just finished renovating and where, until forty-five minutes ago, he and his wife had been raising their young family. This was just one of many surreal and dispiriting sights firefighters would witness over the course of this fire: their own houses, and those of their friends and family, burning down in front of their eyes—in front of their equipment—and no way to save them. It was like doctors in a small-town hospital who were used to dealing with patients one or two at a time, suddenly having to perform triage on neighbors and family members.

Up on Beacon Hill and Abasand things were happening that were unthinkable in traditional firefighting: crews fled neighborhoods leaving their hoses still attached to hydrants. In most cases, these failures occurred because members were trying not just to evacuate civilians, but to keep track of each other, too. It is hard to imagine an environment better engineered to induce disorientation and forgetfulness. And yet, as far as anyone knew, every civilian and first responder made it off those burning hilltops, even if some of their equipment didn't.

Their crews may have been intact, but there was nothing in their experience to prepare them for defeat on such a scale. As the Bronto operator Evan Crawford put it, "We got our ass handed to us." Crawford lost his own house in Beacon Hill, and Abasand was a similarly lost cause. "The command [in the REOC] basically wrote off Abasand," he said.

It would be a big write-off. With the humblest condo going for around $200,000 and detached homes fetching upward of half a million, $500 million was a conservative estimate for Abasand alone.

There were many ways to quantify the damage the fire was inflicting on Fort McMurray; one of them was "a million dollars a minute."

Even as the crews retreated, wind-driven embers from Abasand were landing on the medical center downtown, a mile to the east. Several houses nearby caught fire and burned down; several others were saved with garden hoses—by professional firemen. The size and speed of the flames, combined with the smoke and ember barrages, had the same frightening, disorienting, and disabling effects as a psyops attack. And, in the same way that snipers demoralize and destabilize organized resistance, the random appearance of ember-generated spot fires caused firefighters to separate and disperse in their efforts to subdue them. Throughout the afternoon, firefighters were seen racing through backyards with garden hoses, wet mops, and shovels, and sometimes, no tools at all. "I was standing there with one of my other captains," Stephenson recalled, "and we didn't have water trucks [to refill the pumpers]; we didn't have a *fire truck*. So we ran around stomping out little grass fires on the lawns—stomping on them with our *feet*. We had crews of people going through people's backyards just to make sure we didn't leave any small spot fires behind."

A lieutenant named Damian Asher compared these frantic efforts to cats chasing a laser pointer.

It explains why so many firefighters weren't wearing bunker gear: they were running too fast. Many wore only their lace-up station boots, coveralls, and a T-shirt. The fire was a cruel teacher, but its lessons were heeded: in the space of two hours, the Fort McMurray Fire Department had been transformed from a hierarchal, Blue Card–following paramilitary organization into loose cells of hit-and-run guerillas, mixing fluidly with other departments and crews as they fought street to street and house to house. They were learning on the fly what the Slave Lake crew already knew well enough to teach in a course, and had, since 2011, been preaching to anyone who would listen.

14

We look at the present through a rear-view mirror.
We march backwards into the future.

—Marshall McLuhan, *The Medium Is the Message*

As Wayne McGrath made his way off the hilltop in Abasand, followed by evacuating firefighters, he encountered the same horrific conditions that the Ayearsts and their neighbors were enduring a mile to the south in Beacon Hill. Videos taken by evacuating residents that afternoon look as if they were shot at midnight, inside tunnels of fire swirling with embers. The audio—screaming, swearing, praying, begging, and crying—is difficult to listen to. And so, in their way, are the long moments of stoic silence, broken only by the hum of the engine and the crackling of the fire outside.

It has been suggested that one reason so many of us are attracted to disaster movies—beyond voyeuristic catharsis—is because they offer ways to visualize, and perhaps prepare for, such events ourselves. By midafternoon, right about the time a lot of people in Fort McMurray start thinking about a Double Double at Tim Hortons, the city had become its own disaster movie, with a uniquely dissonant tempo—the astonishing swiftness of the fire's progress set against the excruciating slowness of evacuating traffic. Whether it was figuring out how to reassure a panicking spouse or how to stop the next block from burning down, one way or another, everyone was improvising.

Combustive energy had drawn people to Fort McMurray in

steadily increasing numbers over the course of a century, and combustive energy was driving them out again, en masse, in a single
afternoon. As the people of Fort McMurray made their escape, it was
through apocalyptic conditions that recalled the seventh plague in
the Bible's Book of Exodus: "So there was hail, and fire mingled with
the hail, very grievous, such as there was none like it in all the land
since Egypt became a nation."

There was none like it since Canada became a nation, either: the
exodus of May 3 was the largest, most rapid displacement of people
due to fire in North American history. It took the form of an unbroken ribbon of vehicles crawling in ranks, like army ants, northward
and southward out of the city while fire raged along the highway, in
some cases right up to the breakdown lanes. Visible in every rearview
mirror was a monstrous plume where their city should have been, as
if the city itself had erupted. Many who saw this sight speculated that
the entire city was lost. The fire plume, which was growing steadily
larger, was actively changing the region's meteorology. No longer simply a ground-level interface fire, it had become a force of Nature. As
temperatures rose past 1,000°F, the air at the smoke column's center
rose ever more rapidly, driving upward, like smoke up a hot chimney.
As this superheated air rose higher and faster, it created a vacuum into
which cooler air was drawn from all sides at greater and greater velocity. Operating like a recirculating fountain, storm systems this large
also generate powerful downdrafts along their outer edges, which, in
the case of a wildfire, can cause it to burn even more intensely, like an
atmospheric turbocharger.

Smoke columns behave like fountains in other ways, too: suffused
within that swirling vortex, inconceivable in the face of so much fire,
was a colossal amount of water—not just from moisture bound up
in the forest, but also from melting ice, broken water lines, and fire
hoses. In order for fuels to burn as explosively as they did in Fort
McMurray, any residual moisture had to be removed by evaporation.
All that water has to go somewhere, and it does: what looks from
a distance like "smoke" is really a combination of soot, combustive
gases, toxic chemicals, and steam. Hundreds of thousands of gallons
of water vapor were being carried skyward though the smoke column,
ten, twenty, thirty thousand feet above the fire, where it condensed

and then froze. There, miles above the city, hurricane-force down-drafts hurled fusillades of black hail back to earth, just as they had done in ancient Egypt. Reduced to their most elementary ingredients, these carbon-infused ice pellets were all that remained of the trees and houses so recently devoured by the fire.

As the fire intensified, ash and glowing spruce needles grew into firebrands the size of work boots, and then branches, treetops, fence panels, and entire garden sheds—all flying through the air, on fire. Some of these were carried thousands of feet into the smoke column, just as they would in a tornado. Pilots flying over large wildfires have reported charred tree branches bouncing off their windshields at twenty thousand feet. A photo taken from an airplane window late on the night of May 3 shows a vast and luminous smoke cloud where the city had been while, high above, the northern lights blaze across the sky. In another age, this might have been an omen worthy of formal record, but that night, it was just one more illumination from the twenty-first century, captured in this smartphone-crowdsourced record of apocalyptic visions.

Other anomalies appeared as well, and, from this vantage, they sound more like details from the Old Testament or Greek mythology than events reported from one of the twenty-first century's wealthi-est industrial centers. Among them was a fire-borne thunderhead. Known to meteorologists as a pyrocumulonimbus cloud, or pyroCb, these massive formations can be two hundred miles wide and reach into the stratosphere. A fully developed pyroCb, like the one shroud-ing Fort McMurray on May 3, is so huge and energetic that its behav-ior is influenced by the coriolis effect—the rotation of the earth. In the Northern Hemisphere this will cause such a system to spin coun-terclockwise, just like a hurricane. Because of their size, particularly their height, pyroCbs are Nature's most efficient delivery system for high-altitude pollutants, including carbon monoxide, hydrogen cyanide, ammonia, and vast amounts of carbon and other particu-lates. Once these smoke columns reach the lower stratosphere, between thirty thousand and forty thousand feet above the earth, the aerosols and particulates within them can be carried around the world on the jet stream, which circles the poles like a high-speed conveyor belt.

Breakthroughs in aerosol-sensing satellite technology have revolutionized scientists' understanding of these phenomena. As recently as the 1990s, hemisphere-spanning aerosol clouds generated by enormous wildfires were mistakenly attributed to volcanic activity. In part because they were so rare, wildfire-generated pyroCbs have only been formally identified and studied as such since 1998, the dawn of this new era of twenty-first-century fire. One of the most exhaustively studied pyroCb events to date occurred during the Chisholm Fire, the same one American satellite data analysts initially suspected might be a nuclear bomb test. The plume it generated obscured an area of more than fifty thousand square miles, roughly the size of Greece. While they remain an atmospheric rarity, pyroCbs have become significantly more common over the past two decades, occurring around the world, in places they have never been observed before.

In addition to hail, pyroCbs can also generate their own lightning. "Pyrogenic lightning" has been described since ancient times, but almost exclusively in the context of large volcanic eruptions. While ember-generated fires are relatively easy to predict (they appear downwind, typically less than five miles from their source), fires caused by lightning can be ignited virtually anywhere within a fifty-mile radius of a pyroCb, where they are accompanied by all the hazards associated with electrical storms—tower strikes, power outages, and electrocution. By 4:00 p.m., as tens of thousands of citizens were making their slow escape, Fort McMurray was experiencing the same "darkness at noon" phenomenon associated with apocalyptic events recounted throughout the world's histories and mythologies. With the forest already primed to burn, a pyroCb, combined with wind-driven embers and lightning, changed this fire from a localized conflagration into a perpetual motion machine of destruction operating on a regional scale. Given the long-term forecast, this fire could burn as long as the fuel held out, and, in these conditions, the boreal forest was nothing *but* fuel.

As residents fled, a skeleton crew of firefighters, first responders, and volunteers numbering in the very low hundreds was left behind to fight for the city's life. Within the superheated miasma that had enveloped Highway 63, life-and-death dramas were unfolding. Jet Ranger helicopters were seen flying over downtown with buckets of water and retardant to target crucial infrastructure, including the hospital and city hall, where the morning press conference had wrapped up barely three hours earlier. Vince McDermott, a journalist for the local paper, *Fort McMurray Today*, tweeted what could have been a reporter's line from *Armageddon*: "All I can hear right now in downtown are sirens and helicopters." "Hear" was the operative sense, because, even in the heart of town, many blocks from the fire, it was now physically painful to have your eyes open, and the simple act of breathing was growing difficult. Along with the smoke came heat, and it pushed downtown temperatures into uncharted territory. Jill Edwards, the business manager at KAOS Radio, a local Christian station, was evacuating from downtown, when her car thermometer registered 109°F. Radiant heat from the fire had pushed the local temperature almost twenty degrees above the high, which—even without the fire—exceeded the forecast, topping out at 91°F, shattering the previous record for that date. As Arizona heat baked the city, the downtown on-ramps for Highway 63 seized in gridlock. Buffeting winds caused by the fire interacting with the region's complex topography sent smoke and embers swirling in all directions. Upwind or downwind, nowhere was off-limits to ignition. In this way, the fire had created optimal conditions for its spread that bore an uncanny similarity to a successful virus, or a monopoly: once these entities reach a certain breadth and density of distribution, there are no longer any bad opportunities; every situation can be turned to advantage.

Out on the highway, drivers were making some more painful discoveries: an $80,000 truck equipped with a touchscreen, crew cab, four-wheel drive, and a 6.4-liter Hemi engine is only as good as the fuel in its tank. Most pickups can hold twenty-five gallons or so—more than half a barrel's worth, and more than enough to get to Edmonton, five hours away. But many fuel tanks were running on empty that day. With each successive mile, a steadily growing number of vehicles were pulled off onto the shoulder and median. Surrounded by fire, with an infinite supply of bitumen just below them, those engines had nothing left to burn.

Because the scale of the event was poorly understood and in flux, many evacuees left town under the impression that they would be returning soon, while others feared that the entire city would be lost. Some who had friends in the fire department, or smartphone-linked alarm systems, discovered as they drove that their homes were on fire. An unintended consequence of the evacuation was the collapse of the 911 system. With virtually everyone evacuating, the city effectively lost its eyes and ears. It hardly mattered: the most likely reason for calling ("I'd like to report a fire") was now moot, and those in a position to respond—roughly three hundred firefighters, police, assorted first responders, and volunteers—were already overwhelmed. Most of their partners and families were evacuating, and, if they were able to contact them at all, it was to speak for what more than a few believed might be the last time. This was true even in the relative safety of the REOC at Hall 5, which wouldn't be safe for long. At lunchtime—both a moment and a lifetime ago—terminal good-byes had been the last thing on anyone's mind.

Even at this late hour, there were residents of Fort McMurray who were unaware of the fire. Because it was a workday, many of them missed the press conference and the later updates. David Smith, an instructor in environmental studies at Keyano College's downtown campus, was working in his basement office that afternoon. It was quiet as a cave down there, and while he was sequestered in the monk-like solitude of his windowless room, time and events had been accelerating in a way that they rarely do anywhere, and never do in the basement of Keyano College. Smith had no idea what awaited him outside until around three in the afternoon, when an evacuating colleague knocked on his door just to make sure no one was left behind. The surface world he emerged into was barely recognizable. Smith is a trained ecologist, well versed in the boreal fire cycle, but he had never imagined he would be part of it. His first thought was his dogs.

The abrupt end to business as usual was manifesting itself in prosaic ways that now felt almost poignant. Squeezed in between evacuation notices, the municipality's Twitter feed posted this message: "Due to the forest fires, today's Council Meeting and Sustainable Development Committee meeting have been cancelled." There was nothing sustainable about any of this, and the idea of a meeting—of any kind—was suddenly inconceivable. The post came with a new

Seismic lines across the boreal forest, northern British Columbia

Syncrude mine and upgrader with haulers in the foreground
The pale rectangles in the upper right are sulphur piles.

Light pillars, northern Ontario

Fire 009 from Fire Hall 5 at 6:00 p.m., May 2

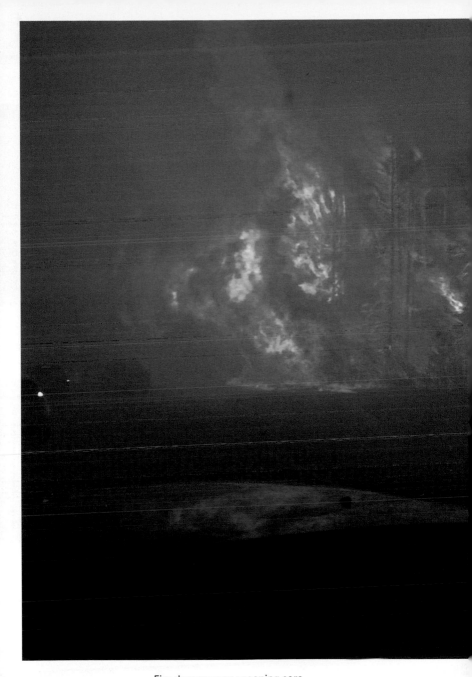

Fire dragon over escaping cars
Paul Ayearst took this photo through his truck windshield.
His wife and daughter are in the cars ahead.
Beacon Hill, 2:49 p.m., May 3

Alpine Court in flames
Note truck wheel burning and melting.
Abasand, May 3

Backhoe tearing down houses to slow the fire's progress
Prospect, May 5

Fire hose wetting down houses
Note how the hose stream is bent, vaporized, and "inhaled" by the fire.
Prospect, 12:44 a.m., May 5

Burned D8 Caterpillar bulldozer. Prospect

Total destruction. Prospect

Redding, California
August 2018

Transmission tower with high-
tension cable and pickup truck (top)

Truck and tree (above)

Cast-iron frying pan (right)

Steel tractor seat (bottom)

Revirescence
Abasand, June 8, 2016

hashtag: #ymmfire; only minutes old at the time of its posting, #ymmfire would remain active not for days, but for years.

Because of the massive smoke pall over the city, and the fact that the recently expanded Fort McMurray airport was in grave danger of being overrun by fire, there were no flights in or out. Instead, charter flights were running through the bitumen plants' own airfields, some of which are capable of handling passenger jets. As fast as workers were flown out, evacuees from the city took their places. Meanwhile, six of the hottest, driest, windiest hours of the day were still to come. The relative humidity—second only to wind in terms of its importance to fire behavior—was still dropping. It would bottom out at 12 percent, a level of desiccation typically found in kiln-dried lumber.

⁓⁓⁓

On Sunday, May 1, when Fire 009 first posed a serious threat, Chief Darby Allen had made a call to Dale Bendfeld. Bendfeld, fifty-ish, lean, crisp, and calculating with short, sandy hair over a high forehead, was a former RCMP officer who also held the rank of lieutenant colonel in the Canadian Armed Forces, where he served in Afghanistan, Europe, and the Middle East. In 1998, he helped lead the response to a devastating ice storm, the costliest disaster in Quebec's history. In 2003, he led a military team responding to the Okanagan Mountain Park Fire, then one of Canada's most destructive modern fires, in which 240 buildings were lost in southern British Columbia. In 2010, he managed security strategy for the Winter Olympics in Vancouver. In 2012, he was hired as Fort McMurray's executive director of municipal law enforcement and protective services. It was an unusual title because Fort McMurray is an unusual place, a complex blend of foreign, federal, provincial, corporate, and municipal interests with assets as sensitive as a military installation. It was understood that the city and the bitumen industry were vulnerable to terrorists as well as natural disasters, and Bendfeld had expertise in both.

Each day, based on field reports, weather forecasts, and computer models, educated judgment calls had been made by Darby Allen, the wildfire manager Bernie Schmitte, and their deputies and advisers, including Bendfeld, Darryl Johnson from Forestry, Jody Butz from the fire department, and Chris Graham from Alberta Emergency Management, among others. The situation was dynamic in the extreme, and the interdepartmental politics and consultation were time-consuming; at times, hundreds of emails, phone calls, and texts were flowing through the REOC every minute. And yet, in the end, it was as if their accumulated discussions and decisions were in response to a different threat than the one actually confronting them.

Allen, Schmitte, Bendfeld, Butz—all of them took the threat of this fire seriously and were still blindsided by what it did. David Staples, a reporter with the *Edmonton Journal*, spoke to these men immediately afterward, and they confirmed what Jamie Coutts had observed in the REOC at lunchtime—leaders who seemed unable to meaningfully grasp the enormity of the danger facing them. "At times," Staples wrote, "they headed to the REOC's lookout tower to get their own view north, east and west. Only then did they have their own, 'What the f*ck is going on!?' moment. At one point, as they walked to the lookout, Bendfeld turned to Allen and said, 'We are so fired.'"

"Mate," Allen said, "we are so done."

In spite of the REOC being equipped with state-of-the-art communications (and a clear view of the fire), and staffed with experienced members from every relevant agency, communications had broken down in fundamental ways. One of Allen's lieutenants, Jody Butz, a youthful deputy chief you'd want on your rugby squad, described this moment to the CBC's Laura Lynch: "We were starting to receive reports, not from Forestry," he told her, "but from citizens on social media that they could see flame from the Shell station in Beacon Hill."

A forensic analysis of the fire response prepared for Alberta Forestry in 2017 was blunt: "Rather than learning about the wildfire's imminent incursion into Fort McMurray through the [Incident Command] structure, the RMWB Operations Chief [Darby Allen] discovered the wildfire was in the community through public reports over social media."

The RCMP found out the same way.

This kind of failure, baffling as it may be, is not unique to the REOC in Fort McMurray; it is as old as human judgment. Nassim Taleb, a statistician, risk analyst, and author of *The Black Swan: The Impact of the Highly Improbable*, calls it the "Lucretius Problem." Named for Titus Lucretius Carus, a Roman poet and philosopher, it refers to a bug in human perception observed by Lucretius in the first century BCE and described in Book VI of his epic poem, *De Rarum Natura* (*On the Nature of Things*):

> *Yes, and so any river is huge if it be the greatest man has seen*
> *who has seen no greater before, . . .*
> *and each imagines as huge all things of every kind*
> *which are greatest of those he has seen . . .*

Two thousand years later, in a companion to *The Black Swan* called *Antifragile*, Nassim Taleb paraphrased the Lucretius Problem this way: "The fool believes the tallest mountain in the world will be equal to the tallest he has observed." But "fool" seems awfully harsh for such a persistent human failing. The TV journalist Reid Fiest experienced it himself when he saw the fire enter Beacon Hill: "I didn't quite believe it yet," he said. "I had never been in a situation like that." The firefighter Evan Crawford had the same reaction: "I thought, 'What the hell is going on here?' I never thought it would hit us like it did."

In essence, the Lucretius Problem is rooted in the difficulty humans have imagining and assimilating things outside their own personal experience. Hundredth-percentile fire weather conditions during the hottest, driest May in recorded history, following a two-year drought in a sudden city filled with twenty-five thousand petroleum-infused boxes and surrounded by millions of desiccated trees, is something no Canadian firefighter or emergency manager had experienced. But this is the nature of twenty-first-century WUI fire, and not just in the boreal forest. Authorities in California, Australia, Greece, Spain, Russia, and elsewhere have found themselves in the same situation—basing their responses on outdated concepts, on what they've already seen, instead of what fire weather is capable of now. The data was there, but the interpretation wasn't, and this—the Lucretius Problem—gave the Fort McMurray Fire an all but unassailable advantage over the people charged with fighting it.

The consequences would be severe. On May 1 and 2, the REOC's primary objective was the same as Slave Lake's in 2011: establish a perimeter around the fire and prevent it from entering the city. By 2:00 p.m. on May 3, it was simply to keep the death toll as low as possible.

A citywide evacuation had never been considered by the city's emergency planners, but as the fire overran neighborhood after neighborhood, and the surging traffic slowed to a crawl, Highway 63's limitations became obvious: there simply wasn't enough road to accommodate all those cars, especially when that road was enveloped in smoke with fire burning right up to the breakdown lanes. Dale Bendfeld's training had taught him to look at risk through a military lens: to prioritize threats, values, and objectives. Drawing on his experience managing logistics in crisis situations, Bendfeld did the math, and the numbers were daunting: five hundred vehicles from Beacon Hill would require at least two or three miles of lane space on their own; one thousand vehicles from Abasand would require twice that. The west side neighborhoods would need ten times more. Given the speed and mobility of the fire, there simply wasn't enough pavement to move all these vehicles out of harm's way—if they all went south. But if traffic was evacuated in both directions, it would double the available road space on 63, and they might have a chance. Where all these evacuees would go once they were safely out of town was unclear, but there was no time to worry about that now. "This gives the most people the best chance of surviving," Darby Allen explained to Marion Warnica, a reporter with the CBC. "Because right now, I'm concerned we'll have vehicles on the road on fire."

Privately, leaders in the REOC wondered whether deaths would be counted in the dozens, or the hundreds, or worse.

There was no middle ground: anyone who didn't evacuate north toward the man camps and Fort McKay headed south as far as the fuel held out, snaking through the eye-blink hamlets of Mariana Lake, Wandering River, and Grassland, wiping out gas supplies, ATMs, and convenience store shelves as they went. The original emergency shelter—the massive rec center on MacDonald Island, surrounded by water and connected to downtown by a narrow causeway—was far too small to accommodate so many evacuees. That afternoon, it was being used as a bus depot to evacuate those left behind. In the coming days, MacDonald Island would transform again—into a marshaling yard for emergency vehicles, and a kind of drop-in center for exhausted emergency personnel.

John Knox, the program director for Country 93.3, left the radio station shortly after 3:30, when a mandatory evacuation was ordered for downtown and the RCMP started banging on doors. By then, 63 South was in imminent danger of being overrun by fire. Despite this and the daunting traffic, Knox opted to go south; through the choking smoke, he saw firsthand how deeply the fire had penetrated his city. Less than a mile south of downtown, directly opposite Beacon Hill on the east side of the highway, was the riverside neighborhood of Waterways. There, clustered around the playing fields of J. Howard Pew Memorial Park, rows of newer trailer homes stood side by side with some of Fort McMurray's oldest buildings. The Royal Canadian Legion was located there, and so was the Athabasca Tribal Council office. "I looked down at Waterways," said Knox, "and I started to cry. It was gone."

The way the traffic rolled so slowly past these simmering ruins, headlights glowing like lanterns in the smoke, Knox might as well have been in a funeral cortege. There was no way to know if everyone had escaped in time—from Waterways, or Beacon Hill, or Abasand, or Thickwood; there was no way to know if cars behind him were on fire, or if the road would be blocked ahead. Knox tried to take a picture as he passed, to somehow capture the scale of this calamity, but the destruction was too broad to frame on his iPad. It had all happened so fast; the fire seemed to be everywhere.

In the umber haze, suffused with a throbbing volcanic glow, hallucinatory inversions were taking place: down in Waterways, on Pelican Drive, the ambient heat had grown so intense that a full-sized metal streetlight folded over on itself like a wilted flower. In Beacon Hill, Plexiglas bus shelters melted like milk jugs. On every block, cars and trucks were burning to the chassis, deforming the way bottles do in campfires, while their wheel rims bled aluminum in many-fingered streams. Through the fire's countless acts of transformative violation, the ordinary was made grotesque; neighborhoods once distinguished by tidy uniformity now looked like suburban Hells rendered by Salvador Dalí.

As it blackened and flattened block after block, the fire was imposing on this highly ordered, type A city a kind of nihilistic anarchy that flouted every value its citizens, companies, and churches held dear: Down with everything! Burn it all! Go back where you came from! Binge and purge! In the face of this surreal undoing, the efforts of its industrious and prosperous inhabitants now appeared futile. This was a city of doers, but now the only thing to do was run away with their children and whatever they could carry. It was frightening, and it was also humiliating. As they fled down the highway in the heart-pounding slow motion one encounters in nightmares, scenes of destruction and defeat confronted them on every hand. A mile down the highway from Waterways, past the smoking ruins of Centennial Park Campground, the exploded Flying J gas station, and the burned-out Super 8, was the recently evacuated neighborhood of Gregoire and, next to it, the Mackenzie Industrial Park. While fires burned in the surrounding trees, a lone ladder truck could be seen spraying down a warehouse with the kind of high-pressure monitor used to fight fires in apartment buildings. But long before reaching its target, that laser-like jet of water—powerful enough to blow through windows and knock down doors—was being swept northward into the heart of the inferno like so much mist in the fire's terrific, all-consuming wind.

Downtown, the six-story medical center was being evacuated and helicopters were water bombing the solid-brick city hall. The streets were choked with cars and trucks; the last time traffic had been this bad was during the boom. By 7:00 p.m., the entire city was under a mandatory evacuation order, and all four lanes of Highway 63 were clogged with southbound traffic. In just a few hours, firefighters had been forced to give up a shocking amount of ground as the fire advanced, virtually uncontested, through the south and west sides of the city. Soon, even the REOC would be forced to evacuate Fire Hall 5.

That evening, before a dark gray cloud that could have been mistaken for an approaching storm, CBC TV's Marion Warnica interviewed Chief Darby Allen. He seemed a changed man, unable to meet the camera's gaze or even, it seemed, his interviewer's. The volatile and rapidly changing scenario Allen had alluded to during the morning press conference had metastasized into something inconceivable a few hours earlier. When it comes to wildfire fighting, most seasoned incident commanders have a story or two about the one that "went over the hill"—that got away. There are consequences for this, but it's usually tallied in acres burned, not in homes, or lives. The magnitude of the losses Bernie Schmitte and Darby Allen had to account for—and also to own—was on another scale altogether, almost beyond reckoning. It was a fire chief's worst nightmare, and it took all of Allen's strength to summon the words without breaking down. "I—I would say it's been the—it's been the worst day of my career," he said, his voice cracking with emotion. It was an unusual way for the director of emergency management to begin an update, but this was not scripted; this was a raw, unsanitized reaction to the worst day—not only of Allen's career, but in the lives of tens of thousands of people whom he was charged with protecting. The chief kept his eyes averted as he wrangled the sobs in his throat into intelligible words. "And I am uh—you know, the whole uh—the people here are devastated. Everyone's devastated. The community is gonna be devastated. This is going to go on; this is gonna take us a while to come back from, but we—we'll come back."

There was, at that early stage—with the fire raging unabated, the city completely obscured by smoke, and many believing, not unreasonably, that Fort McMurray was lost—no way to know who would come back, or when, or even if. But duty obliged the chief to say it, and the chief was a dutiful man. At this point, Allen got a grip on himself. Seeming to realize that he had gotten ahead of the story, and, remembering that he should provide details, beginning with what had actually happened, he started again: "We've had a devastatin' day. Um—Fort McMurray has been overrun by wildfire."

15

When thou walkest through the fire, thou shalt not be
burned;
neither shall the flame kindle upon thee.

— Isaiah 4:2

At 10:00 p.m. on May 3, Darby Allen and Bernie Schmitte went back on the air together. Not even twelve hours had passed since the morning press conference, but now it sounded like they were calling from Kandahar or Beirut, rather than from a prosperous Canadian city. Conducted by phone with no video, it included Allen, Schmitte, and several of their deputies, including Dale Bendfeld, who was overseeing the last-minute evacuation strategy. After a brief reprise of his earlier comments, Allen put Schmitte on the line. Schmitte described the extreme fire weather scenario predicted by every forecast going all the way back to NOAA's "Seasonal Fire Assessment and Outlook," and then concluded with a humble admission of defeat: "Basically, the fire behavior was beyond all control efforts." It was just like the Chisholm Fire of 2001, which Schmitte had witnessed, and whose alarming intensity he had been reminded of when he had flown over Fire 009 forty-eight hours earlier.

Allen returned to the call, summarizing the damage thus far: 'We lost about half of Abasand properties. We believe that Beacon Hill is—um—appears to be lost . . . The fire crossed the Clearwater to the north of the city . . . We currently have a free-burning fire in the

Waterways and Draper area . . . We have lost property in the Wood Buffalo area, and those fires are still going."

In other words, the entire city, from side to side, and end to end, was on fire, or in imminent danger of igniting. With the fire's true nature revealed, the status quo obliterated, and any danger of inciting panic now moot, Allen no longer needed to mince words. The news was uniformly bad save for one startling detail: as far as anyone knew, there had been no fatalities or serious injuries. Given the speed with which the fire was spreading, and the haste, scale, and chaos of the evacuation, this was hard to believe. But in Fort McMurray, there are a great many believers—not just in Jesus or God, but in Mary, Allah, Vishnu, and other deities, new and old. Many of those outbound vehicles became informal chapels as their occupants prayed in a Babel's worth of languages for deliverance from this sudden and unexpected Hell. If Allen's information was correct, such an ecumenical mercy could be interpreted by all concerned as a vindication of the power of prayer.

"All I thought," recalled an evacuee named Tina LeDrew Sager, "is 'Dear Lord, if you can get my children safe' . . . I was almost on my knees in the truck."

For safety and logistical reasons, no reporters were present at the evening press conference, so the municipality's press secretary read their questions aloud as they were called or texted in. Just as the morning conference had turned on the fateful question "What's the worst-case scenario?"—the answer to which was now making its way around the world—the evening conference, too, would have its pivotal, veil-lifting moment. It was question number thirteen: "If the fire has gone through the majority of the communities, does that mean that the worst of the fire is over?"

"No," answered Schmitte.

The forecast for Wednesday, May 4, and for the foreseeable future, was for optimal fire weather. In fact, even more extreme conditions lay ahead. The closest rain was weeks away.

As the evening press conference wound down, there was a question about how many firefighters were active in the city, and Allen's answer was surprising: in addition to the thirty-four members scheduled for duty that day, another sixty-five had responded to the all-call that afternoon. This left nearly half the fire department unaccounted for, but there was no follow-up question.* On the Forestry side, Schmitte reported 150 wildland firefighters on duty, and both men reported reinforcements on the way from jurisdictions to the south and across Canada. Assistance from the Army and Air Force had also been requested, but they were two days away, at least. At the rate things were going, that could be too late.

Although it had been dark since the early afternoon, night had truly fallen now. From the air, the city was shrouded in smoke mottled with orange, a blemish on the atmosphere as lurid as a bruise. It was impossible to know what was going on down below—how much of the city was gone, how much was still at risk, what would be left in the morning. The city was effectively empty—a first in its 150-year history—and civilian life had ceased. Homes, shops, offices, and schools; churches and restaurants; the Showgirls Night Club, the Heritage Museum, and the Boomtown Casino—in one single, frantic afternoon, everyone had dropped what they were doing and made for the door. Left behind were still life tableaux of lives interrupted much like those discovered in the abandoned city of Pripyat following the meltdown of the Chernobyl nuclear power plant. In both cities, there were those who understood how truly grave things were, but many others had left under the wishful impression that they would be gone for only a day or two. Surely, firefighters would get the situation in hand by then.

All night, trucks and crews roamed the ghostly, smoke-bound city, lights flashing but otherwise unannounced save for the growl and whine of their engines and heavy tires. With no traffic, and no one left to warn, there was no need for sirens anymore. Besides, what was the hurry? Visibility was at best a block, often far less, and speeding under those conditions is dangerous. Night was hard to distinguish from day, and, as exhaustion and sleep deprivation took their toll,

---

* This number did not include Site firefighters like Lucas Welsh, or volunteer fire department members like Jamie and Ryan Coutts.

time and events began to blur inside the disorienting miasma. Meanwhile, the fire—this invading energy—had commandeered, not just the physical laws of the land, but the bylaws of the city. Traffic signals with their bright colors and civic rhythms—Stop-Go; Walk–Don't Walk—seemed, in the stultifying gloom, to be relics of another civilization. Nothing moved on the streets now; even the ravens had fled. The firefighters, police officers, heavy equipment operators, and water truck drivers who remained behind in this murky, post-human limbo had seen the movies, and they recognized this place. They called it Zombie Land.

~~~~

By midnight, thousands of temporary bitumen workers had been flown out on passenger jets from petroleum company airfields as well as from the Fort McMurray airport. Taking their places in the camps were roughly twenty thousand evacuees from the city. The rest—more than sixty thousand people, including Paul and Michele Ayearst, Wayne McGrath, and Shandra Linder—fled south. They were now scattered from Lac la Biche to Calgary and beyond—in gyms, hotels, sports arenas, campsites, and the homes of family and friends. Total strangers were opening their doors and wallets, and the Red Cross was launching a nationwide response. As workers left the city and the fire intensified, bitumen production was rapidly scaled back. Mining, SAG-D, and upgrading operations were all impacted, and total plant shutdowns appeared imminent.

"As the closer camps filled up," recalled Chris Vandenbreekel, who, at three in the afternoon, was still broadcasting from Mix 103, "evacuees were being pushed farther and farther north. At that point, our station manager knew his house in Abasand was toast, and he's trying to console his wife, who had come down to the station with her dogs, which were now running around the station. He was saying to all of us, 'We need to go,' but I lobbied to stay and keep the lines open. There was some shouting then, and there was panic in downtown—car accidents, people racing to fill up with gas."

A half an hour later, the RCMP came through downtown, ordering anyone who hadn't evacuated already to leave immediately. Vandenbreekel left the station with the transmitter on autopilot, playing Bruno Mars, Rihanna, and Justin Bieber, interspersed with PSAs urging listeners to evacuate immediately. Seeing gridlock to the south, Vandenbreekel opted to head north, hoping to reunite with his wife. Alone in his Jeep, creeping up 63 in the slowly flowing traffic jam with an anxious eye on his fuel gauge, Vandenbreekel was struck by how many vehicles were pulled over, some out of gas, others because their drivers were simply in shock, standing and staring back at the mushroom-shaped cloud blotting out their city. From a distance, Vandenbreekel said, "it looked like an atomic bomb had hit Fort McMurray."

Vandenbreekel rolled into Gray Wolf Lodge on fumes. The camp was designed for a thousand workers, but by evening, more than double that number of people were packed in there. This was happening at every camp, and the people running these facilities were not used to seeing crying, bewildered civilians, trailing pets on leashes and children with Dora the Explorer suitcases, wandering the potholed gravel parking lots where regimented busloads of somber, work-weary men normally came and went. The men and women receiving these evacuees became the accidental recipients of countless stories of panicked escape. Through their doors, and through their eyes, passed thousands of faces altered by terror, grief, and loss, and those expressions became theirs, too. Many present were in shock; far too much had happened far too quickly for people to process. One engineer who was put in charge of reactivating a mothballed camp north of the city broke down in tears repeatedly as he set about accomplishing in a matter of hours a job that would normally take several weeks. He did it, but it very nearly broke him, and he is a different man today. In the camps, as in the city, hundreds of people worked harder than they had ever worked in their lives. Twelve-hour shifts blurred into twenty-four and forty-eight hours.

It was only when they arrived—at man camps, campgrounds, shelters, hotels, or the homes of family and friends to the south— that these refugees realized that they had brought the fire with them: everything around them—clothes, pets, car interiors—reeked of smoke. Many evacuees were short on basic supplies, from clothes and drinking water to diapers and toothbrushes, but they were greeted with great sympathy and kindness wherever they stopped. Improvised welcome centers appeared offering water, snacks, toiletries—whatever could be spared. Those who'd had the time to grab a few belongings made some choices they would have trouble explaining later. Among more plausible belongings were found snow pants, low-fat cheese, a mounted bear's head, chocolate Easter eggs, a propane tank, a samurai sword, empty bottles, bounce boots, an encyclopedia of gardening. The rest was left to the whims of the fire.

As the news spread from coast to coast at a speed that felt faster than real time, a spontaneous wave of solidarity surged back toward Alberta and Fort McMurray, bearing a flood of donations from all over Canada. Everyone understood the outsized role Fort Mac played in the country's economy, and most citizens, no matter where they lived, had some personal connection to the place.

Paul Ayearst, who along with his wife and daughter had fled Beacon Hill that afternoon, had not stopped thinking about his nineteen-year-old son, who was still stranded at a man camp north of the city. Anything was possible with this fire, and he wanted his boy out of there. Ayearst was determined to gather his family together, but the city was now a disaster area, off-limits to anyone who wasn't emergency personnel. With the highway aggressively patrolled by local and federal police, there was no way Ayearst was going to get through there on his own.

All through the afternoon and evening of the 3rd, heavy equipment was being moved around on trailers as new areas were identified for firebreak clearing. Heavy equipment is Ayearst's business; he knew many of the operators and truck drivers, and he also knew how to blend in to a convoy. South of the city, he dropped into one of these like a small boat into a current. When an RCMP officer looked at him, he gestured toward the trucks ahead as if to say, "I'm with them," and, in this way, he flowed through the first roadblock, unimpeded. He didn't get far; more roadblocks awaited, and he didn't want to push his luck. In order to avoid them, Ayearst turned off onto Airport Road. Just past the airport is the small community of Saprae Creek; the houses are bigger out here, and so are the parcels of land. These were on the fire's menu, too, but no one knew this yet and, on May 3, the area was still seen as a safe refuge. Ayearst went to a friend's house, hoping the roadblocks would be taken down later in the evening. "He was nineteen," Ayearst told me. "I wasn't leaving town without my son."

Not far from where Ayearst was resting and waiting, a group of Newfoundlanders had found their way into a friend's liquor cabinet and a party was under way. Given that Saprae Creek was in the fire's path, such behavior might seem callous or crazy, but there are precedents for it. Hurricane parties are well known on the Gulf Coast of the U.S., but there have been fire parties, too. One of them took place in 1879, when the boreal city of Irkutsk, in Siberia, lost two-thirds of its five thousand buildings to a catastrophic fire. The British explorer and missionary Henry Lansdell happened to be passing through at the time, and he marveled at the mood of evacuees as their isolated city smoldered by the river. "The people's demeanor," he wrote, "was in strange contrast with their pitiable condition; for many, having saved their samovars, were drinking afternoon tea, and all sides were joking and laughing at their comical situation."

A front-page article in *The New York Times* from June 16, 1886, described a far more chaotic scene in Vancouver, British Columbia, even as the nascent city was burning to the ground:

> During the confusion which prevailed, when rowdies and roughs saw that every one was leaving, they entered the saloons which had been left entirely unprotected and commenced drinking. Many a one was seen staggering along the streets with a keg of beer on his shoulder and as many bottles of liquor as he could appropriate. Men were seen sitting completely hemmed in by the fire and apparently oblivious to their surroundings drinking liquor.

No one knows for sure how many died in the Great Vancouver Fire; charred human remains were still being discovered twenty years later.

Paul Ayearst, however, was on a mission; he kept his focus, and he kept to himself. Shortly before midnight, he ventured out again; as he had hoped, the roadblocks were down. With the highway empty save for the occasional rescue vehicle, Ayearst made his solitary way northward, toward the McKay River Lodge where his son was staying. Despite the hour, the fire was still actively spreading and his headlights didn't reach far in the heavy, rolling smoke. As he crossed the bridge to the west side, the throbbing glow of the fire hove in and out of view, illuminating a world he did not recognize.

It was well after midnight by the time Ayearst located his son, and it was 2:00 a.m. by the time they were southbound again on 63. Just past downtown, by the burned-out Shell station, was the turnoff to Beacon Hill; they took it and made their way slowly up the hill. It had been exactly twelve hours since Ayearst had made his escape with his wife and daughter, and he wondered if there was anything left. His wife had forgotten her purse in the rush, and she was hoping he might be able to get it for her. Ayearst's route into the neighborhood took them past his wife's childhood home. Miraculously, it was still standing, and he became hopeful. "But as soon as I go by the school, we see the devastation. It was just ashes."

Ayearst pulled into his driveway that was no longer a driveway but an empty space among many empty spaces. He and his son got out of the truck, and they peered into the simmering void where their house had been. There, in the basement, rising up through a foot or two of ash, was a row of jack posts, often used in place of studs or columns to support the upper floors of a house. Despite being made of heavy steel and rated to support four tons each, they had warped in the ferocious heat. Nothing else remained. Ayearst's mother had been the family historian, and, before she died, she had entrusted her eldest son with all the information she had gathered, going back for generations—"stuff from the 1800s," Ayearst said, "passed on to the oldest of all the families." He took his charge seriously and had stored the precious documents in his safe, but not even the safe was safe from the fire. Nothing was. Every artifact of his existence was gone.

All around them, a strange phenomenon was manifesting itself. "It was pitch black," Ayearst recalled. "Three o'clock in the morning, and there was no wind; it was just dead calm. Eerie. The smoke was hovering above the ground here." He gestured just above his head. Beneath was a scene impossible to reconcile with the vibrant neighborhood that had stood there half a day earlier. "At the end of the night," he said, "you put your campfire out—you just got the glowing embers, a few little logs flickering—that's what every basement looked like."

In every direction, beneath that low smoke ceiling, were glowing holes where their neighbors' homes had been.

Up there, in the dark and smoke and crushing quiet, Ayearst and his son could have been the last people on Earth. Their truck could have been the last truck; their memories could have been the last memories. They were totally alone, and, for a time, they cried for what was gone and for the emptiness left behind.

16

It is a possibility that we may lose a large portion
of the town.

—Alberta Emergency Management executive director
Scott Long, May 4, 2016

Officials say #FortMcMurray fire now exceeds
10,000 hectares [40 square miles]. "All efforts
to suppress it have failed."

—@CBCAlerts, May 4, 2016, 9:06 a.m.

At around 9:15 on Wednesday morning, May 4, the CBC tweeted again: "Alberta Premier Rachel Notley says '1,600 structures' affected." "Affected" was a gentle way of saying "burned," but no one could be sure of the actual number, or how it might have already changed. Because of the smoke and active flames, it was difficult to assess the fire's impact, but 1,600 was a big number—three times the number of structures burned in Slave Lake in 2011, and more than six times the number lost in the Okanagan Park Fire of 2003, until now the most destructive fires in modern Canadian history. That this number would grow was both a fear and a certainty.

At 10:00 a.m., there was another press conference, this time with video. "This is a nasty, dirty fire," Chief Darby Allen said, staring down into his notes. A lot had changed since the previous morning. In the meantime, it had become Allen's lot to be the bearer of catastrophic news, and his audience was growing by the hour. Fort McMurray was now trending internationally. Nowhere else in the world had (or has) such a large modern city been forced to evacuate so suddenly due to a wildfire.

May 4 happens to be St. Florian's Day, also known as International Firefighters' Day, because Florian is the patron saint of firefighters. It would take a miracle to stop Fire 009, especially in this weather, but another miracle had already occurred. The most striking statistic to come out of the morning press conference concerned the citizenry of Fort McMurray: "We have successfully evacuated 88,000 people," Allen said, his throat catching as, once again, he fought back tears. "No one is hurt, and no one has passed away right now. I really hope we get to the end of this and we can still say that."

Allen's relief can only be imagined. Later, he would share his deepest fear with the CBC's Marion Warnica: "I really believed at the end of Tuesday, if we wake up at first light and we've got 50 percent of our homes left, and we've only killed a few thousand people, we'd have done well."

Given the speed of the fire on May 3, and the infinite variables presented by tens of thousands of human beings all rushing for the same exit in powerful machines, the odds against a 100 percent successful evacuation were enormous. Even on ordinary days, injuries and accidents are common events in a city the size of Fort McMurray. So far, no one had been reported missing, but no one knew for sure what, or who, might be found after sifting through the hundreds of ash-filled basements. Meanwhile, the fire was zero percent contained, and burning at will.

Allen had seen the forecast, and he knew they were in for another terrible day. Downtown, nighttime temperatures had resembled Phoenix in August. In spite of the blistering heat, another inversion had settled over the city, cloaking it in smoke and making aerial surveys impossible. As the unusually high morning temperatures soared back into record territory, the relative humidity would again bottom out at a desertic 12 percent. Meanwhile, the morning winds, light and from the south, would swing northwest—the opposite of the previous day, building as they did so and driving the flames back through town.

Winds were expected to remain strong well into the night, with gusts exceeding thirty-five knots. Lightning was also predicted. It was, in short, a disastrous forecast. What Wayne McGrath had experienced in Abasand—the fire doubling back to take another run at his home— would be occurring on a citywide scale. "This fire," Allen said gravely, "is a moving animal . . . There are certainly areas within the city that have not been burnt, but this fire will look for them, and it will find them, and it will want to take them."

It was an unusual way to talk about a chemical reaction, and it was at this point that the fire, four days old and growing exponentially, completed its transformation from an objective hazard to an independent entity with ambitions of its own. Allen wasn't being fanciful; this was how it felt to be in this fire's presence—a hungry and motivated adversary intent on maximum mayhem. Satellite images did nothing to discourage this interpretation: the fire had quadrupled in size since the previous day. It now resembled a gigantic crab traveling northwest across the landscape with its pincers wrapped around the unburned portions of Fort McMurray—downtown, along the west bank of the Clearwater River, and Timberlea-Thickwood, on the west side of the Athabasca River. That map, ominous as it was, offered the most graphic proof of the firefighters' success, which, given the magnitude of their losses, was hard to appreciate from street level. Roughly three hundred municipal, Site, volunteer, and Forestry firefighters, operating across multiple fronts, often miles apart, in brutal heat, toxic air, and negligible visibility without rest or reinforcements, had managed to keep the fire out of the city's most valuable and populous areas. As grievous as the damage was, and as ferocious the fire, there was, on May 4, still a city left to defend.

Whether this would be the case at day's end was an open question, and the fire was already preparing an answer. By the time the press conference was over, Fire 009 was once again entering crossover, blowing up to Rank 6, and doing exactly what the chief had said it would: probing the city from every side; capitalizing on every undefended front and lapse of attention; and broadcasting embers like incendiary confetti.

With the fire entering its second of day of burning inside city limits, it was revealing, in new ways, what a truly extraordinary creature it was. The Great Baltimore Fire of 1904, which destroyed 1,500 buildings and damaged a thousand more, was brought under control in thirty hours—by firefighters using nineteenth-century equipment. The Great Chicago Fire of 1871, which killed more than 300 people, destroyed nearly twenty thousand structures, and left a third of the city's 325,000 citizens homeless, burned for only a few hours longer. Fires, even devastating urban fires, tend to follow a pattern: arriving under exceptionally hot, windy conditions, they move through in a hurry, wreaking havoc as they go. Like a bad storm or a swarm of locusts, the worst is over in a matter of hours. Fort McMurray was different; it might even be unique in the annals of urban fire. It burned day and night with no let-up for days on end, leaving firefighters no time to rest or recover. By the time the city's newest housing development ignited at around midnight on May 4, the fire had already been burning within the city for thirty-six hours, as long as the Great Chicago Fire—one of the worst conflagrations of the Petrocene Age—burned in total. Only a handful of urban fires have burned longer, and they are legendary. The Great Fire of London in 1666 burned continuously for five days. The Great Fire of Meireki in Edo, Japan (present-day Tokyo), burned for three days in March 1657 and destroyed two-thirds of the capital city. A combination of drought conditions, high winds, and wood and paper house construction led to an appalling death toll estimated at 100,000 people, a third of the city's population.

On May 4 and 5, desperate battles were still being fought for individual streets, neighborhoods, and infrastructure all across Fort McMurray, and active firefighting continued for many days after that. Due to a combination of abundant fuel, shifting winds, and record-setting temperatures, the fire, which would soon be described as a "siege event," attacked the city from every point on the compass. One reason it was able to do this is because the atmosphere never cooled down, and that is what is different about the twenty-first century. The highs are certainly higher, but it is the lows—in all seasons—that are, in their way, more disturbing. A typical spring night in Fort McMurray used to be in the 40s; in May 2016, nighttime temperatures barely dropped below 70°F. In Canada and northern Europe, nights during which the temperature stays above 68°F are referred to as "tropical." Twenty years ago, Toronto might experience one or two tropical nights in the month of July; in 2020, it counted fourteen. During the same month, Phoenix, Arizona, sustained an average twenty-four-hour temperature of 99°F for the entire month, a new record. Not only can fire sustain itself much more easily in warm temperatures, but ice, snow, rain, and dew will all melt and evaporate more quickly (as will lakes and rivers), making all fuels easier to ignite.

On the west side of Fort McMurray, the fire had made its way up through Wood Buffalo and Dickinsfield, where firefighters, backed up by dozer groups and water bombers, were still fighting a pitched battle to keep it out of Timberlea and, particularly, out of Birchwood Trails, the west side's forest park. Meanwhile, the fire was also threatening the entire south side of the city. At half past noon, with the fire well into its daily crossover, Mix 103's Chris Vandenbreekel tweeted the following alert from his new location north of the city: "We are receiving confirmed reports, accompanied by video footage, of fire crews hosing down Fire Hall No. 5 near Airport Road." With the REOC itself under threat, Darby Allen, Bernie Schmitte, and the rest of the emergency leadership would have to relocate. This meant that much of their communications apparatus would have to be dismantled. It would take hours for the REOC to reassemble and reconnect in the new location at Nexen Long Lake, a Chinese-owned SAG-D facility twenty-five miles southeast of town on Highway 881. That same afternoon, on the same road, a southbound SUV collided head-on with a tanker truck bound for Fort McMurray with desperately needed fuel. The teenage driver and passenger in the SUV were killed, and the tanker exploded, igniting another forest fire. Miles overhead, Fire 009's gyrating plume had organized itself into another colossal pyrocumulonimbus cloud that pierced the stratosphere and generated lightning that started still more fires, some of them twenty-fives miles away. In the Alberta boreal that day, it seemed as if combustion was the default response—to anything.

By late afternoon, Highway 63 had also been blocked by fire, again, this time at the critical intersection with Airport Road. Within hours, communities to the east and southeast of Fort McMurray would be forced to evacuate. This included Long Lake, to which the REOC had just relocated. Shortly after 6:00 p.m., a propane depot in the Mackenzie Industrial Park, north of Airport Road, caught on fire. Nearby, a new Catholic school, still under construction, burned to the ground. Out by the airport, a hotel with 170 rooms was destroyed. As the wind shifted northwest, it pushed the fire back into Abasand, wiping out almost everything that had been spared the day before.

Even with the addition of recently arrived firefighters from nearby hamlets, there were still barely 300 city, Site, and volunteer firefighters spread out across the city, along with 160 RCMP officers patrolling shopping districts and neighborhoods. In addition, dozens of heavy equipment operators and water truck drivers were working the WUI along with 100 wildfire fighters who were reinforced by seventeen water bombers and ten helicopters. Many more firefighters, trucks, and aircraft were being mobilized from farther away but had yet to arrive. With the REOC down, these exhausted crews, already spread dangerously thin, were further isolated as they faced new and ongoing battles in a steadily strengthening fire. Stands were being made all across the city, from the neighborhood of Prairie Creek, just south of Airport Road, all the way to Timberlea in the northwest.

Several significant battles would be fought throughout the day, and one of these was for the airport itself. Fort McMurray International was the main base for all the aircraft fighting Fire 009, and, by late afternoon, the fire was gnawing at surrounding buildings, blanketing the runways with smoke and flying embers, and forcing air operations to move farther south. Another pivotal battle was unfolding on the opposite side of town, in Birchwood Trails. Through aggressive intervention, crews had managed to hold the fire at bay all through the previous day, even as it tried to follow those greenway "fuses" between the west side neighborhoods of Wood Buffalo, Dickinsfield, and Timberlea. But now the fire had breached those barriers and had the advantage; radical measures would need to be taken to stop it. The park was too big and the terrain too convoluted for catguards, so the only viable option was to carpet the forest with fire retardant—a wildland firefighting tactic.

Unlike military bombing, which can be done from miles up, air tankers operate at low altitudes, more like crop dusters than B-52s. The lower the release, the less evaporation of water, or diffusion of retardant, will occur. For this reason, drops are typically made just two or three hundred feet above the target, which leaves a perilously narrow margin for error, especially when flying through the thick smoke and unstable airs of a wildfire. Even under ideal conditions, the combination of a heavy load of liquid and a low approach means that even the smallest human error or mechanical glitch can carry an aircraft into the ground. Or, in this case, a neighborhood. Birchwood Trails was completely surrounded by houses, and a number of them were already on fire.

There are techniques for avoiding accidents like this, and Paul Spring, a local helicopter pilot and the owner of Phoenix Heli-Flight, who oversaw air operations throughout much of the fire, has mastered them. "When we actually come through and do the dump," he explained, "we're in level flight so, if things don't go right, we haven't set ourselves up for disaster. And you don't bomb into hills," he added. "And you don't come in downwind. There are a lot of dos and don'ts because if you fuck it up, you die."

These dos and don'ts mattered on May 4 because the wind was rising, embers were flying, and Birchwood Trails was essentially a cluster of small hills nestled into the larger hill of the west side plateau. There was, in addition, a sense of desperation that afternoon: after all the losses suffered over the previous thirty hours, losing Birchwood Trails felt like it could mean the end of the city. If the fire got a foothold in there, it would be a devastating blow to property, but also to morale. While the airport was critical infrastructure that Fort McMurray could not afford to lose, Birchwood Trails was the green heart of the residential west side. On Wednesday afternoon, it felt like the hinge on which the fate of this benighted community swung.

A water bomber—whether it's an airplane with a belly tank or a helicopter with a Bambi Bucket—cannot make the drop unless the pilot can see the target. Typically, a smaller plane—the "bird dog"— will scout the target area and radio its observations to an air boss like Paul Spring who is circling high above in a helicopter, overseeing the big picture. Once the okay is given, the bird dog will guide the bombers in to the target while sounding a loud, chirping siren to alert ground crews that a drop is imminent. Since airdrops of water were useless against this particular fire, retardant—a mix of water, fertilizer (ammonium phosphate salts), and red dye (iron oxide) for visibility—would be used. This mixture adheres to whatever it lands on, creating a viscous layer that smolders rather than burns and can stay in place for weeks. "You don't want to get in front of a load of retardant," Spring told me. "It'll kill you. It'd be like being hit with a ten-thousand-pound loogie. I've seen some low dumps where they've just rolled the trees up in a pile."

For these and other reasons, retardant drops are generally reserved for open country, and Birchwood Trails was anything but that. It was the green hole in a donut made of twenty thousand houses. To be effective, and safe, the drops would have to be perfect, and that meant the pilots would need good visibility. But there was no visibility. In its place was anxious back-and-forth between the battalion commander on the ground and Deputy Chief Jody Butz, the manager of firefighting operations in the REOC—wherever it was. Butz had ordered the tankers, but there was no sign of them. After what seemed like far too long, the firefighters finally heard the bird dog's alarm through the smoke, followed by the low drone of the tankers. They waited for the wet clouds of retardant to descend over the forest, but they never materialized, and then the sound of the tankers faded away. As low as the pilots were flying, they couldn't see the target through the smoke. Butz told the battalion commander, Captain Mike Woykin, that he needed to make a target visible to the pilots, but Woykin's crews were not equipped for this. Emerging, once again, was a dangerous and increasingly common WUI situation: municipal and Forestry—two complex organizations with dramatically different skill sets, cultures, equipment, and tactics—trying to co-manage a wildfire in an urban setting. Who do you call if Central Park or Stanley Park is on fire in explosive conditions? How do you deploy them once they arrive? Who's in charge—the city or Forestry?

The tankers were already setting up for another pass, so the munis used the only targets they had—their helmets, which they began heaving into the trees as high as they could. After a nerve-racking wait, the twin engines of the bird dog once again penetrated the smoke, along with its chirping siren. It passed directly over them, perfectly positioned with the much larger tankers rumbling behind, creating an exaggerated Doppler effect. The ground crews braced themselves for the drop, but once again, the sound receded. The helmets hadn't worked. The wind was swinging into the northwest, and, as forecast, it was accelerating. Embers were landing everywhere now, and more spot fires were igniting in the woods. Firefighters had resorted to chasing them down on commandeered golf carts, but there was no way to catch them all.

Butz was on the phone making it emphatically clear to Woykin that there would be no drop without a visible target. Somehow, they were going to have to pull one out of a hat before the tankers circled back. This would be their final pass, Butz said; the tankers were almost out of fuel. It was then that a resourceful firefighter remembered a piece of standard equipment—the rescue blanket. Six feet square and bright red, it was their last best chance. After the firefighters spread it out in a little clearing, the air boss was notified, and the ground crews retreated to wait. Once again, they picked up the whine of the bird dog, and its siren, followed by the heavy tankers—so loud and low this time it sounded as if they were about to land. Moments later, billowing red clouds of retardant burst through the smoke— four in a row, right on target.

It was a badly needed victory, and, after that drop, the fire made no significant inroads into Birchwood Trails. The fire, however, had many other options. Firefighters made stands where they could, and deployed sprinklers where they couldn't, but the fire seemed to be everywhere, exploiting every weakness. It was flowing into the city like smoke through a picket fence while the embers rained down. And then there was the lightning.

Thirty-six hours in, it no longer seemed to matter whether the sun was up or down; at midnight on May 4, the fire was burning with the same fury it had at midday on the 3rd. Fire resembles living things in many ways, but there is a crucial difference: as long as there is sufficient wind, fuel, and dry weather, it never gets tired, and it never sleeps. The people fighting this fire, however, were running out of gas, and so were their trucks. They had now been operating in crisis mode for a day and a half straight, and there was no end in sight. The intensity of physical activity is hard to measure, but a Fitbit tracker on the wrist of the Suncor firefighter Lucas Welsh offers some idea. From when he went on duty at lunchtime on the 3rd until he returned to Site late that first evening, his heart rate stayed at or above 150 beats per minute, with the only dips occurring during location changes in his truck.

While reinforcements had begun to arrive from as far away as Calgary, eight hours to the south, local crews who knew the neighborhoods and the lay of the land were now beyond exhausted. Not only had they barely slept, they were also running out of food. Because the city's water supply had been compromised due to the sudden demand, potable water was becoming an issue, too. "The first two days, I ate granola bars and potato chips," Welsh recalled. "That was all we had. Day three, they started getting us some food, but I probably lost ten or fifteen pounds that week."

This is where the KnoxBoxes found an unintended usefulness. KnoxBox is an emergency access system that enables police and firefighters to enter commercial properties in the case of an emergency. Anyone with the key or code can enter a participating building at any time, day or night. In a number of cases, firefighters, and also police, used their KnoxBox keys to get into closed supermarkets in order to raid the bread, deli, energy bar, and soft drink aisles. The men were honorable, leaving notes, tallies, and IOUs behind, but without these borrowed calories, their own fires were in danger of sputtering out. In terms of social norms, a kind of zero gravity had been achieved in Fort McMurray, and those who remained behind occupied it in new ways. Looting a supermarket, aided and abetted by officers of the law, was one more surreal inversion; some unoccupied homes were used this way, too. But it was just as well that they put those cold cuts to use: gas and electricity had been shut off across most of the city, which meant that roughly twenty-five thousand as-yet-unburned refrigerators and freezers—from superstore walk-ins down to the humblest mini fridge—were on borrowed time. Unless power was restored, a wholesale meltdown was going to occur. In a fifty-gallon tank by the seafood section of Save-On-Foods—dark now, and warming rapidly in the hundred-degree heat—the lobsters were left to their own devices.

Eight miles above the city, the fire looked like the Mount Saint Helens eruption—a billowing roil of smoke so dense with wood ash and petroleum soot that it appeared almost pyroclastic. Deep inside this fire-powered storm, scattered bands of firefighters battled on, all but invisible in the gloom.

17

If you see me running toward the Clearwater,
you might want to follow.

—John "Toppy" Topolinski

Vyto Babrauskas is a pioneering physicist based outside of Seattle, where he serves as a kind of one-stop shop for understanding modern house fire behavior. In addition to developing a wide variety of tests, tools, and industry standards for assessing flammability and fire behavior in man-made environments, Babrauskas has published hundreds of articles and several textbooks, including *Fire Behavior of Upholstered Furniture and Mattresses* and his thousand-page fire bible, *Ignition Handbook*. In an effort to better understand how and why Fort McMurray's houses burned as they did, I wrote to him through his website, doctorfire.com. After pasting in Lucas Welsh's observations about Wood Buffalo's five-minute burn times, I asked if he could explain the physics behind this kind of ultra-rapid combustion scenario.

Babrauskas responded with surprising speed, and also brevity: "Yeah, that's a difficult question," he wrote just hours later. "The best analogy is the Hamburg firestorm."

I hadn't been sure what to expect, but it wasn't that. The Hamburg firestorm is an infamous event in which Allied forces studiously and systematically carpet-bombed Germany's second-largest city at the height of World War II. Over eight days and nights in late July and early August 1943, hundreds of British and American bombers dropped thousands of tons of high explosive and incendiary bombs across the city. The Allies' objective was twofold. The first was physical—to destroy the strategic river port and cripple its shipyards, factories, and refineries. The second was psychological—by targeting workers' housing with incendiary bombs, the raid's architects hoped to kill, terrorize, and demoralize the citizenry, the real engine of the Nazi war machine.

Military code names often contain symbolic references to their missions, but most aren't as pointedly vengeful as the one given to the Hamburg bombing campaign. They called it Operation Gomorrah, but only because they couldn't get away with Operation Sodom. Both refer to scenes of apocalyptic punishment from the Book of Genesis in which "the LORD rained upon Sodom and upon Gomorrah brimstone and fire . . . And he overthrew those cities, and all the plain, and all the inhabitants of the cities." In order to accomplish this, the Allies took a methodical approach, employing the finest minds available. Given that oil is fire on demand, a sideline in incendiaries is a natural for petrochemical companies. This is why Operation Gomorrah was equipped and advised by the Standard Oil Development Company (the chemical research arm of present-day Esso/ExxonMobil). Working in conjunction with the U.S. Army's Chemical Warfare Service Technical Division, they spared no expense, hiring German expat architects, Hollywood set designers, and staff from the Harvard Architectural School to faithfully re-create typical German workers' housing—inside and out.

During the spring of 1943, test structures were built and bombed at Harmondsworth in the United Kingdom, at Standard Oil's test facility in Elizabeth, New Jersey, and, most elaborately, at the top-secret Dugway Proving Ground in the Utah desert. There, after erecting buildings that replicated several common German construction styles—right down to sofa stuffing and the placement of babies' cribs—they were bombed, repaired, and bombed again. It was a military version of the Underwriters Laboratories living room experiment, and it provided the Allies with the information necessary to cause maximum damage and loss of life—not by blowing the houses up one at a time, but by burning them down en masse. Generating a firestorm was not a serendipitous by-product, it was the overarching goal: "Fire severity in German structures . . . is mainly a function of the combustible furnishings," states a declassified government report on the Dugway tests. "For this reason a thorough study of typical furnishings was made so that proper fire severity would be reproduced."

It was diabolical, but this was total war; the eight-day assault on Hamburg would be payback for the equally merciless London Blitz.

On July 27, the fourth night of Operation Gomorrah, eight hundred heavy bombers released nearly 5 million pounds of bombs and incendiaries over southeast Hamburg in less than an hour. The bombs came in waves: first to drop were the high explosives. Designed to flatten whatever they landed on, their attendant shock waves blew out windows and doors for hundreds of yards around while the rubble and craters they made hampered rescue and firefighting efforts. With the buildings broken and ventilated, and escape routes blocked, the incendiaries came next. Containing phosphorous and thermite, these bombs were much smaller, designed to puncture roof tiles and explode in the upper floors and stairwells of houses. "The bombs," wrote one survivor, "often came to rest in beds or on the floor beside wardrobes, chests of drawers, behind bedframes, and ignited these fuels from which the fires then developed."

As it happened, July 27 was exceptionally hot and dry, in the mid-80s, and there hadn't been a proper rain in weeks—conditions similar to those in Fort McMurray. As ton upon ton of thermite and phosphorous bomblets descended on the city, the tight, high-walled streets were transformed into canyons of fire. Within half an hour, the tightly packed six-story stone apartment buildings in the working-class district of Hammerbrook were roaring like blast furnaces. So ferocious was the heat that streets and alleys took on the characteristics of giant bellows, drawing in the surrounding air with such force and velocity that residents, particularly the lighter women and children, were sucked back into the very buildings they were trying to escape. Others managed to flee their roasting, smoke-filled bunkers only to become mired in molten road tar, where they, too, were burned alive. High above the city, smoke and fire spun in a terrible gyre that only drew the surrounding air in faster, causing the fire to burn so hot it caused bricks to melt, and objects one hundred yards from a flame source to ignite due to radiant heat alone.

Flashover was now occurring on a massive scale; fire trucks and ambulances, blocked by craters, rubble, and liquefying streets, burst into flame spontaneously. As updrafts surpassed one hundred miles per hour, witnesses observed trees being torn from the ground and civilians being swept into the air, igniting like firebrands before disappearing into the annihilating vortex overhead. Down below, the flames coursing through the now-roofless and -windowless buildings mimicked the behavior of Japanese anagama "climbing" kilns, which are constructed in a series of ascending chambers through which rivers of fire race for hours on end, generating temperatures well over 2,000°F. According to an official with Hamburg's Fire Protection Police, the only things that withstood the terrific heat were the steel safes built into the walls of banks.

Twenty thousand civilians were killed that night, many of them immolated; many more baked or suffocated in basement bunkers. After eight days of relentless bombing, the death toll rose to forty thousand. Sixteen thousand apartment buildings were destroyed and nearly a million people were rendered homeless. In time, the bombing of Hamburg would come to be known as the "Hiroshima of Germany." Despite the devastating impact of Operation Gomorrah and the similarly destructive raids that followed, the Nazis fought on for two more years.

How, I wondered, could there be a connection? Fort McMurray, after all, began as a random forest fire. What could it possibly have in common with an unprecedented act of state-sanctioned arson? I wondered if Babrauskas had misunderstood my question. His email included declassified documents analyzing the behavior of the Hamburg firestorm, one of the most extensively studied such events in history. This is where the answer must be, I thought. But when I looked for similarities between Hamburg and Fort McMurray, I saw few, other than the fact that, in both cases, the fires ignited in residential areas under conditions that were unseasonably hot and dry. I couldn't get past the haunting fact that the Hamburg fires were set intentionally by people who had studied how these neighborhoods would burn and then ignited them for maximum effect. This is where it helps to think like a fire rather than a human being, and this is what the people who study fire have learned how to do, even if they might not describe it that way. A fire doesn't care if it was set by accident or intention; in the end, all that matters to the fire is fuel, weather, and topography. This is the primary fire triangle, and it is in triangles—there are several of them—that students of fire are trained to think.

When I applied the fire triangle to Hamburg, it was clear that fuels played a key role, but as volatile as they were, they wouldn't have erupted into a firestorm—a whirling, self-perpetuating system with hurricane-force winds and glass-melting temperatures—without weather (the near-drought conditions afflicting northern Germany that summer), combined with topography (the funneling, chimney effect of narrow streets lined with bomb-ventilated apartment buildings). Working my way around the triangle, it was clear that both Hamburg and Fort McMurray had the weather, topography, and in-place fuels to support major fires, but not without an unusually powerful initial push.

It was this—the driving force behind the Hamburg fires—that got me thinking about Fort McMurray, not in terms of weather, fuel, or topography, but in terms of incendiaries. After all, house and apartment fires are relatively common, and so are forest fires, but what propels them into catastrophic territory? This is when it dawned on me that, in Fort McMurray, the incendiaries didn't need to be delivered by plane because they were already in place—thousands upon thousands of them—not just in town, but all through the surrounding forest. However, as explosive as these "incendiaries" were, they took such benignly familiar forms (houses and trees) that you would never associate them, or their potential for devastating synergy, with a World War II firebombing raid.

Unless you were a fire.

It's a frightening thought: a literal forest of freestanding, fifty-foot firebombs, but this is what black spruce are, ready to burst into flame with the touch of a match. With their exceptionally flammable sap, resinous cones and needles, and pitch-infused wood, you have a naturally occurring incendiary that would impress any war department. Add drought conditions, noonday heat, and a stiff wind, and you'll get something closer to a blowtorch. In May 2016, it was so hot and dry in northern Alberta that poplar and aspen trees were behaving the same way. In light of this, it could be said that Fort McMurray bombed itself.

It is not only the military and petroleum companies that take an interest in incendiaries; insurance companies do, too. They pay particular attention to house fires and the damage they can do, and a common cause of such fires is dry Christmas trees. The insurance industry has determined that a burning Christmas tree can generate about four megawatts of energy for every second of burn time. A megawatt is a million watts, or ten thousand 100-watt light bulbs. A typical Christmas tree is only about five feet tall; a typical black spruce tree is fifty feet tall. Fort McMurray was surrounded by hundreds of thousands of these incendiary towers. The result was an insurance agent's nightmare.

With their wind-driven sparks, each burning tree was capable of propagating more fires as abundantly as dandelion seeds. By the time this total-conflagration system hit Fort McMurray's interface zone, it was already projecting radiant heat comparable to that experienced in Hamburg. While Hamburg's century-old houses filled with legacy furniture were extremely flammable, Fort McMurray's modern equivalents were even more so: in addition to their kiln-dried wooden frames, every single house and garage contained a built-in stockpile of petrochemical fuels and accelerants, many of them created by the same companies that design and manufacture incendiary weapons. Add to this every gas grill and vehicle, every set of summer (and winter) tires, every plastic trash bin—the list of flammables is virtually endless. Whatever form their products take, there is no getting around the fact that, wartime or peacetime, petroleum companies trade in fire. Like the boreal forest, their products are made from materials capable of burning with great intensity—even if they aren't advertised that way.

It raises a grave question: What role does the petroleum industry play in promoting and approving building materials that are supposed to shelter families from harm?

The firestorm engineers at Standard Oil and the U.S. Army's Chemical Warfare Service had left nothing to chance, right down to analyzing the moisture content in the timbers of the houses they were intending to ignite. In Hamburg, the moisture content in those hundred-year-old houses was calculated to be between 10 and 15 percent—matchstick-dry and, as it happened, identical to the relative humidity around Fort McMurray on May 3.

⁓⁓⁓⁓

The fire's assault on the Prospect subdivision began in earnest at midnight on May 4. In order to fight it, firefighters would once again have to redefine what a house was. These brand-new detached houses and condominiums, suffused with petroleum products and volatile chemicals, were no longer homes, or property, or "values." Nor were they, as Lucas Welsh had discovered, units of time. During this all-night battle, these dwellings were approached simply as accelerants, the same way one would approach a can of gas, or a spruce tree. Under different circumstances, they might even be called incendiaries. In Slave Lake, at the tender age of fifteen, Ryan Coutts had been given a crash course in WUI fire, and he understood the cognitive leap firefighters were having to make: "A house is just another kind of tree to a forest fire, right? Once it hits the city there's nothing you can do to stop it."

The past two days had been frightening and humiliating, but they had also been instructive, and firefighters were adapting to their new adversary in numerous ways. One of these adaptations was around the machinery they deployed against it. Despite their strenuous efforts to get ahead of it, the fire was still setting the terms of engagement, and firefighters had to come up with a new approach if they were going to have any hope of slowing it down. Municipal firefighters are trained to fight structure fires with water, but when this fire got into a neighborhood, juiced by its terrific momentum and all those petrochemicals, water had little effect.

Because of this, Fort McMurray's firefighters were forced to resort to firefighting techniques normally reserved for wildfires. Water bombers cannot fly at night, but bulldozers can operate around the clock, and they did—all along the northwest edge of the city.*

---

* Since 2020, night vision technology has been used by some water bomber pilots in California.

Most, but not all, of Timberlea had been saved, thus far, by a moat of hundred-yard-wide firebreaks, combined with relentless firefighting. Just to the north of Timberlea, separated by a thin strip of spared forest, lay an enormous new development comprising nearly two square miles of raw, bulldozed earth. This was Parsons Creek North, and it exemplified the aggressive optimism of this city-in-progress. While it contained only about two hundred finished houses, clustered together in a section called "Phase 1," space had been cleared for two thousand more. Once this expansion was completed, only five miles of forest would remain between the city limits and Suncor's South Tailings Pond, a body of toxic effluent so huge it could absorb all of Parsons Creek North with room to spare.

It was through the narrow band of forest between Timberlea and the treeless no-man's-land of Parsons Creek that the fire found its way eastward to Prospect Drive. Almost completely surrounded by forest, this guitar-shaped enclave contained about five hundred houses and condos jammed together, gutter to gutter, in concentric rows of zero lots on streets named after various types of stone. Looping around this new development, outlining the body of the guitar, was Prospect Drive, which began and ended at Confederation Drive, the main link to the nearby highway. This subdivision was so new that recently finished houses stood among unbuilt lots. At its center sat a pair of hundred-yard-long condominium complexes barely a year old, surrounded by freshly paved parking lots and broad swaths of undeveloped earth.

When the fire arrived, energized by twenty-five-knot winds, it hit the west side of Prospect Drive so hard that one hundred houses around Siltstone Place burned to the basements in less than an hour. Ryan Pitchers, the former infantryman who had been showing his pumper truck to a class of kindergarteners when the fire first entered the city, was in Prospect that night. He showed me a picture of a ferocious fire, inside of which the telltale angles of man-made structures were barely visible. "That's conflagration," he said. "You're just not stopping this. It was about three minutes for a house to go. Three minutes."

I wanted to be clear on what he meant: "From something with a roof to—?"

"To nothing."

Flashover occurring outdoors on a neighborhood scale is an extreme rarity at any time; for it to be occurring in the middle of the night, during peacetime, in northern Alberta, is unheard of. "You look at a house and all that's left is foundation," Pitchers said. "Nothing recognizable. That's just the absolute intensity of the heat that we were dealing with. It was so hot that it burnt everything down to ash."

This is normal in an incinerator, or a crematorium, but it is not normal in a house fire, or even a forest fire. What was happening to the houses in Prospect had more in common with artificial conflagrations like the Hamburg firestorm than with any homegrown fire. What became clearer with every passing minute was that Prospect was no longer a neighborhood to be saved, it was a liability to be managed. But these houses were not shacks, and they weren't trailers; most of them had taken well over a year to build and had a median sale price north of half a million. Some of them had tiled bathrooms, wine racks, and multicar garages with lights, alarms, and appliances that could be programmed by smartphone. They were, in 2016, pinnacles of North American middle-class aspiration—even if your neighbors were only six feet away.

When those houses were being designed and built and sold, no one considered the possibility that they could burn like a refinery fire, or that the same apparatus used against such fires would be brought to bear on Prospect Drive. "Aircraft rescue fire fighting" vehicles, also known as ARFF or "crash" trucks, are standard equipment at major airports and also at oil refineries where explosive petrochemical fires need to be accessed and extinguished quickly. Because of this, ARFF trucks are built to accelerate faster than ordinary fire trucks and also to go off-road (because many plane crashes occur off the runway). With their low centers of gravity, enormous tires, and beveled snouts, ARFF trucks can plow through brush and debris to access a fire, even on steep hillsides. With large tanks of both water and foam, and a maximum pump pressure of three thousand gallons per minute, they can suppress a large fire more quickly and decisively than any other land-based firefighting machine. ARFFs were the most powerful pieces of firefighting apparatus in the bitumen industry's considerable arsenal, but in Prospect, they were having only a temporary effect.

The fire would always make up the ground and keep advancing. Neither water nor foam alone was going to stop this fire. They had to find another way.

Members of the Slave Lake crew were on scene in Prospect, and for them it was like 2011 all over again: high wind, cyclonic barrages of embers, and a neighborhood giving way to fire like a beach town to a tidal wave. The Slave Lake Fire Department learned many painful lessons when their town burned, and they were given the chance to apply them in Fort McMurray. Ronnie Lukan was a Slave Lake volunteer whose day jobs included building houses and leasing heavy equipment, a skill set that would prove indispensable that night in Prospect. Forty-ish, irreverent, with boundless energy and the rapid, upbeat patter of a salesman-raconteur, he is a man of action temperamentally suited to catastrophe. ("That's why we brought him," Ryan Coutts told me.) What Lukan and the Couttses understood that most people didn't was that, in firestorm conditions like Slave Lake in 2011, and Fort McMurray in 2016, everything is a potential bomb: "If we leave that F-150 sitting on the street," said Lukan, "it's an oilfield town so, what's in that truck? Is there explosives in that truck? Is there propane bottles? Is there acetylene? We don't know, and if that thing blows up behind us, we gotta worry about that with guys on the ground. So, the first thing—we shove that truck into the basement and it does two things: it eliminates two sources of fuel, and it's helping us get leverage to knock down this two-story house."

But in order to shove a pickup truck into a basement, or knock a house down on top of it, you need heavy equipment. There is more heavy equipment in Fort McMurray than just about anywhere in North America, but most of it is up at Site, and Lukan needed it at the intersection of Prospect Drive and Siltstone Place. "There's probably $20 million worth of engines, pumpers, and everything else sitting on this street," Lukan told me, "and they're absolutely getting their rear ends handed to them. REOC wasn't responding to the incident commander when we were there, so I just kept prying on this guy. I was told numerous times to stop doing that, but what he was doing was losing. We're all competitive guys, and we knew we were losing. Finally, he said, 'Let's try it.' They got us a couple excavators—400s or 450s—big excavators."

Others were making this leap, too, and the excavators (backhoes) were soon joined by two D8 Cats with fifteen-foot blades. The strategy was radical: use the backhoes and bulldozers to cut firebreaks—not through the surrounding forest, which was already in flames, but directly through the neighborhood itself. Removing fuel ahead of the fire was the only way to break its relentless momentum. This meant knocking down intact, unburned houses, most of them brand-new—and not one or two at a time, but by the block, just as a bulldozer would knock down a stand of trees. One of the features of this fire that made it so hard to fight was how high the flames were; not only did this mean more fire and heat, it also meant more wind and flying embers. By flattening the houses, it would lower the flames dramatically, thereby making them easier to subdue. It's an unorthodox mode of thinking that only makes sense when you're using forty-ton machines to remove all available fuel in the face of a rapidly expanding conflagration in an urban setting—something very few firefighters have ever had to do, or even consider doing.

Lukan explained the process this way: "So, the D8 [Caterpillar] guy shows up and says, 'What can I do?' Well, first thing you're gonna do—we know you suck at pushing down houses [a bulldozer tends to knock a house off its foundation rather than flatten it] so, see those four houses at the end of the street that are burning? You're gonna grab this F-350 pickup, that Mercedes SUV—you're gonna take those vehicles and you're gonna clear the street and shove it all into those burning basements so we don't have to worry about that a block away. Then, you're gonna come back, push up all the garages [along with whatever's inside them]. Then, when the backhoe comes, we're saving six minutes per house on knockdown time. Our guys are safe because now we're fighting a fire that's in a basement with a fifteen-foot flame, instead of an eighty-five-foot flame."

But safe is a relative term. When your world is on fire, houses are no longer houses but fuel; vehicles are no longer vehicles but bombs; and water is not for dousing fires but for cooling heavy machinery and saturating demolished homes. Because of his experience in contracting and heavy equipment, Lukan understood that, while you can't use a bulldozer to crush a standing house, you can use it to push other things—like pickup trucks—through the walls to knock out structural members. Failing that, you can use the Cat's blade edge to tear off the outer walls and break the corner posts. Then, the backhoe, using its dexterous bucket like a giant hand, can reach in and rip out the staircase, the spine of most multistory homes. Once this is done, the rest will come down in a matter of minutes—almost as fast as the fire was destroying them.

It is one thing to demolish a house, it is another to demolish an inhabited home. With the exception of invading armies and, occasionally, police, nobody does this. The experience for the machine operators was a surreal one, but by 2:00 a.m. on May 5, surreality was a defining feature of this fire. "Tearing into somebody's house—it's fucking sad," said Jim Rankin, a thirty-three-year-old backhoe operator who grew up in Fort McMurray. "Your house is your life, more or less, and you're seeing people's life." As Rankin and others disemboweled these living homes, random assortments of personal belongings would cascade into view. "You see everything," he said, "baby pictures and baby cradles. Their couch, their fridge—I could have jumped out and grabbed ten beers while I was working." As picture windows burst, entire rooms—furniture, carpets, toys, and TVs—tumbled through the broken walls. They may have been state-of-the-art smart homes, but they were dead before they knew it: with the beams broken and the walls gone, modems, Xboxes, and Alexas could be seen dangling by live wires, their indicator lights blinking dutifully, awaiting the next command in the advancing fire's glow. There was, in Prospect that night, an atmosphere of *Fury Road* apocalypse—the living world receding before an onslaught of fire and machines. It is a feeling anyone who works up at Site is already familiar with.

Jim Rankin's wingman, and the interface between him and the firefighters orchestrating the demolition, was Chris Hubscher, a childhood friend, also raised in Fort McMurray. While Rankin is dark-haired, quiet, and compact, Hubscher is tall, fair, and athletic. He is also married and, tattooed across the knuckles of both hands in flamboyant biker script, was a reminder: "STAY TRUE." Both men grew up around heavy machinery and were operating before they could drive. Earthmoving was second nature to them, and the basics were summed up by the prophet Isaiah: "Every mountain and hill shall be made low: and the crooked shall be made straight, and the rough places plain." Hubscher's father, a longtime supervisor for a local heavy equipment company, and a mentor to both men, had this to add: "You've got to be smarter than the dirt."

Because they look like such blunt instruments, it is hard to see how surgical one can be with a backhoe or a bulldozer. They may not be helicopters, but an intimate, almost biomechanical connection develops between a skilled operator and his machine. Even through all that horsepower, hydraulics, and steel, an experienced backhoe operator can sense when he's hooked a three-quarter-inch gas line three feet down, just as one might detect a plastic drinking straw through heavy gloves. Likewise, a knowledgeable Cat operator can read the Braille of the earth through his treads and blade. "You feel it through the ass of your seat," Rankin told me.

"The pivot point of the tracks," Hubscher added, "that's where you know you have a transition that needs to be balanced out. You'll feel the impact going over a rock, or anything. There's no forgiveness with that amount of steel, it'll just shake your body."

In the earthmoving business, where smoking is common, a cigarette pack is used as an informal unit of measure known as a "deck"—ten centimeters (roughly four inches). Solely by eye, and by the feel of the ground through the seat of his pants, an experienced operator can grade a playing field from end to end to within the width of a pack of Players. Demolishing houses in the face of a rapidly advancing fire was new to Rankin and Hubscher, but years of honing their skills on multimillion-dollar bitumen contracts enabled them to adapt to these extraordinary new demands. "We were trying to make a fire barrier on either side of the street, right?" Hubscher said. "So, instead of going into the burning houses, go four or five houses up the street, tear it down, and carry on, 'cause then at least everything's flat and, if the fire gets there, you can control it."

Simply put, the objective during the small hours of May 5 was "to crush and drown." The heat was terrific, and so was the noise. On top of the racket and roar of detonating houses, the din of machinery was almost deafening: the shriek of powerful water pumps charging the hoses soared over the nerve-jangling, metal-on-metal clatter of truck engines running at high idle. The bulldozers, working hard and fast, were even louder—not just the arrhythmic revving of their huge engines, but the groan and squeal of their steel treads as they ground across residential streets and over curbs, pulverizing everything in their path. Equipped with spotlights and the startling agility of alligators, these machines exuded a menacing and predatory focus: once targeted by those piercing beams, nothing could resist them. As the D8s and backhoes moved from house to house through the heavy smoke and fire glow, laying waste to everything they encountered, they looked and sounded like tanks invading a city. Punctuating this cacophony, as it did in every other burning neighborhood, was the steady but random concussion of exploding fuel tanks and transformers. That night, Prospect looked, smelled, and sounded like a war zone; military veterans present that night compared it to Afghanistan—right down to the booby traps. While Hubscher was helping to unload a Cat from a lowboy trailer, he received a call from a colleague: "Chris, get away from that house! There's oxyacetylene in the garage!"

Warnings of this kind highlighted the strange intimacy of this disaster: many of the people fighting this fire knew whose houses were burning, and what they did for a living. Welding tanks like this— the same kind that were in Wayne McGrath's garage in Abasand— are made of heavy steel and pressurized to two thousand pounds per square inch. They can weigh over a hundred pounds. "They kind of take off like a missile," said Hubscher, "and that one went right through the door and shot across the street." It explains why Ronnie Lukan and Jamie and Ryan Coutts were so eager to neutralize all possible sources of ignition.

While backhoes and bulldozers were tearing down houses and plowing them into their own basements, firefighters were arrayed behind them with hoses and truck-mounted "deluge guns" turning those basements into cisterns. In the process, they also sprayed down the heavy equipment along with the men inside them, who were often no more than a boom or blade length from raging fire. Neither the machines nor their operators were designed for this kind of heat, and there was real concern about the hydraulic hoses melting, the men themselves getting radiant heat burns, or worse. Not only was falling into a basement a constant hazard, the front lawns, saturated by continuous streams of water, had become dangerously soft—"tender," in earthmover's parlance—and were turning into quagmires.

When one of the eighty-thousand-pound D8s bogged down in the liquefied mud of a front yard on Siltstone Place, its thrashing movements and roaring engine gave the impression of a mammoth caught in a tar pit. As its treads churned, only embedding the bulldozer further, its spotlights jerked this way and that like wild, rolling eyes. With sufficient time, a skilled operator could self-rescue, but there was no time, and the operator wasn't up to the task. With the fire bearing down, heavy chain and cable were brought in and hooked up to another Cat, but all efforts to pull it out failed; both the welded steel "schedule 78" chain and the two-inch cable snapped under the strain. Like a captain losing his ship, it is bad form for an operator to lose a million-dollar machine, and this one happened to be brand-new. He stayed at the controls as long as he could, but these men were machine operators, not firefighters, and none of them had protective gear. "You could feel the heat coming from the burning house next door," said Hubscher, who assisted with the rescue effort. "It was like your face was melting."

At the last moment, with the Cat tilted at a crazy angle and flames lapping at the windshield, the operator leaped from the cab. Almost immediately, the solid steel machine was engulfed in flames, melting the radio, igniting the seat, and burning through the cab while the mud around it hissed and steamed. The aftermath was grotesque, the Cat a total loss. Half buried in the front lawn of a smoking, ash-filled house foundation, it was half-charred, windows gone, hydraulic hoses melted down to the steel fittings. It would not be the only bulldozer lost to this fire.

Anyone who wasn't operating a machine, managing a hose, or actively supervising was running through backyards with portable extinguishers putting out ember fires. Out on the street, batteries of pumper trucks sent water flowing freely across every surface; between the surge and flicker of the fire and the metronomic flash of fire engine lights, the streets appeared to pulse with electric blood. And yet, despite the torrents of water being directed toward the fire, crews and trucks were forced, again and again, to fall back as row after row of homes succumbed—Prospect, Siltstone, Shalestone. Each one meant another hundred houses gone. By the time Rankin and Hubscher went to work at the southern edge of the fire, near the intersection of Siltstone where it looped back into Prospect, the focus had shifted: "They came in and started spraying our iron, trying to cool it down," Hubscher said. "They weren't even spraying the houses anymore."

As the fire advanced, localized whirlwinds of smoke, embers, and debris roamed the streets, and the air rippled in the withering heat. At one point, when Hubscher moved his blistered truck to stay clear of the fire, he lost contact with Rankin in the backhoe. Fearing for his friend's safety, Hubscher began running back to where Rankin was working. "Ashes and embers were going in circles," he said. "It was just like a fire tornado—just cycloning—couldn't see nothing, just hot embers." As he tried to push through it, he found it difficult to breathe and, despite weighing close to two hundred pounds, he had the feeling his legs might be blown out from under him. He couldn't see where Rankin was, and, in that moment, Hubscher got a sense of the disorienting terror felt by residents of Hamburg, Dresden, and the dozens of Japanese cities destroyed by Allied firebombing raids. "The fire was creating its own weather system," Hubscher said. "The wind would just switch on us, and the fire would pick—it was like it was choosing which house to burn. Flames were licking the clouds. It was like you were in Hell."

It wasn't long after this that Rankin went into a basement. That night in Prospect, the machine operators were closer to the fire than any firefighter. Shielded only by windshield glass, the heat was ferocious. While "scalloping" a house (tearing it down with a scooping motion of the backhoe bucket), Rankin got too close and one of his treads slipped over the foundation and into the basement. These machines weigh nearly forty tons, and, with the boom and bucket already overextended, Rankin had no leverage to push himself back out. "He called me on the radio," Hubscher told me. "He said, 'Chris, I'm stuck in a basement.' That's when my heart just—"

"Okay," interjected Rankin, "I wasn't worried for my own safety. I'm thinking, 'We're gonna lose a machine.'"

Hubscher *was* worried for his friend's safety, and with good reason. The fire was literally next door—so close that, before a Cat could pull him out, it had to push up a dirt berm between the backhoe and the advancing fire just to keep it (and Rankin) from igniting off the radiant heat. Meanwhile, Hubscher dragged in another two-inch cable. In order to connect it, he had to crawl under the precariously balanced backhoe, which had a half-demolished house leaning over it. Hubscher had just finished hooking up the cable when, out of nowhere, he was struck by a typewriter. Some unseen shift had caused it to tumble out of one of the upstairs rooms. The typewriter, an old manual, hit Hubscher so hard that he lost his footing and slipped into the splintered and nail-studded debris under the backhoe.

"I looked at it for a second," Hubscher said, "and I was like, *Really?*"

It was a strange, suspended moment: Hubscher noting the heavy gray body of the obsolete machine, its chrome space bar, how the keys appeared to float, independent of their inner workings. And then events resumed. "The cuff of my jeans caught a nail," he said. "Otherwise, I would have been at the bottom of the rubble."

Hubscher used the rescue cable to save himself, and then he signaled to the Cat. He was bruised, but the rescue was successful. "It was just operator error," said Rankin. "I was rushing. I was rushing, and I shouldn't have been." But one look at photos from that night, and it becomes clear why rushing was the only sane approach.

Jim Rankin estimated that he tore down about thirty houses that night, and moved or buried at least twenty vehicles, and he wasn't the only one working. The total loss represented tens of millions of dollars, but this was simply the scale of what was at stake. What mattered was that the strategy worked: by removing fuel, they successfully broke the chain of ignition, thereby saving all the houses and businesses south of Siltstone Place. A similar firebreak was bulldozed through on the east side of Prospect Drive, between Shalestone and Gravelstone Way, and there, too, the line held. With the exception of the isolated condominium complex at the center of the neighborhood, virtually everything else within the Prospect loop—about five hundred houses and garages—burned to the foundations. It was, at best, a bittersweet victory. The fire wasn't out, not even close. It meant merely that, by superhuman and mechanical effort, this particular front of the fire, on this particular night, had been denied a source of fuel—for now.

The effort took its toll: some men were crying; others stared into the hypnotizing flames as if they were shell-shocked. A firefighter named Jerron Hawley fell to his knees, wavering on all fours so long that his friend, a paramedic, wondered if he might be having a "jammer"—a heart attack. He wasn't, but exhaustion of a profound and mind-altering kind was setting in across the city. Some members hadn't slept since the night of May 2. After gathering himself, Hawley took stock from a vantage on some high ground with his comrades. "It was on fire as far as we could see," he recalled in a memoir he wrote with two fellow firefighters. "It was never-ending."

"It" was the horizon around Fort McMurray. The city was now a smoking island in a sea of flame. At that moment in Prairie Creek, miles away at the south end of town, the forest was torching like an Australian bushfire, releasing hundred-foot fire dragons that swirled in the night air as if it were midday. Surrounded as they were, the exhausted and demoralized men wondered aloud how long they could last, if and when the REOC would give the order to evacuate, or if they would be trapped there. Hawley recalled one of the men trying to reassure the others. "They aren't just gonna let us die here," he said. "We're not gonna die here tonight, fellas."

First light was still two hours away.

By sunrise on Thursday, May 5, it was starting to seem as if the fire had taken up residence in Fort McMurray. And why not? In his translation of *Beowulf*, Seamus Heaney refers to fire as "the glutton element." It is in its nature to exploit its resources to the fullest, and this fire's ability to do so was even more impressive than Suncor's or Syncrude's. Since May 1, it had burned through an area more than ten times the size of Manhattan—about 350 square miles. Firefighters, while not "used to it," exactly, were growing familiar with this fire and, after three days of nonstop assaults on the city, the fire was given a name. It started with Chief Darby Allen, and it stuck. "The beast is still up," he told viewers in a video message posted late in the evening of the 5th. "It's surrounding the city. And we're here doing our very best for you." The name caught on because it fit: a beast is something enormous and unusual that behaves capriciously and mercilessly, inciting terror and wreaking havoc as it comes and goes at will—like Jaws, or Grendel, or Godzilla.

All wildfires are assigned a number based on the order in which they occur, and major wildfires will be named for their place of origin. In special cases, such as the Great Miramichi Fire (1825), the Great Fire of Boston (1872), or the Great Vancouver Fire (1886), the fire will be named for the place it destroyed. But May 5, 2016, marked the first time in post-contact North American history that a fire was endowed with the qualities of a living thing and named accordingly. This fire had become personal, not simply because of the way it entered people's lives and upended them, but also because of its overwhelming power, the way it attacked the city day and night, from all directions, and the way it appeared to single out targets for destruction in ways that were at once random and intimate.

Malevolent creatures tend to arrive from dark and hidden places: Godzilla and Moby-Dick emerged from the deep sea; Wells's Martians came from outer space; Grendel lived in a swamp; man-eating tigers and leopards lurk in the jungle; the devil comes from Hell. The Beast of Fort McMurray was born in the forest. What makes a beast a beast in our minds is repetition: in each case, real or imagined, these beings appear repeatedly, impacting different people and places in the larger community. This is standard behavior for monsters, but atypical of fires, and this, even more than its speed, size, or ferocity, is what set this fire apart: the way it *persisted*—for days—much as monsters do in the stories we tell about them. It is through these repeated encounters that victims and adversaries assess their attacker's character, establish a relationship, build a narrative, and arrive at a name. Naming and repetition are how myths and legends become anchored in the collective mind, the name serving as a kind of mnemonic avatar for the larger, more complicated whole. We may not read the Bible or the Koran, but we've all heard of Satan, and we all have an image that comes to mind.

Earlier that afternoon, Acting Captain Mark Stephenson had

made this leap on his own. Stephenson, who had been ordered to abandon Abasand (and his home along with it) on May 3 and had barely slept since, was now on one of the fire's new front lines—Signal Road, a quiet residential street on the southern edge of Thickwood. All told, there were perhaps half a dozen trucks and twenty men stationed along its length, and they would be taking the rapidly advancing fire full in the face. There were hydrants to tie into, but they could no longer be trusted; the city's water supply was already under terrific stress.

With a hose connected to a pumper truck, Stephenson and his comrades were doing the same thing the Slave Lake Fire Department had done in 2011: spraying down the forest. Signal Road ran along the southern edge of the west side plateau, which meant the fire was coming at them from the valley below, and firefighting is that rare form of combat where having the high ground is actually a disadvantage. Fire is the only thing in Nature that gains energy and speed by traveling uphill, and, with ample fuel and a rising wind, the roar this fire made was ferocious. Like a wild beast.

Stephenson, an enormously powerful man who resonates to mythic tales, looks as if he would be at home on a battlefield in Middle Earth. Considering what he did to his own garage door, if it came to a bet on who would prevail, the money would be on Stephenson. But this fire was different; it had beat him once in Abasand, and now he was going to face it again. There was no guarantee the outcome would be any different. This was the first fire any of these men had fought where water had no meaningful effect, but water was the only weapon they had, and they still believed in it.

Even after their humbling defeat across the river, the men did not fully comprehend the otherworldly power and voracious *intention* this fire was capable of bringing to bear, until it confronted them on Signal Road. "The fire was coming up the hill," Stephenson recalled, "and it was like it took this big breath—the wind was so strong going into the trees that the pant legs of my coveralls were flapping—and all of a sudden, the whole front erupted at once. A hundred meters of forest—up, *poof,* gone."

What had, a moment earlier, been a line of trees was now a towering wall of flame. It was like some kind of atmospheric magic trick.

Stephenson was twenty yards away. Judging by its effect on the fire, his firehose might as well have been spraying gasoline. When asked if he would ascribe a character to this fire, Stephenson didn't have to think for long: "You know that big fire monster in *The Lord of the Rings*?"

I thought he was referring to Smaug.

"Not the dragon," he said. "The fire demon."

One of J. R. R. Tolkien's most terrifying creations is the Balrog, a huge and diabolical embodiment of fire itself. Deep inside the Mines of Moria, the Balrog confronts the wizard Gandalf while his companions watch in helpless terror: "The dark figure streaming with fire raced towards [him]," Tolkien wrote in *The Fellowship of the Ring*, ". . . and the shadow about it reached out like two vast wings . . . Suddenly it drew itself up to a great height, and its wings were spread from wall to wall."

"Yeah," said Stephenson, "just flame—everywhere."

It was the fire's third day inside Fort McMurray.

PART THREE

RECKONING

～～～

I gave them fire

. . .

But I have no device to free myself
from this disaster.

—AESCHYLUS, *Prometheus Bound*

18

Dissonance
(if you are interested)
leads to discovery.

—William Carlos Williams, *Paterson*

The Middle Ages saw a slowly dawning awareness that there was "something in the air," something invigorating, distinct from the invisible entirety moving through and around us. The first time the word "atmosphere" (literally, "vapor sphere") appeared in the English language was in 1638, but it was in reference to the moon, which goes to show how poorly the concept was understood. Any insight into Earth's atmosphere, its life-giving properties, or what it might be composed of would have to wait for another century. Its revelation, fueled in part by coffee, appeared in the Western consciousness at the same time as the American Revolution and a prototype for the spark plug.

That there were elements of the intangible void surrounding us that could be altered, added, or isolated was discovered almost simultaneously by a Swedish-German apothecary named Carl Scheele and a "furious free-thinker," Unitarian minister, and polymath from Birmingham, England, named Joseph Priestley. Both men had the exhilarating and increasingly rare experience of finding something new under the sun, a feat Priestley performed repeatedly, discovering nine new gases, among his many other remarkable accomplishments. But what to call this invisible spirit-like essence that seemed to energize fire and animals alike? Scheele settled on the German *brandluft*— "fire air." Priestley, building on the work of his contemporary Joseph Black, went with a Greek-derived English variant: "dephlogisticated air," a mouthful that meant virtually the same thing. Both Scheele and Priestley understood that there was a close relationship—a kind of codependency—not just between oxygen and fire, but between oxygen and life itself.

In the summer of 1771, Priestley began a series of experiments with live mice, putting them into an inverted bell jar, which he sealed by immersing in a shallow pan of water. The mice died quickly in this environment, sometimes in a matter of seconds. But why? "Once any quantity of air has been rendered noxious by animals breathing in it," Priestley wrote, "I do not know that any methods have been discovered of rendering it fit for breathing again."

He followed this with an astonishing corollary intuition:

> It is evident, however, that there must be some provision in Nature for this purpose, as well as for that of rendering the air fit for sustaining flame; for without it the whole mass of the atmosphere would, in time, become unfit for the purpose of animal life [and also for fire].

Like his remote colleagues, Priestley was, in his off time, using secular means to fathom the deep space of their immediate surroundings—the invisible, intangible mysteries previously relegated to the clergy and to God. If his neighbors didn't consider these inquiries actively heretical, then many considered them pointless; after all, there was, literally, "nothing to see here." Priestley summed up this paradox in his observation on Isaac Newton: "He had very little knowledge of air, so he had few doubts concerning it."

Despite the fact that there is no tangible barrier between the human eye and the remotest visible star, Priestley perceived that our atmosphere was not only malleable but finite: if it wasn't a closed system, it was a very restricted one, much like his glass bell. Wondering how other living creatures might fare in "putrefied air," he introduced a mint plant into a sealed jar recently vacated by one more suffocated mouse. The mint plant lived—for days, and then for weeks, in what had been lifeless, "phlogisticated" air. Priestley continued, this time introducing a mint plant into a jar where a candle had smothered. Again, the mint continued to grow. That the same conditions which killed mice and smothered candles would have no adverse effect on a plant left Priestley to speculate on what was happening inside those sealed jars, in plain view yet out of sight. After a week, he placed a mouse inside the jar with the mint. Instead of dying immediately like all the others, this mouse lived long enough for Priestley to observe it, retrieve it, and put it into a second sealed vessel filled with mouse-polluted air, where it promptly died. Then he put a candle in with the mint and it, too, burned longer than the candles had in empty jars. "This observation," he wrote later in his seminal work, *Observations on Different Kinds of Air*, "led me to conclude, that plants, instead of affecting the air in the same manner with animal respiration, reverse the effects of breathing, and tend to keep the atmosphere sweet and wholesome, when it becomes noxious."

Using equipment cobbled together from his wife's kitchen and the potting shed, and methods a resourceful twelve-year-old could master, Priestley was, solely by the brute force of his curiosity and powers of deduction, systematically linking the responses of fire, plants, and animals in a series of connections that led, inevitably, toward oxygen and, by extension, to carbon dioxide.

Priestley claimed that the motive behind his provocative experiments was "exciting the attentions of the ingenious." Of all the ingenious thinkers active in the late eighteenth century, it is hard to name one whose attentions were more excitable than Benjamin Franklin's. By a wonderful coincidence, Franklin visited Priestley's home in June 1772 while he was refining his mint experiments. Following his visit, he wrote to Priestley, "that the vegetable creation should restore the air which is spoiled by the animal part of it, looks like a rational system, . . . The strong, thriving state of your mint, in putrid air, seems to show that the air is mended by taking something from it, and not by adding to it."

Franklin, running on a potent mix of intuition, raw data, and his caffeine-powered encounters with Priestley, was already halfway to photosynthesis: it *was* a "rational system," and the mint had indeed "mended" the air "by taking something from it"—carbon dioxide. Narrowly eluding Franklin's grasp was the fact that the mint further mended the air by adding something: oxygen. Priestley, meanwhile, was close enough to smell it. Beneath those small glass domes, so much like our own atmosphere, Priestley was, tumbler by tumbler, cracking the safe that held the secrets to life on Earth. Not only had Priestley managed to simulate Earth's atmosphere in microcosm, he understood—250 years ago—that it was both contaminable and restorable, that human intervention could render it lethal or life-sustaining.

~~~~

Our atmosphere envelops the cosmic sand grain of Earth just as Priestley's glass bells enveloped his mice, just as bitumen envelops a grain of bituminous sand, just as our skin envelops our own bodies: relative to what they are covering, each of these insulating layers is gossamer-thin. The vertical distance from sea level, where most humans live, to icy suffocation at Camp 4 on Mount Everest is less than five miles—a mere .06 percent of Earth's eight-thousand-mile diameter. Put another way, your skin is ten times thicker, relative to your body, than the habitable portion of the atmosphere is relative to Earth. This gaseous membrane is all that separates us from the lifeless oblivion of deep space, and neither we nor fire could survive without it.

That our atmosphere is malleable and sensitive to changes may be hard to grasp in the abstract, but it is easy in microcosm and, in this sense, we're as sensitive as Priestley's mice: if you are traveling in a car and a fellow passenger releases methane, you will know in seconds. Likewise, bacon, wood smoke, roses, or gasoline. The same is true for heat: the reason Arctic bush pilots travel with candles is because, in the event of a crash, it only takes one or two of them burning in the fuselage of a small plane to make a night at -40°F survivable. Our atmosphere is acutely sensitive to subtle changes, and so are we.

Despite the fact that we are protected by a formidable combination of ozone, gravity, solar radiation, magnetic fields, and life-enabling gases, our atmospheric "living room" remains as fragile as a fish bowl—and as easily contaminated. The idea that our atmosphere could be changed—by us—is not something we have ever, in our entire history, had to consider seriously until a single lifetime ago, which is about as long as we have had to seriously consider the automobile.

The Petrocene Age has enabled ordinary people to command energy in ways kings and sultans could only dream of, and with an ease hitherto unimaginable. Behind the wheel of a Chevy Silverado, a one-hundred-pound woman can generate more than six hundred horsepower as she draws a six-ton trailer at sixty miles an hour while talking on the phone and drinking coffee, in gym clothes on a frigid winter day. Prior to the Petrocene Age, only a king or a pharaoh could have summoned such power, and its equivalent would have required hundreds of enslaved people and draft animals. Today, with cheap and plentiful oil at our disposal, everyone's an emperor. Every time we get in a car, on an airplane, or on a ship, we are traveling with a vast invisible retinue that multiplies our potency even as it multiplies our emissions. During this first century and a half of the Petrocene Age, as we have harnessed, democratized, and amplified fire on demand, we have also unleashed some unintended consequences: a by-product of becoming a petroleum-based society—in other words, a fire-based society—has been the superheating of the atmosphere.

Fort McMurray, founded at the dawn of the Petrocene Age, has grown into an unlikely flashpoint in this collision between the rapid expansion of our fossil fuel–burning capacity and the rigid limitations of our atmosphere. Here, in this city's fire and the events leading up to it, can be seen the sympathetic feedback between both the headlong rush to exploit hydrocarbons at all costs, in all their varied forms, and the heating of our atmosphere that the global quest for hydrocarbons has initiated, and that is changing fire as we know it.

Reckoning with the negative aspects of oil and gas is a responsibility that duplicitous marketing, short-term governance, superb engineering, and a certain amount of willful blindness have enabled us to keep at bay for a century. In addition to being extraordinarily flammable, petroleum is lethally toxic, both in its liquid and vapor forms. In light of this, it is almost spooky how comfortable we are traveling with powerful, poisonous bombs positioned directly behind our children's car seats. There is a palpable dissonance between this and the auto industry's recent preoccupation with "safety" that only intensifies when you consider a car's emissions. Exhaust fumes, like the atmosphere they flow into, are mostly invisible and easy to keep out of mind, but if that Silverado's tail pipe were directed back into the vehicle, the driver and all her passengers would be dead in minutes. If the Silverado's exhaust were piped into the driver's living room, she and her family would be dead in an hour. But somehow, when we run our cars "outside," in our shared atmosphere, all the soot and toxic gases magically disappear.

When we cast a vote, we do it believing that it will combine with others and add up to something transformative, and it often does, even if we don't like the result. Combustion works the same way: every fire we light, whether we see it or not, is a vote for the transformative power of carbon dioxide. Every ton of coal, every barrel of oil, every tank of gas is a genie; once the command is given—once that fire is summoned forth—CO_2 is released and its heat-trapping properties are activated. Once in the atmosphere, CO_2 will persist for centuries. Meanwhile, methane, the main ingredient in "clean" natural gas, retains heat at least *twenty-five times* more effectively than CO_2. A by-product of fracking, gas flaring, bitumen processing, livestock raising, heating, and home cooking, methane (CH_4) remains active in the atmosphere for years after its initial release.* Earth's atmosphere may be huge and invisible, but it is also as finite as a room: what happens in it stays in it. Unless we send it into deep space by rocket, nothing we make, or emit, ever truly goes away.

This is hard to remember, or even believe, when we gaze skyward through our domed, transparent ceiling, past our lone moon and lonelier sun, toward the legions of luminous pinpricks beyond it. From our tiny vantage, it's nearly impossible to truly apprehend that we exist inside a closed container—together with every other living thing, every fire, and every molecule of our cumulative emissions. And it is even more difficult to accept the possibility that humans could conjure up an insult large enough, or noxious enough, to impact the

---

* Recent mapping and measuring data suggests that methane emissions—both anthropogenic and natural—have been grossly underestimated.

Earth as seen from Saturn's orbit,
900 million miles away (Cassini Probe, NASA)

integrity of something so apparently limitless and vast. If you think
about it, it is amazing that anything survives here. Compared to the
vastness of our solar system, our galaxy, or the universe, Earth's atmosphere is too insignificant to register—a soda bubble drifting in the
dark. Seen in this way, it becomes unnervingly clear that we are here
by a strange and precarious grace.

We evolved close to the ground, in small communities, and
we remain parochial by nature. When it comes to our atmosphere,
this hardwired trait really shows. Even though both the causes and
effects of greenhouse gases are as clear as the causes and effects of
poor hygiene, many remain surprised by the impacts they are having, while many others cannot or will not accept that human activity
plays a leading role. The comparison of manufactured greenhouse
gases to personal hygiene is not an arbitrary one. Both are closely tied

to the Petrocene Age. The first time a Western doctor recognized the connection between handwashing and patient survival wasn't until the mid-nineteenth century. In 1847, the Vienna-based obstetrician Ignaz Semmelweis noticed that births attended by midwives seemed to result in fewer infections than those attended by physicians. Wondering why this might be, Semmelweis observed that, while doctors in his hospital handled all manner of sick and septic patients, including cadavers, midwives focused solely on mothers and babies. Ruling out other possibilities, Semmelweis concluded that the doctors' hands were the one consistent link between sick or dead patients and birthing mothers. Semmelweis broke this link by washing his hands in a systematic way before attending births. The incidence of infection dropped dramatically among his patients, and he made handwashing standard practice.

Convinced of the connection though unable to explain it, Semmelweis tried to persuade other doctors to follow his example and wash their hands. Lacking an understanding of microbes (despite access to microscopes), and convinced of their own methodology, Semmelweis's male colleagues dismissed him, resulting in the unnecessary deaths of many more mothers and infants. Semmelweis persisted, but his peers ostracized him and he was forced to leave Vienna. As the years passed, he grew increasingly frustrated, strident, and, ultimately, irrational. In late July 1865, eighteen years after discovering the link between handwashing and natal health, he was lured under pretense to a Viennese insane asylum, where he was beaten and incarcerated. He died two weeks later, allegedly from sepsis resulting from a wound incurred during the beating.

Ignaz Semmelweis would not be vindicated until the 1880s, when, thanks to breakthroughs made by Robert Koch and Louis Pasteur, germ theory gained acceptance in the larger medical community. Since then, handwashing has become second nature to virtually every member of functioning societies around the world. Had handwashing become politicized, or had there been a profit to be made by encouraging skepticism toward the existence of microbes, public health would look very different today. Prior to 2020, we might have laughed at such a suggestion, but resistance to elementary public health precautions during the coronavirus pandemic offers a real-time example of the individual and societal costs incurred by politicizing science and rejecting solid evidence. Carbon dioxide proliferation is even more preventable than germ spread, and the failure to control it has induced its own "pandemic fever." The relationship between the heat-trapping properties of carbon dioxide and the warming of our atmosphere was understood, in principle, decades before Dr. Semmelweis began washing his hands. And it was explored and demonstrated in granular detail well before the adoption of germ theory.

Trucks, buses, and SUVs—the direct descendants of Étienne Lenoir's revolutionary Hippomobile—are what enabled ninety thousand citizens to safely flee Fort McMurray in a matter of hours. And it was the need to fuel such vehicles that drew those people there in the first place. There is in our codependent relationship with petroleum an element of Stockholm syndrome, and the image of shiny new $600-a-month pickups, dead in the median awaiting Samaritans with jerry cans while smoke and flames soared skyward in the background, became an iconic image from this fire. It suggests, on the one hand, a kinder, gentler, Canadian *Mad Max* in which gas has become the only relevant currency and, on the other, the serpent Ouroboros swallowing its own tail. The Ouroboros, you could say, is the literal embodiment of a "feedback loop."

The creative explosion that set this loop in motion represents a segue between the Age of Enlightenment and the Industrial Revolution, both of which were as pivotal for fuel, propulsion, and mass production as they were for art and philosophy. Less well known is how momentous this transitional period was for climate science. Simultaneous with the introduction of the steam engine in the mid-eighteenth century was the dawning awareness that Earth's atmosphere had heat-retaining characteristics analogous to a greenhouse. Solar energy reaches Earth in the form of radiating waves, most of which are reflected back into space. During daylight hours, some of these waves are absorbed by earthly objects, including the oceans, which cool down again at night, and even more so during the winter months—much the way a frying pan heats up on a stove and then cools down after the stove is turned off. But just as a frying pan does not cool all the way down to outdoor temperatures because it's inside an insulated house, the planet does not cool all the way down to the temperature of outer space because it's insulated by our atmosphere—specifically, water vapor and carbon dioxide. This is the "greenhouse effect," and it is critical to life on Earth.

The power of solar heat is, literally, as clear as day, but how that heat is retained, reflected, and, in some cases, amplified, is less obvious. "It is difficult to know how far the atmosphere influences the mean temperature of the globe," wrote the French scientist and climate pioneer Joseph Fourier in 1824. "It is to the celebrated traveller, Monsieur de Saussure, that we are indebted for a capital experiment, which appears to throw some light on this question." In 1824, this "celebrated traveler" had already been dead for twenty-five years. Horace Bénédict de Saussure was a natural philosopher of the same vintage as Joseph Priestley, and equally brilliant. Born into a wealthy family in Geneva, Saussure quickly distinguished himself as a polymath whose expertise ranged from botany and geology to the physics of heat. He was also a mountaineer, recording a number of first ascents in the Alps where he conducted some of the first experiments on atmospheric changes at altitude.

One of Saussure's many lines of inquiry involved a device he invented called a helio-thermometer—in essence, a portable greenhouse equipped with a temperature gauge. On a sunny July day in 1774, Saussure packed his contraption up Mont Crammont, a nine-thousand-foot peak in the Italian Alps. After exposing the glass surface of the box to the afternoon sun, he noted its internal temperature before descending five thousand feet into the valley below. Even though the outside air temperature was more than 30°F warmer there than at the summit, the temperature achieved inside the insulated box (190°F) was almost identical. Saussure deduced that, while the power of the sun was relatively constant, the outside air retained more heat at lower elevations—almost as if there were a pane of glass over it. "The more dense the air," Saussure wrote, "the more it is humid, and the more it is warm . . . Here I agree with [Pierre] Bouguer, in the reflection of the Sun's rays by the surface of the Earth."*

---

* Saussure was building on observations made decades earlier by Pierre Bouguer in the Peruvian Andes.

Fifty years later, Joseph Fourier, a true Renaissance man[*] who, in addition to being a gifted scientist was also a baron, a colonial governor, and a mathematician, expanded on this seminal idea in his 1822 magnum opus, *Analytical Theory of Heat*, an exhaustive tome described by Lord (absolute zero) Kelvin as "a great mathematical poem." By focusing on temperature changes between night and day, and across the seasons, Fourier calculated that Earth was much warmer than it should be if it were dependent only on sunshine and residual heat from the planet's core. Something else, he concluded, must be heating it—or insulating it. The term "greenhouse effect" would not be coined until much later, but now it was only a matter of time.

[*] Fourier has one of the most astonishing CVs of anyone who ever lived.

19

Some say the world will end in fire,
Some say in ice.
From what I've tasted of desire
I hold with those who favor fire.

—Robert Frost, "Fire and Ice"

If one were to ascribe a specific date to the dawn of modern climate science, a strong case could be made for August 23, 1856—though its significance went unrecognized for 150 years. Eunice Newton Foote was an artist, inventor, citizen scientist, and early suffragist from upstate New York whose singular contribution to climate science was lost in plain sight until a retired petroleum geologist stumbled over it in 2010. Without knowing it, Foote conducted and described what could be called the first modern climate change experiment. It was a variation on Saussure's chambered box idea consisting of a pair of sealed glass cylinders equipped with thermometers. After filling one with "ordinary air" and the other with "carbonic acid gas" (aka carbon dioxide), Foote took readings at room temperature before exposing them to direct sunlight. While both cylinders heated up, the one filled with CO_2 grew twice as hot in a matter of minutes. "An atmosphere of that gas would give to our earth a high temperature," Foote wrote in a brief article entitled "Circumstances affecting the Heat of the Sun's Rays," "and if as some suppose, at one period of its history the air had mixed with it a larger proportion than at present, an increased temperature from its own action . . . must have necessarily resulted."

It was on that late summer day in 1856, in Albany, New York, that Foote's account of her experiment was presented at the Annual Meeting of the American Association for the Advancement of Science, where it was read aloud by a male colleague. Attendees would have assumed correctly that Mrs. Foote's experiment was an investigation into climates of the distant past—a topic of growing interest at the time. Few, if any, could have imagined that they were being given a glimpse of the future, too. Foote's conclusions included an ominous warning that would have gone unnoticed by an audience still inhabiting a horse-drawn, steam-driven world: "The receiver containing the gas," Foote wrote, "became itself much heated—very sensibly more so than the other—and on being removed, it was many times as long in cooling."

Humanity is now re-creating Eunice Foote's groundbreaking experiment in real time, only "the receiver containing the gas" is our atmosphere.

~~~~

Climate science came of age in tandem with the oil and automotive industries, and all of them are products of the Petrocene Age. In 1859, just three years after Eunice Foote's landmark CO_2 experiment, "Colonel" Edwin Drake drilled his first oil well in Pennsylvania. That same year, Étienne Lenoir built a prototype of the first commercially viable internal combustion engine. In an uncanny coincidence, 1859 was also the year an Irish physicist named John Tyndall proved once and for all that concentrations of certain gases in the atmosphere had the potential to alter Earth's climate. His findings, which he presented at the June 10th Evening Meeting of the Members of the Royal Institution, effectively confirmed discoveries made by others over the previous century:

> When the heat is absorbed by the planet, it is so changed in quality that the rays emanating from the planet cannot get with the same freedom back into space. Thus the atmosphere admits

of the entrance of solar heat, but checks its exit; and the result is a tendency to accumulate heat at the surface of the planet.

This was the greenhouse effect described in every respect but name. While the concept may sound familiar, Tyndall was the first to prove it, and also to explain in unambiguous terms that water vapor and carbon dioxide are strong absorbers of reflected solar heat in the form of infrared radiation. As similar as Tyndall's findings may have been to Eunice Foote's, he appeared to have no awareness of her research. In fairness, Tyndall was working at a much more advanced level with the most sophisticated equipment available, and doing so in London, a scientific hub a world away from upstate New York. Tyndall was a remarkable man by any measure; in addition to being a dedicated scientist with a long-standing interest in the physics of heat, he was a successful commercial surveyor, an inventor, and a popular writer and speaker with several influential texts to his credit, including *Heat: A Mode of Motion*, which stayed in print for more than fifty years. Like Saussure before him, he was also a pioneering alpinist, logging one of the first ascents of the Matterhorn.

It would take another forty years, but in 1896 (the same year Henry Ford built his first car), a Nobel Prize–winning Swedish chemist named Svante Arrhenius published new research on global warming (and cooling) driven by changes in atmospheric CO_2. At the time, there was growing consensus that epochal climate change in the form of ice ages and interglacial warming periods was driven by fluctuations in carbon dioxide, and that these fluctuations were due, primarily, to volcanic activity. After making tens of thousands of calculations—by hand—in which he accounted for Earth's elliptical orbit, diurnal and seasonal changes, and variations in "nebulosity" (cloud cover), among many other factors, Arrhenius determined that a 50 percent increase in CO_2 from 1890s levels would raise the mean surface temperature of Earth by roughly 6°F. Considering that climate modeling did not exist then, his calculations were astonishingly accurate. He even foresaw that the warming effects of increased CO_2 would vary depending on latitude, a fact now painfully clear to those of us living at higher latitudes.

In that same prescient article, which was inspired by "very lively discussions" at the Physical Society of Stockholm, the thirty-seven-year-old professor also speculated on the impacts of carbon dioxide from coal burning. This was an extraordinary thing to be considering at that time—as intuitively brilliant, in its way, as Joseph Priestley recognizing that mice and plants could alter the quality of air inside a bell jar. Svante Arrhenius often gets credit for this leap of imagination, but he was following the lead of a colleague, the geologist Arvid Högbom, whom he quotes extensively: "This quantity of [CO_2], which is supplied to the atmosphere chiefly by modern industry," Högbom wrote in 1894, "may be regarded as completely compensating the quantity of [CO_2] that is consumed in the formation of limestone."

In other words, Högbom had already determined that industrial CO_2 emissions were replacing any CO_2 being removed by natural processes—chemical weathering, uptake by forests, and absorption by oceans (a process called "buffering").

At the turn of the nineteenth century, the dawn of the automobile age, coal was the dominant industrial energy source (as it would be until the 1960s), but oil and gas were gaining ground rapidly. Even at this early date, new language was being found to articulate the effects of carbon dioxide. "The atmosphere may act like the glass of a green-house," wrote the meteorologist Nils Ekholm, a friend and colleague of Arrhenius, in 1901, "letting through the light rays of the sun relatively easily, and absorbing a great part of the dark rays emitted from the ground, and it thereby may raise the mean temperature of the earth's surface."

Like Högbom and Arrhenius, Ekholm believed that increased industrial emissions would have a warming effect on the climate. By this time, "Rockefeller" was already a metonym for extreme oil wealth, and, in 1906, John D. Rockefeller's Cleveland-based Standard Oil Company, which wielded near-total control over North America's petroleum industry, was challenged in what would become a yearslong antitrust lawsuit. It would end with the massive corporation being broken up into *dozens* of "smaller" companies, including the future energy giants Esso, Mobil, Amoco, Chevron, and Texaco. By then, it was clear that the automobile—powered by gasoline (as opposed to coal oil, steam, or electricity, all of which were viable at the time)—was here to stay.

In 1907, while the Standard Oil antitrust suit was before the court, the British physicist John Poynting self-consciously coined two terms that would inform discussions of climate for the foreseeable future. They appeared in the opening sentences of a response to an article by the wealthy Bostonian travel writer and citizen scientist Percival Lowell. "[Lowell] takes into account the effect of planetary atmospheres in a much more detailed way than any previous writer," Poynting wrote in a prominent British scientific journal. "But he pays hardly any attention to the 'blanketing effect,' or, as I prefer to call it, the 'greenhouse effect' of the atmosphere."

The following year, 1908, Henry Ford introduced his multimillion-selling Model T, and Svante Arrhenius published the English-language edition of his magnum opus, *Worlds in the Making: The Evolution of the Universe*. This audacious work, conceived (like the Model T) with a general audience in mind, was the first popular science book to examine the possibility of anthropogenic warming. After describing the "hot-house" theory of Earth's climate, demonstrated by Saussure in the 1770s and expanded on by Fourier, Foote, Tyndall, and others, Arrhenius went a step further: "The enormous combustion of coal by our industrial establishments suffices to increase the percentage of carbon dioxide in the air to a perceptible degree."

At the time, few in the scientific community approached this issue with any great urgency. To most people who gave the matter any thought at all (including Arrhenius), a tilt toward warmer winters sounded like a wonderful idea.

Neither Arrhenius nor his visionary colleague Arvid Högbom could have imagined a new warming period arriving as quickly as the one currently upon us, nor could they have foreseen that anthropogenic emissions could so rapidly overtake volcanic eruptions, the primary source of excess carbon dioxide. But when it comes to fire, heat, and CO_2, we—*Homo flagrans*—are a "volcanic eruption," one whose intensity has only grown since the Petrocene Age began. In the time since Arrhenius's landmark 1896 article was published (almost, but not quite, within living memory), industrial fossil fuel emissions have increased by twenty-five times, far outstripping Nature's ability to absorb or neutralize them in the short term.

Following Arrhenius's and Högbom's early projections, the pace of atmospheric science began to accelerate in parallel with the rapid adoption of the gas-powered automobile and the concomitant expansion of petroleum use across the globe. In an astonishingly short time, what had once been merely speculative became measurable. One of the first people to calculate the impact of industrial CO_2 in a systematic way was a Canadian-born and British-raised steam engineer and amateur meteorologist named Guy Callendar. By the 1930s, there was already anecdotal evidence that the climate was warming, but Callendar was the first to actually track and graph it. His inquiry arose from an old-fashioned impulse: curiosity. The son of a successful (and wealthy) physicist, Callendar, like the natural philosophers before him, was free to pursue science for its own sake. He had doubts about carbon dioxide's influence on Earth's climate, and he wished to test it. After analyzing a hundred years of temperature records from two hundred weather stations around the globe, Callendar detected a trend. His results were published in 1938, when he was forty years old, just as the automobile was achieving true ubiquity on North American and European roads. In his paper, entitled "The Artificial Production of Carbon Dioxide and Its Influence on Climate," Callendar foretold a future no one was ready to see:

> Few of those familiar with the natural heat exchanges of the atmosphere, which go into the making of our climates and weather, would be prepared to admit that the activities of man could have any influence upon phenomena of so vast a scale. In the following paper I hope to show that such influence is not only possible, but is actually occurring at the present time.

Because there were no computers, and no precedent for this kind of long-term climate analysis, Callendar made his calculations, and his graphs, by hand:

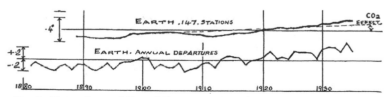

FIG. 4.—Temperature variations of the zones and of the earth. Ten-year moving departures from the mean, 1901-1930. °C.

After a painstaking review of thousands of weather records, Callendar determined that the mean global temperature had risen 0.9°F between 1890 and 1935.

While many scientists and civilians alike acknowledged the occurrence of temperature fluctuations (cooling as well as warming), most attributed them to sunspots or other natural cycles. Callendar, like Högbom and Arrhenius before him, was an outlier when he attributed the fifty-year warming trend to anthropogenic CO_2. It was a lonely position to be taking on the eve of World War II and the petroleum-driven explosion of growth and prosperity that was to follow. Although Callendar's work was published in a respected British meteorological journal, it garnered scant attention outside that niche. At the time, with most climate studies still focused on past glaciations, such a rapid temperature increase would have been seen more as an anomaly than an augury. Who was to say such a trend might not reverse itself in the coming decades?

But it didn't, and, more than a century on, it hasn't. In fact, the similarities to NASA's current data are uncanny—as was Callendar's confident prediction: "The course of world temperatures during the next twenty years," he wrote more than eighty years ago, "should afford valuable evidence as to the accuracy of the calculated effect of atmospheric carbon dioxide. In any case, the return of the deadly glaciers should be delayed indefinitely."

NASA's graph of temperature variation, using state-of-the-art data and equipment, shows a nearly identical pattern.

The significance of Guy Callendar's proposition, all but ignored at the time, was enormous: he was saying, unequivocally, that humans, due specifically to their preoccupation with fire, had become a force of Nature. The term "anthropogenic" did not exist in Callendar's day, but if it had, he likely would have used it. That we have entered the Anthropocene Epoch (the geologic era of humankind's global influence on weather and ecosystems) is now generally accepted among scientists. Precisely when this period began is a matter of debate. Was it 50,000 years ago, with the first evidence of our ability to extirpate populations of Pleistocene megafauna? Was it 12,000 years ago, with the onset of the Holocene Epoch, generally considered to be the age

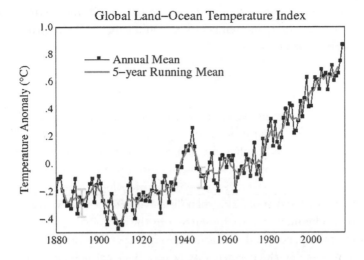

Global Land–Ocean Temperature Index

of modern humans? Was it 10,000 years ago, with the appearance of agriculture and city-states? Was it 2,000 years ago, with the accumulation of global pollution residue from Roman smelting operations? Was it 160 years ago, with the introduction of the internal combustion engine? Was it 75 years ago, with the introduction of the atom bomb? Or was it a million years ago, with the first evidence for the controlled use of fire?

That is for scientists to decide.

What is clear is that geologic epochs are typically divided into ages, and Guy Callendar was the first to document and chart the atmospheric changes being wrought by the Petrocene Age. Like so many other groundbreaking scientists, he was on his own and vindication was not guaranteed. At the time of its publication, Callendar's work was largely ignored, but it could have been so much worse. He could have met the same fates as Joseph Priestley, the oxygen pioneer, driven from Britain and buried in exile, or Ignaz Semmelweis, the advocate for handwashing, harassed out of Switzerland and killed in a mental institution, or, more recently, Alfred Wegener, who, in 1912, forcefully advanced the theory of continental drift, only to be publicly attacked and ridiculed and then perish on the Greenland ice cap before he could be vindicated. Callendar, who died in 1964, would live to see his work accepted, if not universally. In his lifetime, the link between CO_2 and temperature would come to be known as the Callendar Effect.

The "next twenty years" that Guy Callendar believed would vindicate his global warming theory ended in 1958. During this time, a Canadian-born geophysicist named Gilbert Plass had been using infrared spectroscopy to challenge and, ultimately, reconfirm discoveries made over the previous two centuries: long-wave radiation—aka infrared radiation, aka solar heat—is retained by water vapor, and also by industrial CO_2. Starting in the early 1950s, a number of prominent papers and magazines began reporting on Plass's work. *The Washington Post* covered it on May 5, 1953, employing as they did so some now-familiar similes: "Releases of carbon dioxide from coals and oils . . . blanket the earth's surface 'like glass in a greenhouse.'"

The New York Times followed with a similar story a couple of weeks later, using the same imagery, and so did *Time* magazine: "In the hungry fires of industry, modern man burns nearly 2 billion tons of coal and oil each year. Along with the smoke and soot of commerce, his furnaces belch some 6 billion tons of unseen carbon dioxide into the already tainted air . . . This spreading envelope of gas around the earth serves as a great greenhouse."

In June 1953, *Life*, one of the most popular weekly magazines of the day, ran a twenty-page article entitled "The Canopy of Air," which addressed the suspected link between warming temperatures, rapid glacial retreat, and industrial CO_2. Three years later, in 1956, Plass would discuss his findings in *American Scientist*: "It is not usually appreciated," he wrote in the July issue, "that very small changes in the average temperature can have appreciable influence on the climate. For example, . . . a rise in the average temperature of perhaps only 4°C would bring a tropical climate to most of the earth's surface."

What is refreshing, but also unnerving, about scientists is the way they relay their findings—mundane or devastating—in the same measured tones. An increase in average temperature of 4°C, or roughly 7°F,* would end life as we know it. In the twenty years since Guy Callendar published his CO_2 and temperature graphs, the picture had already changed noticeably. After citing Callendar, Plass wrote, "Today man by his own activities is increasing the percentage of carbon dioxide in the atmosphere by thirty percent a century."†

Even though we emit carbon dioxide constantly—from our fires and engines, and from our very mouths—it remains, for most, an abstraction. In recent years, its role has become an article of faith that one can choose, or refuse, to believe. But its impacts are less subjective: "In the last fifty years," wrote Plass in 1956, "virtually all known glaciers in both hemispheres have been retreating." He continued: "*There can be no doubt* that this will become an increasingly serious problem as the level of industrial activity increases" (italics mine).

In his article, which he titled "Carbon Dioxide and Climate," Plass went on to explain how the full exploitation of known coal reserves would drive up CO_2 levels by a factor of ten, pushing the average global temperature into uncharted territory. More disturbing than Plass's conclusions, or his confidence, is the fact that his article was published almost seventy years ago. In this landmark work, we see for the first time the suggestion, albeit oblique, that in order to maintain any semblance of atmospheric equilibrium, humanity would need to moderate its use of fossil fuels (i.e., "keep it in the ground"). Throughout the 1950s, Plass's work continued to be written up in scholarly journals and in well-regarded popular magazines and newspapers. And yet, as graphic, reasoned, and alarming as his message was, these articles passed through the news cycle and into the library stacks, leaving behind the barest ripple.

But unlike Guy Callendar, Gilbert Plass wasn't alone. In addition to researchers in Europe, several scientists at the Scripps Institution of Oceanography in California were also doing cutting edge research on atmospheric CO_2. One of them, Roger Revelle, an oceanographer, former Navy man, and the director of Scripps, became the first person

---

* Currently, we are at roughly 1°C (1.8°F) above average, and already there is abundant evidence of disruption.
† This is now a gross underestimate.

to formally raise the topic of anthropogenic climate change before members of the U.S. Congress.

On March 8, 1956, Revelle was called before the House Committee on Appropriations to discuss research and funding for the upcoming International Geophysical Year. The IGY (1957–58) was as much a diplomatic effort as a scientific one; it signaled a partial easing of Cold War hostilities that saw the archenemies Soviet Russia and the United States working cooperatively with dozens of other countries on a multipronged effort to better understand Earth's atmospheric, marine, and terrestrial systems. The IGY's objectives were magnificently ambitious and included programs for studying everything from polar auroras to the deepest ocean trenches, from the jet stream to the Gulf Stream. The vast array of planned experiments would showcase the latest technology from satellites to bathyspheres, and samples would be taken from anything remotely measurable—from the most ancient glacial ice to the most ephemeral atmospheric gases. In many ways, the International Geophysical Year showed humanity at its best, and it was a great honor to be included in such a historic endeavor.

When Revelle addressed a subcommittee of the Appropriations Committee on that balmy March morning, their questions were focused on polar research. In the U.S., in the 1950s, "polar research" included where to dump rapidly growing quantities of nuclear waste, and how the ice caps might be used to hide nuclear submarines and launch missile attacks. More relevant to Revelle's expertise was how long those ice caps would even last.

"Human beings during the next few decades may, almost in spite of themselves, be doing something that will have a major effect on the climate of the earth," Revelle said to the committee. "I refer to the combustion of coal, oil, and natural gas by our worldwide civilization, which adds carbon dioxide to the atmosphere. In this way we are returning to the air and the sea the carbon stored in sedimentary rocks over hundreds of millions of years. From the standpoint of meteorologists and oceanographers we are carrying out a tremendous geophysical experiment of a kind that could not have happened in the past or be reproduced in the future. If all this carbon dioxide stays in the atmosphere, it will certainly affect the climate of the earth, and this may be a very large effect. The slight general warming that has occurred in northern latitudes during recent decades may be greatly intensified."

Representative Albert Thomas, a right-leaning Texas Democrat, weighed in: "Didn't I read, from what Dr. Gould says, we have been warming up for the last fifty years?"

Dr. Laurence Gould was chairman of the U.S. National Committee's Antarctic Committee, and one can see why his introductory remarks would have caught the Texas lawmaker's attention: "Glacier studies," Gould wrote,

> have given clear indications that we are now in a cycle of warming which began about 1900. It is estimated that if the indicated warming continues for another 25 to 50 years (c. 2000), the ice will melt out of the Arctic Ocean in the summer making it navigable. In addition, the warming cycle, if continued, may melt enough ice tied up in glaciers to add to the sea level sufficiently to affect the lives of millions of people living along low coastal lands. It is conceivable within 20 or 25 years (c. 1980) a peninsula such as Florida might become inundated. Whether this actually happens or not, the slow change of climate has already begun to show a change of storm paths and redistribution of rainfall.

"The reason [for this new warming cycle]," Revelle explained, "may be because we have been adding carbon dioxide to the atmosphere . . . One of the aspects of this IGY oceanographic program is to try to find out what proportion of the total carbon dioxide produced by the burning of fossil fuels goes into the ocean and how much stays in the atmosphere."

"This gas that you speak of, has that had any effect on human life, do you know?" asked New York Democratic representative Harold Ostertag.

"It may be having an effect already," Revelle answered, "primarily through the effect on the weather . . . The increase in the number of hurricanes on the east coast, however, is certainly tied in one way or the other with the general northward movement of the warm air."

It bears noting that these observations and exchanges were entered into the *Congressional Record* in 1956, fifteen years before the first Earth Day, and thirty-five years before the Intergovernmental Panel on Climate Change released its first report. There was no such thing as "climate denial" then, and there were no "climate skeptics." Those present, all of whom were white men, and most of whom were churchgoers born at the very dawn of the automotive age, were open-minded about this alarming new information, and they discussed it with intelligent interest.

In May of the following year, Revelle was summoned to Washington, D.C., again for a progress report on America's participation in the International Geophysical Year. He opened the afternoon session with extensive remarks on what he referred to as "heat balance," and he began in a way that seems surprisingly holistic for a Cold War–era scientist and former military man. The fact that he was also a sailor from Southern California might have had something to do with it. "I think the best way to introduce this subject," he began, "is to point out to you gentlemen something which is not often thought of, and that is that the earth itself is a space ship . . . We have lived here on this space ship of our earth for a good many hundred thousand years, and we human beings are specifically adapted to it . . . Our whole physiology and psychology really depend upon the characteristics of the earth."

Albert Thomas, the congressman from Texas, wasn't impressed. "You are talking like an environmentalist," he said. "I thought that you believed in heredity."*

Revelle, whose affinity for his subject was as soulful as it was scholarly, continued undeterred. "We are certainly shaped by the earth on which we live," he said. "A simple example is that we breathe oxygen. This is the only planet in the solar system we know of that has free oxygen on it—"

---

* I believe Thomas meant "heredity" in the biblical sense: "the law by which living beings tend to repeat their characteristics, physiological and psychical, in their offspring" (*International Standard Bible Encyclopedia*).

"That will not be free long," growled the congressman. "The Federal Government will tax it."

Revelle persevered: "—and this is the only planet that has large bodies of liquid water on it. The fact that water has such a great capacity for storing heat means that it can absorb a great deal of radiation and not change its temperature very much."

Thomas, a veteran of World War I and a forceful politician credited with bringing the Johnson Space Center to his home district of Houston, may have presented like a crusty Texas oilman, but he was listening. "You have announced," he said, "a very fundamental principle of weather when you said that these vast bodies of water are reservoirs for tremendous heat loads."

"That is correct, sir," said Revelle.

After a lengthy exchange about drought and water shortages, particularly in Thomas's and Revelle's home states, Revelle moved on to short- and long-term weather forecasting. "You used the word 'climate' there for the long range," Thomas said, "and 'weather' for the day-to-day basis?"

"That's right," said Revelle.

Eisenhower was president, nuclear Armageddon was a preoccupying concern, and Thomas would soon be voting against the Civil Rights Act of 1957,[*] but on this exceptionally warm May Day afternoon this conservative Texan was talking climate science with a West Coast progressive on Capitol Hill. Revelle went on to say that if industrial CO_2 increased by 20 percent, as predicted, "It would mean that . . . southern California and a good part of Texas, instead of being just barely livable as they are now, would become real deserts."

There followed some back-and-forth about the causes and impacts of drought in ancient Greece and Mesopotamia, and then this remarkably prescient conversation continued.

"From a weather point of view," Thomas asked, "how did that happen?"

"No one knows," Revelle said.

"There is no theory behind it or anything?"

* It passed anyway.

"This carbon dioxide thing that I was talking about," said Revelle, "is in fact a way to test some of these theories."

Thomas tried to grasp it: "Carbon dioxide absorbs the infrared rays that are bouncing back from the earth, and when they are absorbed, that absorbs the heat and therefore, what?"

"It raises the temperature," Revelle said. "It is like a greenhouse . . . If you increase the temperature of the earth, the north latitude belt, which covers most of the western part of the United States and the Southwest, would move to the north. Does this make any sense?"

"Yes," said Thomas.

Then, the discussion turned to ocean currents.

Revelle's warnings to those long-dead congressmen have turned out to be surgically accurate. The 20 percent increase in industrial CO_2 that Revelle predicted in 1957 was achieved in 2004, along with the anticipated atmospheric changes. The kind of disruption Revelle alluded to in the context of drought and rainfall is now referred to as a "phase shift": a dramatic, effectively irreversible change in a region's climate regime. There is abundant evidence that phase shifts are now under way across much of the planet. Fire behavior is just one indicator, but it is a graphic one, and Revelle's home state of California offers a good example: in the 1950s, the state's fire season lasted about four months; today, it is effectively year-round, and the acreage burned during the most severe seasons (1950 versus 2020) has increased *eightfold* (to say nothing of lives and property lost).[*] Meanwhile, the drought Revelle predicted has become a serious and persistent condition—winter and summer, threatening the viability of mountain forests, agricultural lands, and the waterways that link them. As for fire tornadoes, those exceeded even Revelle's powers of prediction.

These historic exchanges between men of such different backgrounds and philosophies were a wonderful by-product of the International Geophysical Year (as well as a reminder of how the U.S. Congress is capable of functioning). While the crucial link between fossil fuel burning and carbon dioxide garnered neither the attention nor the action it deserved, Roger Revelle and his colleagues did

---

[*] In New Mexico, fire weather days have increased by 120 percent since 1973.

secure funding to study it. Given where things stand today, it is sobering to consider that Revelle addressed these matters, accurately and emphatically, more than thirty years before the NASA scientist James Hansen gave his own historic testimony before Congress. Since Revelle's presentations on Capitol Hill, three generations, amounting to 5 billion people, have been added to the world's population, along with billions of fuel-burning vehicles, engines, stoves, generators, and power plants of all sizes. In that time, annual CO_2 emissions have increased fivefold from their already-climate-altering 1950s levels.

20

Tomorrow came and went.

—Cormac McCarthy, *The Road*

t may not have kept pace with car sales, emissions, or population, but climate science continued to gain traction in the scientific community. Like germ theory and continental drift before it, the greenhouse effect, once a provocative idea swirling in the back eddies of scientific inquiry, was drawn into the mainstream, thanks in large part to Gilbert Plass and Roger Revelle. Fragmented though their efforts were, a picture was taking shape across the scientific community. There is a doomed, Cassandra-like quality to these early climate messengers. They had the vision to see what was coming, and the science to support it, but their words, however eloquent, or urgent, seemed unable to penetrate the collective consciousness—to be truly *heard*. Part of the problem was the nature of their message: not only was it abstract in the extreme, it ran absolutely counter to the bullish and triumphal postwar narrative: it was patriotic and unifying to challenge communists and dictators, but not the billowing chimneys of industry or the exhaust pipe of the family car.

More significantly, it runs counter to the colonial and capitalist impulses. Whether it's Britain in North America, Spain in South America, Belgium in the Congo, China in Tibet, or Russia in Ukraine, colonizers are uniformly destructive to the health and well-being of the people and places they deem themselves entitled to occupy and exploit. Roger Revelle's warning to the House subcommittee regarding the potential impacts of fossil fuel development on the atmosphere (talk about colonization) was a more diplomatic echo of the fur trader John M'Lean's observations a century earlier on the Hudson's Bay Company's scorched-earth policy toward fur-bearing animals in the boreal forest.

As new and challenging as these ideas were, sincere attempts were made to introduce them to the public. It may come as a surprise today, but in 1958, the potential for CO_2-driven climate disruption was part of the public school curriculum. In the late 1950s and early '60s, the director Frank Capra (*It's A Wonderful Life*, etc.) collaborated with Bell Telephone (AT&T)—the Standard Oil of telecommunications—on a series of educational films that were aired on national television and distributed widely through American schools. Using a hybrid of animation and live action popular at the time, 1958's *The Unchained Goddess* featured Meteora, an animated Rita Hayworth–like weather deity who becomes infatuated with the domed and bespectacled Dr. Frank Baxter, a legendary (real-life) professor at the University of Southern California. Over the course of the film, Baxter, a truly delightful man, explains the science and mechanics of weather, finishing up with a warning that "man may be unwittingly changing the world's climate through the waste products of his civilization." As Baxter paraphrases Gilbert Plass's research and Roger Revelle's lyricism, we see dramatic footage of collapsing glaciers juxtaposed with fuming smokestacks, bumper-to-bumper traffic, and animations of rising seas inundating the coastal United States. *The Unchained Goddess*, financed and distributed by one of the biggest and most powerful corporations in U.S. history, was seen by tens of millions of young Baby Boomers.

On November 4, 1959, the brilliant but controversial nuclear physicist Edward Teller sounded the alarm again. Teller, the "father of the hydrogen bomb," was, thus far, the most famous scientist to address the topic of carbon dioxide, and he did so—not on television or Capitol Hill, but as a guest of honor at a symposium entitled "Energy and Man." The year 1959 marked the one-hundredth anniversary of "Colonel" Edwin Drake's pioneering oil well in Titusville, Pennsylvania, and the oil and gas industry's leading business association, the

American Petroleum Institute, had joined forces with Columbia University's Graduate School of Business to mark the occasion with an all-star gathering of three hundred top scientists, business executives, historians, economists, and government officials. Held in Columbia's monumental Low Library in New York City, the symposium's objective on that chilly, sleet-spattered Wednesday was to celebrate the past and plot the future of the Petrocene Age.

Among the many powerful men gathered in the library's drafty rotunda that day was fifty-year-old Robert Dunlop. In addition to being director of the API, Dunlop was president of J. Howard Pew's Sun Oil Company. Sunoco was a rarity among the American oil majors in that it wasn't a spin-off of Rockefeller's dismembered Standard Oil. Sunoco might not have been as big as Standard, but the Pews and their lieutenants possessed a similar acquisitive ferocity. By 1959, Sunoco's holdings reached from Venezuela to Canada, where it had a controlling interest in the Sun Company (the future Suncor), a driving force in the effort to develop Alberta's oil sands. For anyone paying attention, it was clear that the Petrocene Age was moving into a new phase: even as oil was displacing coal as the world's "prime mover," and major discoveries were being made around the world, domestic supplies were clearly limited and offshore drilling was restricted, thus far, to shallow waters. The U.S. had begun importing oil for the first time shortly after the war, and, from the point of view of long-term regional self-sufficiency, the tar sands were looking more attractive.

Dunlop addressed the symposium that morning, following a Pulitzer Prize–winning historian, and his tone was exuberant as he celebrated what he called the "the Petroleum Revolution." He began by quoting the editor of *Harper's*, John Fischer, who had contributed an article to the centennial issue of the API's quarterly report. In it, Fischer sounds like the adman Don Draper pitching a client in *Mad Men*. "Many petroleum products reach us in strange disguises," he wrote, "as a plastic grocery bag, for example, or a stocking on a pretty ankle. Though my home has been heated by fuel oil for many years, to this day, I have no idea what it looks like . . . only through the carelessness of a service station attendant [have I] ever seen a splash of gasoline."

To hear Fischer tell it (and sell it), petroleum was subtle and sexy, its consumers happily oblivious as it worked its potent miracles behind the scenes, only visible when its masters (many of whom were in that room) wanted it to be. In the end, Dunlop's talk amounted to a victory lap with some grousing about taxes thrown in (a tune that has changed little in sixty years). The main event that day was Edward Teller, and attendees were keen to hear what he had to say. Teller would have been well known to these men; the inspiration for Stanley Kubrick's 1964 film, *Dr. Strangelove*, he was a genius, and a complicated one. While many of his colleagues involved in nuclear bomb development had been horrified by the destruction of Hiroshima and Nagasaki, Teller, a Jewish refugee who had fled the fascist regime in Hungary, was excited by the prospect of building even more powerful weapons. But Sunoco's Dunlop, a career oilman and confirmed capitalist, hadn't brought Teller to New York to talk about ethics. In his freewheeling thirty-minute talk, entitled "Energy Patterns of the Future," Teller became the first and, probably, the only person to link thermonuclear explosions, climate change, and the tar sands.

Like Pew and Dunlop (and Fred Koch), Teller recognized that the world's appetite for energy was growing rapidly, and that sources beyond conventional oil and coal would need to be explored. Where others saw potential in nuclear power plants, Teller saw a role for nuclear bombs. Teller was a champion of the postwar initiative called Project Plowshare, "the peaceful use," as he described it that afternoon, "not merely of nuclear energy, but of nuclear explosions." It is hard to read such words (or write them) with a straight face now, but the Cold War was a different time, and the godlike power conferred by atomic energy—a kind of ultimate Promethean advance—was irresistible, especially if you were Edward Teller. "A nuclear explosion is cheap and big," he explained to his business-minded listeners. "With the help of nuclear explosives you can blow up several hundred feet of overburden from a shale deposit. We might . . . open up for exploitation gooey or solid substances like tar sands . . . All these are dreams, but dreams which to some extent will come true, if only we are permitted to experiment with the effect of nuclear explosions."

You can almost hear the yearning in his voice. Meanwhile, Dunlop, well aware how intransigent Alberta's tar sands were, was paying close attention. But Teller issued a caveat when it came to exploiting fossil fuels: "I would . . . like to mention another reason why we probably have to look for additional fuel supplies," he said. "And this, strangely, is the question of contaminating the atmosphere . . . Whenever you burn conventional fuel, you create carbon dioxide . . . The carbon dioxide is invisible, it is transparent, you can't smell it, it is not dangerous to health, so why should one worry about it?"

Teller, in addition to being the founding director of the Lawrence Livermore National Laboratory, was a professor of physics at the University of California at Berkeley. He was used to being around smart, educated people, and he knew how to communicate with them. That he felt it necessary to offer a primer on carbon dioxide to this blue-chip audience implies a belief that they needed one. "Carbon dioxide has a strange property," he began. "It transmits visible light, but it absorbs the infrared radiation, which is emitted from the earth. Its presence in the atmosphere causes a greenhouse effect in that it will allow the solar rays to enter, but it will to some extent impede the radiation from the earth into outer space. The result is that the earth will continue to heat up until a balance is re-established . . . It has been calculated that a temperature rise corresponding to a ten percent increase in carbon dioxide will be sufficient to melt the icecap and submerge New York. All the coastal cities would be covered, and since a considerable percentage of the human race lives in coastal regions, I think that this chemical contamination is more serious than most people tend to believe."

Teller cited the recent work by Roger Revelle, but it was clear from the follow-up questions that many, if not all, of the attendees were hearing this information for the first time. Courtney Brown, dean of Columbia's Graduate School of Business, was clearly disturbed: "Would you please summarize briefly the danger from increased carbon dioxide content in the atmosphere in this century?" he asked.

Teller obliged Brown in language a New Yorker would understand: "When the temperature does rise by a few degrees over the whole globe," he said, "there is a possibility that the icecaps will start melting and the level of the oceans will begin to rise. Well, I don't know whether they will cover the Empire State Building or not, but anyone can calculate it by looking at the map and noting that the ice caps over Greenland and over Antarctica are perhaps five thousand feet thick."

How alarmed Brown must have been upon hearing this—not once, but twice—can only be guessed at. To many in that room, it would have been an absolutely shocking concept, one with no precedent since the story of Noah's Ark. What is certain is that such a dire prediction from such an eminent scientist would have been unwelcome news to an oilman, or an investor, and there were many in the audience that day. In the postwar years, energy—in all its forms—was a virtue; it was thanks to its cheap and free-flowing abundance that the United States boasted the greatest economy the world had ever seen. In the 1950s, environmental regulations were almost nonexistent, in part because imposing limits—on energy development, on technology, on growth of almost any kind—was seen as un-American, a kind of secular heresy that could get one accused of being a communist (or, as Representative Thomas said of Roger Revelle, "an environmentalist").[*]

In an irony that captures the willful obtuseness (or profound cynicism) of the Cold War–era petroleum industry, Humble Oil, a subsidiary of Esso/Exxon, spun Teller's warning into an ad campaign in *Life* magazine.

In retrospect, the Symposium on Energy and Man appears to have been a moment when powerful individuals in a position to do something chose to look away. The evidence—and it is abundant—suggests that Sunoco's Dunlop (along with the rest of the API membership) took from Teller's unusual talk only the parts they wanted to hear: namely, that conventional oil supplies could be exhausted in a few decades and that nuclear bombs might be used to access unconventional reserves like oil shales and bitumen. That same year, the California-based Richfield Oil Corporation (ARCO), encouraged by Teller's California lab, was in discussions with Alberta's Oil and Gas Conservation Board regarding "Project Oilsand," a scheme to deploy a nine-kiloton nuclear device in the bituminous sands of Fort

---

[*] In the summer of 1955, the U.S. Congress passed the Air Pollution Control Act, a precursor to the Clean Air Act of 1970. Its focus, however, was on human health and "smog," as opposed to CO_2 and its impact on the atmosphere.

EACH DAY HUMBLE SUPPLIES ENOUGH *ENERGY* **TO MELT 7 MILLION TONS OF GLACIER!**

This giant glacier has remained unmelted for centuries. Yet, the petroleum energy Humble supplies—if converted into heat—could melt it at the rate of 80 tons each second! To meet the nation's growing needs for energy, Humble has applied science to nature's resources to become America's Leading Energy Company. Working wonders with oil through research, Humble provides energy in many forms—to help heat our homes, power our transportation, and to furnish industry with a great variety of versatile chemicals. Stop at a Humble station for new Enco *Extra* gasoline, and see why the "Happy Motoring" Sign is the World's First Choice!

HUMBLE
OIL & REFINING COMPANY
America's Leading **E**nergy **c**ompany

ENCO

Life, February 2, 1962

McMurray. "At a single stroke," crowed Richfield's boosters, Project Oilsand could "double the world's petroleum reserves."

The proposal met with enthusiasm from industry evangelists, including Ernest Manning, the born-again premier of Alberta, but there was real concern from scientists about residual radiation, particularly strontium 90, a known "bone seeker." Much to the relief of almost everyone, the project was abandoned—less because of potential health risks than due to fears that such activities might encourage Russian espionage. But ever since then, Suncor and Syncrude, along with local and federal governments, and petroleum companies around the world, have exploited Alberta bitumen—the world's most greenhouse gas–intensive petroleum product—with all the money, machinery, and manpower they have been able to bring to bear.

~~~~

Mass spectrometers had been around for fifty years when Charles Keeling, a young geochemist out of Caltech, began using them to track atmospheric CO_2 in the mid-1950s. This sophisticated instrument was capable of measuring carbon dioxide in parts per million (ppm), and one of Keeling's more startling discoveries was that CO_2 content rose and fell throughout the year—as if the planet were "breathing." During the spring and summer months, CO_2 levels drop in the Northern Hemisphere as trees and plants leaf out and absorb the gas through photosynthesis while in the Southern Hemisphere, CO_2 levels rise as the leaves fall and decaying vegetation releases it. As the seasons change, the process reverses—up and down, in a steady equinoctial rhythm.

By 1957, Keeling felt confident that he had found a baseline level of atmospheric CO_2—310 ppm—but his instruments were also showing an annual increase.[*] As with Callendar's temperature graph, cautious colleagues suggested that it was too soon to declare a trend, but it was only a matter of time. The Callendar Effect, which tracked CO_2 increases by proxy, through temperature, was joined by the now-famous "Keeling Curve." Both were climbing steadily. The preindustrial baseline for atmospheric CO_2, derived from ancient air sampled from polar ice cores,[†] was approximately 280 ppm. Starting around 1750, when the coal-driven industrial revolution got under way in earnest, that number began to climb, incrementally at first. Two hundred and ten years later, in 1960, Keeling's instruments indicated that worldwide atmospheric CO_2 had risen to 315 ppm.

An increase of thirty-five parts per million doesn't sound like much for an invisible, naturally occurring gas that facilitates plant growth as it bubbles and burps harmlessly through our daily lives. But its effects, though subtle, are cumulative. A 12 percent increase in carbon dioxide over two centuries represents—in geologic terms—a sudden change, especially when you consider that, throughout the previous 1 million years, levels had never surpassed 300 ppm. By

---

[*] Roger Revelle hired Keeling to work with him at Scripps in 1956.
[†] This relationship has since been verified through unbroken ice core records dating back 800,000 years.

1960, the steady upward trend in CO_2 appeared to be in lockstep with global temperature.

These observations, as clear and measurable as they were, directly challenged a deeply held belief across the scientific community that natural systems regulate themselves, that when disrupted—by flood, fire, plague, or volcano—climatic changes will tend inevitably toward recalibration and equilibrium. This is true, but only up to a point; the horizon of recovery recedes in direct proportion to the scale of the disruption. A major volcano (or fire) can inject enough soot into the atmosphere to alter the air quality and lower the temperature for weeks—sometimes for months, or even years. Eventually, natural processes (rain, gravity, air circulation) will draw those particulates out of the air and order will be restored; the skies will clear, and temperature will stabilize.* However, large injections of carbon dioxide and methane take far longer to process—decades, or even centuries—and, because our atmosphere is so contained, everything in it is finite—including Nature's ability to restore balance.

As the twentieth century progressed, a growing number of atmospheric scientists were finding evidence to support Arvid Högbom's speculative theory from 1894: too much industrial CO_2 was being emitted too fast for Nature to maintain, or recover its former equilibrium. In 1960, the Keeling Curve proved it. Today, there are monitoring sites all over the world, and you can track CO_2's steady outpacing of natural systems with the digital precision of a countdown clock.

⁓⁓⁓

Throughout the Western world and its colonies, the 1950s and '60s saw an unparalleled awakening to the plights and the rights of the Other. As colonized peoples fought for and won independence, and as oppressed minorities marched and lobbied for basic rights and

---

* Many economists believe that the same self-regulating processes exist in financial markets, but this logic falls apart when you look at market-distorting monopolies like the Hudson's Bay Company, Standard Oil, Facebook, or Amazon. These centrally controlled giants skew the "ecosystem" much the way gigantic injections of CO_2 from volcanoes or industry do.

representation, the state of Earth itself, and of humanity's impact on it, shifted into sharper focus. Rachel Carson's galvanizing best seller, *Silent Spring*, came out in 1962, making its way, translation by translation, around the globe. In 1965, Martin Luther King Jr. marched from Selma to Montgomery, President Lyndon Johnson signed the Voting Rights Act, and the U.S. military began its systematic bombing of North Vietnam (in the name of liberation). On November 5 of that year, as the Keeling Curve continued to steepen, Johnson received a report from his science advisory committee. It was called "Restoring the Quality of Our Environment."

The introduction wasted no time: "Pollutants," it declared, "have altered on a global scale the carbon dioxide content of the air." The report went on to say that further increases in carbon dioxide would "almost certainly cause significant changes," and that they "could be deleterious from the point of view of human beings." The report's conclusion, borrowing Revelle's words from a decade earlier, was unambiguous:

> Man is unwittingly conducting a vast geophysical experiment. Within a few generations he is burning the fossil fuels that slowly accumulated in the earth over the past 500 million years. The CO_2 produced by this combustion is being injected into the atmosphere; . . . By the year 2000 the increase in atmospheric CO_2 . . . may be sufficient to produce measurable and perhaps marked changes in climate, and will almost certainly cause significant changes in the temperature and other properties of the stratosphere.

Across the industry, fossil fuel producers took this warning seriously. In 1966, the president of Bituminous Coal Research, Inc., issued his own alert in the August issue of the *Mining Congress Journal*:

> There is evidence that the amount of carbon dioxide in the earth's atmosphere is increasing rapidly as a result of the combustion of fossil fuels. If the future rate of increase continues as it is at the present, it has been predicted that, because the CO_2 envelope reduces radiation, the temperature of the earth's

atmosphere will increase and that vast changes in the climates of the earth will result. Such changes in temperature will cause melting of the polar ice caps, which, in turn, would result in the inundation of many coastal cities, including New York and London.

The following year, the American Petroleum Institute, the oil and gas industry's business association, commissioned its own study from the Stanford Research Institute. The SRI was not a partisan think tank (those would come later); their scientists were objective and thorough, and their findings confirmed what the physicist Edward Teller had told the API a decade earlier in New York, only with a level of detail that borrowed heavily from Plass, Revelle, and the White House report. Notable is the authors' self-awareness: they saw a dangerous irony in focusing on the obvious (urban smog was a major issue at the time) "when the abundant pollutants which we generally ignore because they have little local effect—CO_2 and submicron particles— may be the cause of serious world-wide environmental changes."

The SRI report was entitled "Sources, Abundance, and Fate of Gaseous Atmospheric Pollutants." There was an oracular quality to a lot of twentieth-century climate science, and "fate" was an apt choice of words. In careful, conservative language, the Stanford scientists said, in sum: levels of atmospheric CO_2 and small toxic particulates are rising quickly; their most likely source is the burning of fossil fuels; their cumulative impact on Earth's climate, and on human and planetary health, is all but certain to be negative and potentially disastrous. "Significant temperature changes," the report said, "are almost certain to occur by the year 2000." Their concluding advice to the API was crystal clear: industry needs to make understanding and management of its emissions a priority.

What happened in the House subcommittee meetings in 1956 and '57, and in Columbia University's Low Library in 1959, and in the White House in 1965, and at the API in 1968, were critical moments—each one a linking of good science from respected sources with people and institutions in positions to act in meaningful ways. These warning were duly noted, but there were, after all, so many more tangibly pressing matters to attend to—poverty, war, civil rights, leaded gasoline, DDT, acid rain, rivers and canals so polluted they were catching on fire—a daunting list. And, always, there were profits and stockholders to consider.

In-house memos from this period—from Exxon and Shell to GM and Ford—reveal a growing awareness of those industries' impacts on atmospheric stability and human health. While the science on anthropogenic CO_2 was sound, and all but conclusive, its negative effects were not as visible, nor as individually provable, as so many more immediate social and environmental problems. There was wiggle room, especially if you had a vested interest.*

In 1978, James Black, a senior science adviser at Exxon, mailed a packet of documents to the vice president of Exxon's Research and Engineering Company.† The subject heading on Black's cover letter was "The Greenhouse Effect," and with it was a transcript of a presentation he had given to Exxon's Corporation Management Committee. Black's presentation resembled Edward Teller's twenty years earlier in that it was a primer on anthropogenic CO_2 written for oilmen. The difference was that Black buttressed his predictions with the latest climate data, which was now extensive and compelling. But his findings weren't new; he came to the same conclusions that Plass had in 1953, that Callendar had in 1938, that Högbom had in 1894, and that Foote had in 1856.

Science can be frustrating for nonscientists because scientists are trained to be humble and cautious and not to speak in absolutes. Almost everything they say is qualified somehow. It's an approach that doesn't generate clicks, or draw eyeballs, or make money, and it leaves openings for cynical actors. Despite the abundant data—current and historic—at his disposal, Exxon's James Black allowed that he could not be absolutely *certain* that the current rise in carbon dioxide was due to fossil fuel burning (at least not in the same way you are absolutely *certain* you have just stubbed your toe), but in 1978

---

* In 1970, the Clean Air Act passed into law, but its focus was more on emissions directly impacting human health. Carbon dioxide, a gas with "little local effect," was not covered by the new regulations.

† The latest iteration of the Standard Oil division that advised and equipped the firebombing of Hamburg in 1943

he was certain enough to say this: "Present thinking holds that man has a time window of five to ten years before the need for hard decisions regarding changes in energy strategies might become critical." In other words, Black the scientist was saying, "While we're not *certain* there isn't another explanation for the current symmetry between rising CO_2 and rising temperature, we should act *as if* we're certain." This is the precautionary principle; it's why people wear seat belts, get vaccinated, and buy insurance, and, this time, the people in a position to do something listened.

Nineteen seventy-nine was a pivotal year that saw the mapping of an alternate path. That year, Switzerland hosted the first World Climate Conference, whose stated objective was to "foresee and prevent potential man-made changes in climate that might be adverse to the well-being of humanity." It was a hopeful beginning: Exxon, in conjunction with the API and a who's who of oil companies, assembled a collaborative research team with a name that sounds like it was led by Greta Thunberg: the CO_2 and Climate Task Force. Its members were earnest, honest, and accurate, and it was partly due to their substantial investment and genuine concern that the 1970s and early '80s became a golden age of climate science. Hundreds of articles were published in scientific journals exploring the relationship between human activity and warming air, melting ice, and rising seas.

Exxon was a leader in this effort, investing millions of dollars, some of which they used to repurpose an entire oil tanker into an enormous floating climate lab. Federal scientists were calling for action, too. The threat of climate change made the front page of *The New York Times* for the first time in August 1981, and the headline was in all-caps: "STUDY FINDS WARMING TREND THAT COULD RAISE SEA LEVELS." The study was led by NASA's James Hansen, and it predicted virtually everything now coming to pass. The catch, in 1981, was that its effects were not yet visible. In other words, it was the Lucretius Problem: the self-protective tendency to favor the status quo over a potentially disruptive scenario one has not witnessed personally. Fortunately, not everyone is bound by this blinkered perspective, and some of those people worked for Exxon. In 1982, Exxon released graphs projecting carbon dioxide levels through the year 2100, along with the point in time at which resulting climate

Figure 9

Range of Global Mean Temperature From 1850 to the Present
with the Projected Instantaneous Climatic Response to
Increasing CO_2 Concentrations.

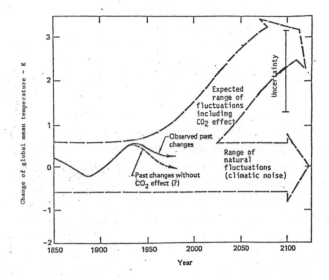

Exxon chart indicating when CO_2 would have a detectable effect
on global temperature (1982)

impacts would break through the "noise" of natural climatic fluctuations and become obvious.

So far, their predictions have been dead-on. As research by Neela Batterjee and her Pulitzer Prize–winning team at *InsideClimate*, by Naomi Oreskes and Geoffrey Supran, and by reporters at the *Los Angeles Times* have now proved, "Exxon knew." So did Shell and Chevron, Ford and GM. The evidence that these and other researchers have uncovered is abundant: "There is concern among some scientific groups," warned an in-house Exxon memo from 1982, "that once the effects are measurable, they might not be reversible and little could be done to correct the situation in the short term. Therefore, a number of environmental groups are calling for action now to prevent an undesirable future situation from developing."

In October of that year, in a talk entitled "Inventing the Future: Energy and the CO_2 'Greenhouse' Effect," Exxon's president of research and engineering openly discussed the need for an active transition away from fossil fuels. A month later, Exxon's leadership distributed another in-house memo on the greenhouse effect. The accompanying text was clear and thorough, and it was supported by a formidable bibliography containing titles that look a lot like recent headlines—only many of them are now half a century old: "Global Climate Change and the Impact of a Maximum Sea Level on Coastal Development" (1971); "Is the West Antarctic Ice Sheet Disintegrating?" (1973); "Understanding Climatic Change: A Program for Action" (1975); "Carbon Dioxide and Climate: The Uncontrolled Experiment" (1977). The bibliography goes on like this for ten pages.

In 1982, Exxon was the world's biggest industrial company. With sixty-five thousand filling stations around the globe and operations in a hundred countries, it fit perfectly Edmund Burke's description of "a state in the guise of a merchant."[*] In May of that year, Jack Blum, a Washington energy lawyer and former Senate investigator, put it in cruder, twentieth-century terms: "There is no wad of cash like this anywhere on earth," he told *The New York Times*. "This is a wad of cash to break banks, even governments." As such, Exxon was in a unique position to set agendas and guide energy policy—and, therefore, climate policy—on a global scale.

In October 1983, climate change made the front page of the *Times* again—its first time above the fold—with this headline: "EPA[†] Report Says Earth Will Heat Up Beginning in 1990s." In language that was reminiscent of Teller's back in 1959, the report stated, "There could be big changes. New York City could have a climate like Daytona Beach, Fla., by 2100."

Forty years ago, EPA scientists effectively predicted 2012's Hurricane Sandy, and the back-to-back deluges that flooded Manhattan again in August 2021.

---

[*] In *Private Empire: ExxonMobil and American Power*, Steve Coll describes Exxon as "a corporate state within the American state . . . one of the most powerful businesses ever produced by American capitalism."

[†] Environmental Protection Agency. A 1983 EPA report entitled "Can We Delay a Greenhouse Warming?" predicted the following: "Current estimates suggest that a 2 degrees C increase could occur by the middle of the next century . . . 5 degrees C increase by 2100."

By the early 1980s, major insurance companies were paying attention, too. "Swiss RE and Munich RE, the two big re-insurers, started attending early global-warming meetings," recalled Don Smith, deputy secretary general of the World Meteorological Organization, "because the prediction was that the 'warming' would be accompanied by more extreme events of all kinds, and they had *already seen* an increase in claims" (italics mine). By then, atmospheric CO_2 was approaching 350 ppm, a 25 percent increase over pre-industrial levels. As the Keeling Curve bent steadily upward, the rate of increase appeared to be accelerating.

Those years, during which the oil and gas industry bravely and intelligently examined itself, represent a pivotal moment in the Petrocene Age. This awareness, combined with the abundant data now available to the API and its members, put the oil and automotive industries in an awkward position: it was clear that business as usual was all but certain to have disastrous consequences, and yet fuel burning—fire—was the core of their business. Without fire (and its silent partner, carbon dioxide), there *was* no business. This vexing quandary happened to coincide with an oil glut, a recession, and a precipitous

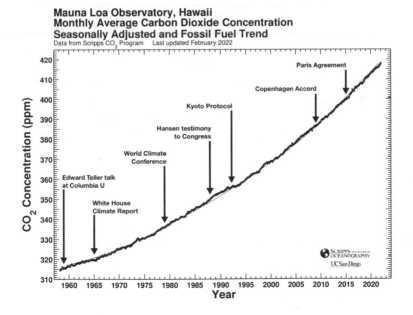

drop in oil prices similar to 2014.* In the midst of all this, with the oil industry desperate to stanch the bleeding, members of the API, led by Exxon, thought it prudent to reevaluate their position on CO_2. In 1984, as scientists from every relevant discipline were comparing notes at international conferences and arriving at similarly dire conclusions regarding the impact of industrial CO_2, the API disbanded its CO_2 and Climate Task Force. Forty years on, the API's decision to turn its back on a century of solid climate science is proving to be the most consequential policy reversal in the history of human civilization.

In 1984, Ronald Reagan was completing the first term of a presidency defined by aggressive deregulation and environmental rollbacks. These policies, strenuously lobbied for by the API, were fueled, in part, by a selective interest in science and an indulgent embrace of the Christian right (who branded themselves "the Moral Majority"), a perennial populist movement that holds science, and expertise in general, under suspicion. For years, Reagan dismissed acid rain and refused to mention the deadly AIDS epidemic. One of his most famous quotes sums up an attitude shared fervently by J. Howard Pew, Robert Dunlop, Ernest Manning, and their many modern counterparts: "Government is not the solution to our problem; government *is* the problem." Reagan's first secretary of the interior, James Watt, was such an ardent advocate for energy extraction, especially on public lands, that a joke began circulating among his admirers:

Q: How much power does it take to stop a million environmentalists?

A: One Watt.

Watt resigned after serving only two years, but the rollbacks con-

* Between May 1980 and May 1982, the price of crude oil fell by almost 50 percent. Prices continued to decline until November 1985, when they went into free-fall, bottoming out at just 20 percent of their 1980 high.

tinued, and so did an erosion of respect for science and scientists.*
This did not negate the fact that the evidence was there, and the
scientists persisted. In 1988, NASA's James Hansen, armed now with
a century's worth of data and a bale of supporting studies and articles, testified before Congress that climate change posed a clear and
present danger to the planet and humanity. For the first time, the
term "global warming" made the front page of *The New York Times*.
The article was accompanied, above the fold, by a jagged, upward-
trending graph that closely resembled the one Guy Callendar had
published back in 1938, only this one included fifty years of additional
data.

The Intergovernmental Panel on Climate Change (IPCC) was
convened the following year. In response, the American Petroleum
Institute—the same organization that had founded and then disbanded the CO_2 and Climate Task Force—formed a new organization, joining dozens of energy, mining, chemical, and manufacturing
companies. They called themselves the Global Climate Coalition, but
the name was deceptive. The GCC's role, officially, was similar to the
API's—to advocate on behalf of industry—but it did so by casting
doubt on climate science and discrediting "alarmists" like James Hansen. Adopting the tobacco industry's playbook (which owes a great
deal to John D. Rockefeller's PR machine),† the GCC disavowed and
undermined decades', and millions of dollars', worth of cutting-edge
climate research. Lobbyists and receptive politicians (including presi-
dents), along with complicit scientists, bankers, economists, CEOs,
evangelical pastors, and conservative news outlets, took the discipline,
caution, and humble openness to alternative theories that are hall-
marks of good science, and weaponized them. Between them, Exxon
and the Koch family alone have spent at least $200 million on think
tanks, lobbyists, pseudoscientific studies, opinion pieces, advertori-

---

* Reagan's first energy secretary, a former Navy dentist and governor of South
Carolina named James Edwards, denied the possibility of climate change and cut
funding for CO_2 research in 1981. Funding was partially restored only after then
Representative Al Gore petitioned Congress, along with Roger Revelle and other
concerned scientists.
† Masterminded by Ivy Ledbetter Lee

als, political campaigns, and advertising, all of which was (and is) designed to obfuscate, minimize, and confuse the issue.

By turning the precautionary principle on its head, the GCC and its allies subverted scientists' absence of 100 percent certainty, pretzeling it into an argument for maintaining the status quo. It's hard work, undermining a century of solid science, but the strategy paid off: by the early 1990s, Republican attitudes toward environmental action of virtually any kind had turned decidedly negative. Meanwhile, energy producers and manufacturers used this extraordinary turnabout as an opportunity to promote even more carbon-intensive products, including plastics (recall the sudden explosion of bottled water in the early 1990s, simultaneous with the first Gulf War). Despite knowing full well the implications for climate, vehicle manufacturers developed and encouraged a market for SUVs, which, in addition to being larger, heavier, and more resource intensive than ordinary cars, were able (thanks to the GCC) to circumvent mileage requirements because of their designation as "light trucks."* Meanwhile, actual light trucks— like pickups—were also marketed aggressively. Ford's F-Series pickup trucks are the best-selling vehicles in automotive history.

Since 1988, when James Hansen testified before Congress and "global warming" began its slow journey into household vocabulary (and partisan politics), the world has changed enormously: more than 3 billion people have been added to the population; atmospheric CO_2 has increased from 350 ppm to over 420 ppm; sea level rise is occurring in real time; the temperature and chemistry of the oceans is changing as fast as the atmosphere; and the human experience of seasons and their attendant weathers is in a state of dramatic flux. One among many ways to quantify these changes is through fire behavior: now, virtually every year, on every continent where *anything* grows,

---

* According to a summary analysis of a report by the International Energy Agency released on November 13, 2020, SUVs are the second-biggest cause of the rise in global carbon dioxide emissions during the past decade. Only the power sector is a bigger contributor.

"The number of trucks and SUVs owned by Canadians has grown 280 percent since they came into high demand during the early 1990s." The average pickup has picked up 1,300 pounds since 1990.

records are being broken for ambient temperature as well as for acres burned and homes destroyed.

One of the few things that haven't changed much in the past quarter century (thanks again to the GCC) is the gas mileage of a Ford pickup—about sixteen miles per gallon for the five-liter, eight-cylinder models. Despite the abundance of fuel-efficient vehicles now on the market, North American sales of Ford's F-Series alone amount to more than *sixty thousand* units per month.

The machinations behind oil, gas, and industry's historic turn away from scientific evidence are in play to this day, to the horror and bafflement of climate scientists and concerned citizens around the world. It is said that time and tide wait for no man. Nor do chemistry and physics: in these squandered decades, generations of dedicated scientists have been vindicated, and everything Plass, Revelle, Hansen, and their earnest colleagues predicted is coming to pass.

If they got one thing wrong, it is the shocking speed with which these changes are unfolding.[*]

Humanity has managed in the past to reduce and even eradicate existential threats on a global scale. International safety standards for cars and airplanes have saved untold lives, as have mass vaccinations and treaties banning dangerous chemicals, certain weapons, and even lethal business practices such as slavery. Global action on the ozone hole is a potent example of how effectively humanity can address even the most remote and abstract threats.[†] But carbon dioxide is pernicious, and its principal source, fossil fuels—fire—encourages excess. What is clear is that the capitalist world has been bingeing on fossil-fueled combustion for a century: Why mandate public transit when you can sell private vehicles? Why encourage fuel-efficient cars

---

[*] The record-breaking "heat dome" of June 2021 that killed nearly six hundred people in British Columbia alone (per coroner), and hundreds of millions of intertidal sea creatures, was, with the exception of the Halifax explosion, the largest mass casualty event on Canadian soil since confederation, and the deadliest weather event by far.

[†] At the end of a presidency many saw as a wholesale dismantling of regulatory policies and protections, Ronald Reagan signed on to the historic Montreal Protocol in 1987, banning ozone-depleting chlorofluorocarbons (CFCs).

when you can sell SUVs? Why take a train at seventy miles per hour when you can fly at eight hundred? Why develop renewables when there's so much oil and gas to burn right here, right now? Why act on CO_2 when you can sow the seeds of doubt and keep those profits (and those dividends) flowing just a little bit longer?

There's a saying in the business world that applies equally to wildfire: "If you're not growing, you're dying."

That's the oxygen talking.

Again, if you find yourself skeptical, don't inhale again for thirty seconds and notice how you feel. "Growing," in this case, is a synonym for burning (more product, more money, bigger markets, but also life and forward motion), and it captures perfectly the no-tomorrow "ethos" of wildfire, and of wildfire economics. The free market, operating by the laws of wildfire economics, is, in essence, a license to burn, and oil has been the ideal medium. Petroleum—fire on demand—has allowed and encouraged more people to burn bigger, brighter, and faster than anything else in history.

Most people, given the opportunity to earn more, spend more—"burn" more—will do so, reflexively, and banks, real estate companies, and auto dealers have taken full advantage of this impulse. After Peter Pond showed up in the Athabasca country with the promise

The Onion ✓ @TheOnion · 8h

Silicon Valley Leaders Sit Down With Wildfire At Investment Meeting After Being Impressed By Its Rapid Expansion trib.al/DXK56hF

of iron, rum, and firearms, exponentially more beavers were killed. Despite no prior experience of European money, debt, or "profit," most eighteenth-century Dene hunters embraced the concepts readily when given the chance. So did the subsistence-farming and -fishing Orkney Islanders hired by the Hudson's Bay Company to run its trading posts. Meanwhile, the Company's upper-class British governors, already familiar with the intoxicating thrill of superabundance, were eager to transcend the limitations imposed by their now-treeless and -beaverless island home. A candle flame will not be content with its wick if it can spread to the curtains and ignite the rest of the house as well.

But binges are, by their nature, finite; eventually, you have to come down. Scientists, it could be said, are the designated drivers of our society: sober, responsible, and vigilant, their job is to tell us when we've had enough, and try to get us safely home. Since the 1950s, when Gilbert Plass and Roger Revelle were telling anyone who would listen that business as usual would have catastrophic consequences, climate scientists have been waving at us from the wings, politely saying, "Don't you think you've had enough?" Nature, in its way, has been waving, too: where there's fire there's smoke, and where there's smoke there's carbon dioxide. Our civilization, our engines, and our markets are not the only things that have been supercharged by fossil fuels. Our atmosphere is a weather engine, and it has been supercharged, too.

21

It will sometimes burst from out that cloudless sky,
like an exploding bomb upon a dazed and sleepy town.

—Herman Melville, *Moby-Dick*

There was no name for the thing Tom Bates captured on video at
the height of the notorious Australian bushfire season of 2002–3.
Prior to that sweltering afternoon, there was no such thing as "pyro-
tornadogenesis," because such a phenomenon—a tornado generated
by a wildfire—was not known to occur on planet Earth. Wildfires in
both hemispheres often whip up small twisters known as fire whirls,
but as impressive as they are to behold, and as dangerous as they
are to be near, they are relatively small and short-lived events—more
like dust devils than full-blown cyclones. What Bates saw and filmed
from a suburban rugby pitch just outside Canberra, in southeast Australia, was different. It occurred during a historic week of lightning-
caused fires that killed four people, injured more than four hundred,
and destroyed five hundred homes west of Australia's capital city.

On January 18, Bates and his neighbors in the Kambah neighborhood, about five miles southwest of downtown, were on high
alert because local fires had advanced to within a mile and a half of
their neighborhood. Looking northward that afternoon, toward the
flames, Bates observed a large funnel cloud over Mount Arawang, one
of several low, tree-covered peaks in the area that are laced with walking trails and surrounded by suburban homes.

Tornadoes are not unheard of in the region, but this one appeared to be rising up out of the fire itself, like an atmospheric Balrog. It was four in the afternoon, the ambient temperature was nearly 100°F, and the air was so dark with smoke that it appeared to be nighttime. The year 2003 was pre-smartphone, but Bates had the presence of mind to get his video camera and record what would come to be a new kind of fire. "I've never in my life seen anything like it," we hear Bates say as the funnel takes shape above the burning mountain. He struggles to describe what he is seeing, not because he lacks the words, but because no Earthling has ever witnessed what he is witnessing now: "Holy *shit* . . . Ho-ly *mackerel* . . . It's a big fireball. It's gotta be rippin' poor bastards' houses up there." Then, right before our eyes, Mount Arawang appears to detonate. The blinding flash, combined with the funnel cloud whirling above it, gives the impression of a nuclear blast. "Ho-ly *Jeezus*," Bates gasps. "This is bad news . . . It's like a big fireball tornado."

It is now clear that this monstrous thing, which Bates has just named, is headed directly toward him. Australians, like Canadians, seem to have a gift for understatement, and, as the wind begins to hiss and roar through the camera mic, we hear Bates say, "This is rather frightening." A moment later, tin roofing and other debris from the homes surrounding Mount Arawang begin clattering to earth all around him. Sticks and gravel are now flying in horizontal gusts. "I'm getting pelted with stuff. It's stinging the daylights out of me," he says, shortly before the video ends. "It's just like being sandblasted."

It was estimated later that, during the single blinding burst that caused Mount Arawang to briefly disappear, an area of roughly three hundred acres ignited in less than a tenth of a second. Tom Bates had managed to document the most dramatic instance of exterior flashover ever observed. The Canberra fire tornado of 2003 was rated an EF-3 on the Enhanced Fujita Scale, with horizontal winds of 160 miles per hour, roughly equivalent to a Category 5 hurricane. As the first documented example of its kind, it was a milestone—another harbinger of twenty-first-century fire. But the Chisholm Fire, two years earlier in Alberta, had offered a preview. A funnel cloud was observed during that fire as well, and the resulting forest damage showed evidence of cyclonic action.

It took years of analysis for Australian fire experts to fully understand what Bates and his neighbors witnessed on that terrible January day. The term "pyro-tornadogenesis" did not enter the literature until nearly a decade after the event. A fire tornado, fire scientists would come to understand, is the delinquent offspring of a pyrocumulonimbus thunderstorm. While you can have a pyroCb thunderstorm without a fire tornado, you cannot have a fire tornado without a pyroCb. In this sense, a fire tornado is (so far) a wildfire's most dramatic terrestrial expression (there are other extraordinary things that wildfires can do now, but they take place in the upper atmosphere). Both fire tornadoes and pyroCbs are generated by high-intensity wildfires burning in hilly terrain on exceptionally hot days that have been further energized by incoming high-pressure systems and, some believe, by massive infusions of superheated steam from rapidly burning forests. These events have the capacity to further amplify an already ferocious fire in shocking ways that human beings have no power to defend against.

Once this new, warmer, CO_2-enriched atmosphere had proved itself capable of conjuring up a fire tornado, the question in 2003 was: Could it happen again? Australia is vast, drought prone, and, in places, heavily wooded, a combination that has generated the largest bushfires and the longest, most destructive fire seasons anywhere on Earth. It is fair to say that Australia rarely has a "good" fire season, but some are worse than others; the devastating fire season of 1973–74 blackened an area the size of France and Spain combined (nearly half a million square miles). The Black Saturday Fires in 2009 were some of the worst ever. February was so hot and dry that year—even for southern Australia—that fire officials in the state of Victoria declared the weather forecast "uncharted territory." "There are no weather records," said an official on ABC television, "that show the kind of fire conditions [predicted] tomorrow." The ambient temperature in Melbourne that day—February 7—was 116°F, a record that broke the previous high (set in 2003) by 5°F. The searing heat was attended by gale-force winds; residents compared the experience of going outside to standing in front of a giant hair dryer.

The Black Saturday Fires, concentrated in the hill country northeast of Melbourne, destroyed more than two thousand homes and obliterated several small towns. One hundred and seventy-three people were killed. These fires, started variously by faulty power lines, lightning strikes, and arsonists, were, as of that year, the most lethal and destructive bushfires in Australia's dramatic fire history. While none of them generated a full-blown tornado, a fire service pilot estimated head fire heights at a hundred yards, and a number of victims perished in their cars, overtaken by flames even as they fled at highway speed. But there was another killing energy released by those fires that moved even faster—at the speed of light. So otherworldly were the fire conditions on Black Saturday that animals and people were killed by radiant heat alone, from hundreds of yards away, as if they had been felled by a death ray.

Afterward, a Royal Commission was ordered to investigate the disaster. One of the recommendations made was for a new fire danger category, because "Extreme" was deemed insufficient to express what had occurred on Black Saturday. The new, more dire classification is "Catastrophic," or "Code Red." On its "Fire Danger" webpage, in a box labeled "What you should do," the Rural Fire Service for the state of New South Wales has posted a list of directives. The directive for "Catastrophic" fire could not be more stark: "For your survival, leaving early is the only option."*

This is not planet Earth as we found it. This is a new place—a fire planet we have made, with an atmosphere more conducive to combustion than at any time in the past 3 million years.

In 2009, the year of the Black Saturday Fires, the Keeling Curve hit 390 ppm, a 40 percent increase in atmospheric CO_2 over preindustrial levels. By then, temperature records around the world were being broken on an annual basis, as fire seasons lengthened along with the lists of damage done and fatalities caused. Two thousand seventeen, the year after Fort McMurray burned, appeared to be a turning point. That year, atmospheric CO_2 hit 405 ppm, a 45 percent increase over pre-industrial levels. It was not yet April before 2,300 square miles of grassland had burned across the Great Plains, from

---

* In 2013, Australia's Bureau of Meteorology had to add two new colors (pink and purple) in order to accommodate new temperature extremes previously capped at roughly 122°F.

Kansas to Texas, killing thousands of cattle and at least seven people. That summer, every country in Europe experienced wildfires, including Ireland and Greenland—a first for that continent. More than a hundred people were killed in Spain and Portugal alone when the first pyrocumulus clouds ever observed there supercharged seasonal wildfires into firestorms. That same year, New Zealand experienced unusually intense wildfires while Chile and British Columbia, two huge coastal territories in opposite hemispheres, suffered the worst fire seasons in their respective histories. California, too, had one of its worst ever, including what was, then, the most destructive fire in state history: the Tubbs Fire in Santa Rosa, a catastrophic blaze that destroyed nine thousand structures, killed forty-four people, and generated winds strong enough to flip cars.

In July 2018, 150 miles north of Santa Rosa, a wildfire near the idyllic Northern California town of Redding introduced something altogether new to the Northern Hemisphere. Flying into Redding that August was like flying into New Delhi: down below, as far as the eye could see, spread a lake of smoke. White clouds floated in the brown miasma like marshmallows in cocoa. Somewhere under there was a city of 92,000 souls, but it was impossible to tell from the sky. Poking through the murk, sixty miles to the north, was the peak of Mount Shasta, fourteen thousand feet high and yet oddly devoid of snow—a recent summer phenomenon. This was Northern California, a hundred miles south of Oregon, but it was as hot and dry as Nevada. Beneath the speeding jet, the smoke pall leeched southward for more than two hundred miles—all the way to Mendocino, where it merged with yet another gigantic fire, growing into what would be, if only briefly, the largest wildfire in California's flamboyant 150-year history. As the plane descended through the smoke layer, the convoluted country came into focus, valleys and ridges in endless combination, most of them burned down to mineral soil. The skeletons of trees, their bark burned through to the now-lifeless cambium layer, groped at the air like black hands.

The Carr Fire, as it came to be known, ignited on July 23 near the hamlet of Whiskeytown, fifteen miles west of Redding, due to sparks thrown by a trailer wheel with a flat tire. Three days later, the fire roared into the city. The temperature that day was similar to Black Saturday, 2009: 113°F (tying a local record that was 13°F above the average high). Like so many California wildfires, the Carr Fire was driven by high winds drawn in from the coast, drought conditions, and a century of fire suppression in the state's forests. On July 26, the Carr Fire forced the evacuation of forty thousand people in a matter of hours. Sixteen hundred homes, businesses, and other structures were destroyed. Five people died.

"It was unreal," said a middle-aged woman by the airport baggage carousel, returning home after evacuating three weeks earlier. "It was like doomsday."

Downtown, it was hard to tell; a fresh westerly was blowing, and it hid a lot of the evidence. A taxi driver explained that it was the first day without street-level smoke since the fires had ignited nearly a month earlier. It was still a presence, shrouding the surrounding ridges, but the sky was blue directly overhead, the shadows crisp on the pavement. So benign and ordinary was the scene that it seemed as if news accounts might have been exaggerated. They weren't. "Everybody," said the taxi driver, "knows someone who's lost a home."

You didn't have to go far to see that whole neighborhoods were missing.

Every day, the world ends for someone, somewhere. Lately, this seems to be happening more often, not only because of fire, but because of the titanic energies unleashed by it. It took fifteen years of steadily worsening wildfires, but on July 26, 2018, the question asked by Australian fire officials—will the Canberra fire tornado ever be repeated?—was answered.

On the other side of the world, eight thousand miles from Canberra and ten minutes from downtown Redding, is Lake Keswick Estates, a compact neighborhood of modest, mostly single-story homes. Like a lot of suburban Redding, it is built in the WUI, the wildland-urban interface. Many of the residents there had been sure that the Keswick Reservoir, half a mile to the west, would stop the Carr Fire's wind-driven advance. Seventy-three-year-old Sarah Joseph was one of them, and she had to gather herself before describing what leaped across the water shortly before 8:00 p.m. on that hundred-degree evening. "It looked like a tornado," she said, "but with fire." Like Tom Bates in Canberra, she was describing something that no one in her world had seen before. It arrived so quickly that Joseph had only minutes to gather up her cat, some photos, and a change of clothes before fleeing for her life.

Sarah Joseph managed to escape the fire and so, miraculously, did her home. But this is in keeping with the whimsical nature of tornadoes: their violence is capricious, meted out selectively, but at random, like a sadistic child crushing ants. All around Joseph's home, entire blocks lay in ruins. A half mile to the east, on a broad, forested slope that felt almost rural save for the steady crackle of high tension lines overhead, Willie Hartman stood knee-deep in the ruins of her home. Hartman is a slight but sturdy grandmother with white hair and a warm demeanor, and, a month on, she was still coming to terms with the fire that had transformed everything, as far as the eye could see. Behind her, the metal porch railing drooped like a garden hose. Spotting a charred skeleton of furniture, she murmured, half to herself, "The lawn chair's in the house."

So was the mailbox. Nothing was where it should have been, or even *what* it should have been. The Hartmans' living room, which had ceased to exist, once had a picture window of double-paned glass, but it melted. You could see it outside, a vitrified river flowing downhill toward her daughters' homes, also burned to the foundations, many of their contents borne away on the incinerating wind that spun out of the fire and into their neighborhood minutes after skipping over Sarah Joseph's home.

There is video, and it is terrifying: surging up out of a cluster of burning neighborhoods is a whirling vortex a thousand feet across, seething with smoke and fire. Its outer bands appear almost taut, undulating and distorting as if something were inside, trying desperately to get out. During its brief existence of approximately thirty minutes, the Redding fire tornado sent jets of flame hundreds of feet into the sky, obliterated everything it touched, and generated such ferocious thermal energy that its smoke plume punctured the stratosphere. The damage at ground zero, a three-hundred-yard-wide, half-mile-long swath of scoured earth, annihilated homes, and blasted forest that ended just south of the Hartman family compound, was hard to comprehend.

In the rising light of dawn was revealed the aftermath of an atmospheric tantrum so violent it looked as if the Hulk and Godzilla had done battle there. A pair of hundred-foot-tall steel transmission towers were torn from their concrete moorings and hurled to the ground, where they lay, crumpled like dead giraffes. A four-ton shipping container was ripped to pieces and heaved hundreds of yards across the landscape. All the houses south of the Hartmans' were gone, stripped to bare slabs. Trees were torn limb from limb. In the branches of those that survived, where plastic bags might flutter, ten-foot pieces of sheet metal roofing were twisted like silk scarves. A camshaft, a flywheel, a kitchen sink, an oven door, and countless other objects were scattered through the charred forest. There was no glass anywhere. Grass, bark, and topsoil were gone.

What tornadoes do best, it seems, is obliterate context. In their wake is left a catalog of violation so thorough and yet so arbitrary that it causes an existential derangement of a kind that makes you want to check your own hands to make sure the fingers are still there. Not far from one of the bare house slabs was a pickup truck wrapped around a tree; another was simply torn in half. Other vehicles were strewn across the barren ground, their roofs crushed as if they had been rolled repeatedly, the bodies burned down to the springs. Most of the doors, hoods, and trunk lids had simply been torn off; any that remained were blown inside out and pocked so thoroughly by flying gravel and debris they looked as if they had been attacked with hammers and shotguns.

Draped over the landscape and snaking through the trees were endless strands of two-inch transmission cable from the fallen towers. Built to carry fifty thousand volts, even a short section was too heavy to lift. Elsewhere, buckets, barrels, handtrucks, and stovetops were scattered willy-nilly, so badly damaged they were hard to recognize. Some of these things were wrapped around tree trunks so tightly they could only be removed with heavy tools. I saw a metal folding chair driven into a tree like an ax blade, and a steel tractor seat crumpled like a paper plate. Cast iron frying pans lay hundreds of yards from the nearest house, bent, with their handles torn off and holes punched through the bottom. Nothing, no matter how sturdy or how small, was left intact. Even the stones were broken.

Larry Hartman, Willie's husband of forty-seven years, is a large, congenial man with a hydraulic handshake and a gift for problem-solving. Finding himself with a dozen bear-hunting dogs that needed regular exercise, he devised a mechanical carousel with twelve chain leashes that now lay upside down in a heap of unrelated wreckage. When asked what he would have imagined happened here if he hadn't witnessed it himself, he regarded the utter ruination all around him, the spaces where outbuildings and other landmarks of his life no longer were. "A bomb," he said. "Like Hiroshima."

When you compare photos of the hypocenter of that historic nuclear blast with the excoriated ground just south of the Hartmans' property, they are hard to tell apart. One of the Hartmans' daughters, Christel, used to hunt bears with her father, and she inherited his formidable handshake. Christel recorded video of their evacuation on her phone, and it shows a fire surging over the hill, the way so many wildfires arrive in the WUI these days, but this fire is burning higher than the transmission lines. Peering into her phone, I could see the towers' latticed silhouettes ghost in and out of the flaming wall like skeletal giants. *War of the Worlds* came to mind. "It made a roaring sound," said Christel, "like a man." She demonstrated and then said, "Only ten times that." Across Quartz Hill Road, a few hundred yards from the Hartmans', an elderly woman and her two great-grandchildren were burned to death in their trailer.

Captain Dusty Gyves, a twenty-year veteran with Cal Fire, California's 130-year-old state firefighting agency, was shocked by what he saw five hundred yards southwest of the Hartman compound. After being lifted into the air, a two-ton pickup truck was subjected to forces so violent that it looked, said Gyves, "like it had been through a car crusher." And then incinerated.

Inside that truck was a thirty-seven-year-old fire safety inspector named Jeremy Stoke who had cut his vacation short to help out with the evacuation effort. A husband and a father of two, Stoke was well liked, and there was a memorial to him on Buenaventura Boulevard where he was wrenched from this world. Flowers, a flag, and a nightstick had been assembled around a humorous portrait of Stoke holding a pistol, along with dozens of ball caps, T-shirts, and shoulder patches representing police and fire departments from all over California. Among the offerings were several lids of Copenhagen tobacco, a bottle of sunscreen for his cleanshaven head, and a handwritten note saying, "Rest easy, brother. We will take it from here."

Stoke was one of five people killed that day, but when survivors tell of their escapes, it seems a miracle there weren't many more. A local dentist, surprised by the flames in the nearby gated community of Stanford Hills, ran for her life through the woods. Disoriented, with no idea where to go, she and her husband followed the animals—deer, rabbits, and squirrels—as they fled downhill, toward a bend in the Sacramento River. Several of her neighbors were rescued by a patrolling helicopter. Another neighbor, a retired police detective named Steve Bustillos, was one of the last to evacuate their neighborhood. Bustillos is a compact, powerful man with dark eyes and spiked gray hair. While he did not fully understand what was brewing that evening a mile to the west, he had been in enough dangerous situations to know that he'd better not let his guard down. Steve's wife, Carrie, who is tall and slender with wavy dark hair, had gone to visit family in San Jose, so Steve was on his own at the house, watching satellite coverage, tracking social media, and keeping an eye on the horizon. Neither Bustillos nor his neighbors ever heard an evacuation order for Stanford Hills; nor did patrolling firefighters appear to issue a warning. There were about fifty homes built along their serpentine, ridgetop cul de sac; prices started at a million dollars and went up from there. Many of the owners were retired, but they were all successful professionals—proactive planners who made a point of staying informed and looking out for their neighbors. Warning or no warning, most of them had been preparing for a possible evacuation. Steve and Carrie Bustillos had a pile of duffels stacked and ready in the garage.

"You could see the plume off to the west," Steve Bustillos recalled. Like his neighbors, he assumed that the Sacramento River, which flowed out of the Keswick Dam eight hundred feet below the ridge, would be wide enough to stop the fire's advance. Between the river, a half mile west, and transmission lines to the south and east, he figured that Stanford Hills was safe from most wildfires. This might have been true under ordinary circumstances, but at 6:00 p.m., the nearby Mule Mountain weather station registered a temperature of 111°F with 7 percent humidity.

"Then," Bustillos said, "around 7:40 p.m., we saw the fire just move across the river. It was weird; it just hopped over."

These effortless transitions—across rivers, highways, and firebreaks—seem to be happening more and more often. Steve's neighbor Kate Baker, who lives at the end of their cul de sac in a palatial home with its own vineyard, knows something about firebreaks because her husband is in the heavy equipment business. Baker had spoken with a local dozer boss who worked the devastating Tubbs Fire in 2017. The dozer boss told Baker that he had overseen a fleet of D9 Cats building firebreaks down there, and that every firebreak had failed. "I'm zero for six," he said. As it happened, he had also been building firebreaks on Buenaventura Boulevard, between Stanford Hills and the Hartmans'. Now, he was zero for seven. Tactics that worked twenty years ago, or even ten years ago, no longer seem to be as effective.*

Once the fire jumped the river, there was nothing left to stop it, so after making sure an elderly neighbor got away safely, Steve Bustillos finished loading his truck. As he did so, he noticed a shift in the wind. "Air is flowing past me from the other direction"—toward the fire—he said. "It was like a vacuum. I was kind of dumbfounded. I didn't really know what I was looking at. What I saw was: when you open an oven, or a barbecue—the heat waves. There was a wall of them as high as the power lines. There was no flame or anything and then, when it hit the back of the houses across the street, just a wall of fire ignited. I watched the palm trees next to me all the way to the backyard. Now, it's full-on blowtorch-type heat and the wind is strong enough that I have to lean into it to stand upright. I can hear the air rushing out of my house, whistling out my garage doors."

Bustillos was having the same sensation that Mark Stephenson had on Signal Road in Fort McMurray—as if the fire were taking an

---

* This is but one example of what the futurist Alex Steffen terms a "discontinuity." A corollary to the Lucretius Problem, it is a situation wherein expertise and past experience cease to be useful guides to future problem-solving.

enormous breath before exhaling like a dragon. The huge, rotating plume of a pyrocumulonimbus cloud, such as the one looming over Redding on July 26, is the engine of an igneous storm, and, like all combustion engines, it needs oxygen, which it inhales from the surrounding countryside. Park Williams, a professor in the geography department at UCLA, explained what happens next in *The Atlantic* in 2018. "Sometimes, that channel of upward-flowing air can collapse in one small spot," he told the journalist Robinson Meyers. "Then the hot air in the atmosphere plummets through the weak point. You get a very fast wind moving down toward the ground, and when it hits the ground, it spreads like jelly slopping across the floor."

This flood of explosively hot air is what Bustillos saw—first as heat waves and then as spontaneously combusting trees and houses. This was exterior flashover in real time. (When I described this scenario to Park Williams, he said Bustillos was lucky to be alive.) Despite his careful preparation, it now looked as if Bustillos had missed his escape window. "My house isn't on fire yet," he said, "but everything else is, and there's smoke and chunks of ash. I wouldn't be able to get anywhere because of the heat and the smoke. I'm thinking I can try to drive through this, or wait and see what burns through and see if I can drive out then."

When asked if he was as calm then as he appeared to be recalling it. Bustillos answered, "Oh yeah."

Carrie interjected, "You know what he used to do for a living, right? Homicide detective? They amp down when we amp up."

"I used to work the busiest sides of San Jose," Steve said. "Every Thursday, Friday, Saturday night, it was just full-on chaos: stabbings and shooting and car chases. When you do that stuff—instead of getting all worked up, you slow it down and you start to work through what I call 'the Process': How'm I gonna solve the problem? You can't do it when you're all flustered. You take a step back; your breathing changes, your perception; you kinda look at the big picture. So, that's what I'm doin': I'm thinkin' this isn't the time to drive; we'll see what happens. Because things are fully engulfed, and I don't know how long they're gonna continue to be fully engulfed."

Steve recalled that the wind subsided somewhat then, but this was relative; the fire was burning on both sides of his cul de sac, and the heat was generating its own local winds. Meanwhile, the tornado, now fully developed and over three miles high, was just a half mile to the north. It was moving eastward at the speed of a street sweeper, grinding through a sparsely inhabited stretch of forest between Stanford Hills and Keswick Lake Estates, where Sarah Joseph had recently evacuated. There is only one road out of Stanford Hills (a common feature of suburban subdivisions), and that is Buenaventura Boulevard, which happened to run perpendicular to the tornado's path. Steve called Carrie and reassured her that he was okay and was going to drive out. By now, he was all but certain their house wouldn't be standing when they returned. He stayed on the phone with Carrie as he got in the truck and headed for the entrance gate. It was a few minutes past 8:00 p.m.

Steve's truck, a GMC Sierra Duramax with a camper shell, was a big one, and it was filled with many of their personal belongings—everything from clothes, photos, and their daughters' keepsakes to computers, cameras, and Steve's guns. Fully loaded, the truck weighed close to five tons. It never occurred to Bustillos that weight would determine whether or not he lived through the next five minutes. Turning left out of the gate, Bustillos made his way slowly through the smoke, past half a dozen fire trucks idling by the gates of the Land Park subdivision, another gated community. This was a puzzling sight, given all the houses on fire, but there wasn't time to speculate, and he hung a right onto Buenaventura.* Heading north now, Bustillos had half a mile left to live. Just ahead of him, somewhere in the smoke, the fire inspector Jeremy Stoke and his Ford 150 pickup had just been heaved through the air. A hundred yards farther on, a bulldozer operator was being flayed with burning gravel. Less than a mile to the west, another bulldozer operator had just been burned over and killed.

But in the gathering dusk and smoke, safe inside his big truck, it looked to Bustillos as if the fire had already passed through: the forest floor was bare, the trees smoking and devoid of leaves. About forty feet above the ground hung a ceiling of thick smoke. He still had Car-

---

* They were sheltering there because fire conditions were so dangerous ("Carr Incident Green Sheet," p. 11).

rie on speaker as he passed a lowboy trailer by the side of the road, and then, on the opposite side, a tractor trailer. Ahead of him on the right were a couple of bulldozers, brought in by the dozer boss who was zero for six in Santa Rosa. Then, Bustillos looked to his left. Over the speaker, Carrie heard her husband of thirty-two years say, "Oh shit."

"And then the window blew out of the driver's side of the truck and there was just ash and debris," Steve told me. "I'm calling it 'fire and brimstone' 'cause that shit was just rolling around in the cab with me. I'm scooping it off of me, and then I looked over my shoulder, and everything in my shell was fully engulfed already."

At this point, Bustillos stopped the truck and devoted himself to the small, whirling fires igniting inside the cab. Embers, sticks, and stones were blowing through his window with terrific force, peppering his face and shoulder; the heat was unbearable. He grabbed a canvas carry-on bag and tried to protect himself. "As I was doing that," said Bustillos, "the truck was lifting. I had both feet on the brake. I'm still seat-belted in. Then, the truck kind of came to rest. The hair on the side of my head is singeing. I'm sitting there wondering, 'What am I gonna do next?' And I look to the right—at the right front fender—and I'm seeing full-on flames. Something's ruptured and the diesel fuel's on fire."

Bustillos understood then that he was going to have to abandon his truck, but this meant exposing himself to the full force of the burning tornado. He was dressed only in shorts and a T-shirt. "My electric locks don't work," he said. "So I use the button, cracked the door, put my foot down in the door, took the seat belt off, and then I started lookin'. I'm lookin', and I see a dozer off to my left and I say, 'Okay, that's where I'm goin'.' I grab the bag I was using as a shield, and I grab a backpack. I make it over to the front end of the dozer, and it's running. I don't want to get run over so I look up—all the windows are blown out, and nobody's sitting in it. Stuff is flying around; it's just pelting me—little chunks all on fire."

Left behind in the truck was Steve's cell phone. At that moment, Carrie was sitting on a curb in San Jose, 250 miles away, still on the line. She could tell by the sounds that the truck was on fire. "I had about two minutes of open mic," she said. "I heard all the popping of the truck—all that fury."

By then, the tires and fuel tank were exploding. Steve's loaded pistols were firing off at random. "I just kept talking to him," said Carrie. "I said, 'Hey, if you can hear me, you're gonna be okay; we're gonna get through this.'"

They had been through so much already—not just Steve's dangerous job, but Carrie's battle with cancer three years earlier. "And then the call failed," she said, "and I knew the phone was destroyed."

Carrie had no way of knowing if Steve was still in the truck or not. In the killing heat and wind, beset by flying gravel and embers that stung like hornets, Bustillos maintained his composure and used the dozer's blade as a kind of blast shield, moving around it as the wind direction changed. He then crawled in between the blade and the dozer's treads, burrowing into the dirt like an infantryman under fire, trying to keep the bag and the backpack between him and the worst of the flying debris. "I hunkered there," he said, "and this is where I kind of lost time—because of the adrenaline and everything going on. At some point, all of a sudden, it calmed down and the temperature dropped—I'm gonna say, three hundred degrees—I mean, the difference between opening your oven and getting blasted to shutting the oven."

With the tornado apparently past, Bustillos took a moment to gather himself. Then, he crawled out from under the bulldozer and stood up. He was severely burned, but so full of adrenaline that he didn't notice. "I'm looking across the road," he said, "and I see this foil blanket up over this guy, and he's running, shouting, 'Get me outta here! Get me outta here!'"

It was the bulldozer operator. Because California's forests are so fire-prone, bulldozers are equipped with safety glass and fire-retardant curtains, and their operators are supplied with fire shelters. The windows and curtains had failed, but the fire shelter—a kind of foil-lined cocoon—had saved this operator's life. Moments later, a Forestry truck appeared; it was burned and pocked with dents as if it had just emerged from a war zone. The driver, a Forestry supervisor, spotted the dazed and smoking men and stopped. Bustillos climbed into the truck unbidden. He found himself behind a young firefighter who didn't appear to register his presence. Staring straight ahead, the young man was in shock. The supervisor drove Bustillos and the dozer operator back to the cluster of fire trucks by the Land Park subdivision. From there, Bustillos and the operator were taken downtown to a nearby trauma center. En route, Bustillos borrowed a phone and called Carrie. Mercifully, she'd had only a few minutes to fear the worst. After a quick examination, Bustillos was helicoptered to the burn unit at the UC Davis Medical Center in Sacramento. In addition to being bruised for reasons he could not fully recall, Bustillos looked as if he had been rolled in red-hot gravel.

About half the houses in Stanford Hills burned to the ground, but the Bustillos home survived. It was an irony, at once merciful and cruel: in their truck, now a smoking shell sitting perpendicular to Buenaventura Boulevard, were all of Steve and Carrie's most precious belongings, including jewelry, passports, and a significant amount of cash. Anything that hadn't burned had melted, including Steve's firearms.

Bustillos and the bulldozer operator compared notes later. If Bustillos had driven down Buenaventura a few minutes earlier, he would have met the same fate as Jeremy Stoke. The dozer operator, like Stoke, had borne the full brunt of the tornado. According to Bustillos, the bulldozer, which weighed more than fifty thousand pounds, was dragged across the ground. The operator told Bustillos that the only way he could get it to stop was by engaging the ripper tine—a massive hook used for tearing up packed earth and pavement.

Forensic analysis of the scene on Buenaventura concluded that the tornado's wind speed was somewhere between 140 and 165 miles per hour, and that "peak gas temperatures likely exceeded 2,700°F"— the melting point of steel. In other words, Bustillos had endured the equivalent of an EF-3 tornado, combined with a blast furnace. He and the dozer operator had been inside the same thing that the Australian Tom Bates had filmed from the Kambah Park rugby pitch in Canberra.

EF-3 is considered "Severe" on the Enhanced Fujita Scale. However, judging from the totality of the destruction between the vehicles on Buenaventura and the downed transmission towers five hundred yards to the east, much of the damage seems consistent with an EF-4 ("Devastating: Well-constructed houses levelled; structures with weak foundations blown away some distance; cars thrown and large missiles generated"). Given the size of the trees uprooted and broken in half, and the way their bark was stripped off with the surviving branches abraded into points and facets, the damage was consistent with a Category 5 hurricane. The addition of metal-melting heat seemed gratuitously biblical.

It raises the question: What should you call something that behaves like a tornado but is made of fire? Many wildfire scientists bridle at the term "fire tornado"; they prefer "fire whirl," but "fire whirl" seems inadequate to describe something that can do what the Redding fire tornado did while building a weather system seventeen thousand feet high. In 1978, a meteorologist named David Goens devised a classification system that placed fire whirls of this magnitude in the "Fire Storm" category, along with the caveat that "this is a rare phenomenon and hopefully one that is so unlikely in the forest environment that it can be disregarded." But Goens wrote that more than forty years ago, and a lot has changed; for starters, the possibility of a fire tornado can no longer be disregarded. The language, too, is evolving to accommodate these recent changes: "Natural fire never did this," explained the Cal Fire veteran Dusty Gyves after surveying the damage in and around Redding. "It shouldn't moonscape."

It is alarming to consider that this annihilating energy came out of thin air, born of fire and fanned by an increasingly common combination of triple-digit heat, single-digit humidity, high fuel loads, dying trees, and the confusion of battling winds that swirl daily through the mountains and valleys of the North American West, and so many other fire-prone parts of the world. That this phenomenon may represent something new under the sun has become a subject of earnest debate among fire scientists and meteorologists. With the exception of the 2003 Canberra fire tornado, there is no record of a fire tornado of this magnitude occurring in a residential setting.

⁓⁓⁓

Heading west out of Redding on Route 299 takes you past Whiskeytown, where the Carr Fire started, and into steep country dotted with old mining towns where "gulch" becomes a common suffix. None of the hamlets in these funnel-shaped valleys had burned, due either to the whims of the fire or to heroic stands taken by Cal Fire, Forestry, and volunteer fire departments. The surrounding forest was a different story. Most of the trees that grow across these ridges and canyons are adapted to fire, but there are limits. After the Carr Fire, everything from ridgetop to canyon floor was burned as far as the eye could see. It wasn't just the smaller and more flammable pine, oak, and manzanita; mature redwoods stood now like broken columns in a blasted temple. Some of the stumps, ten and twenty feet high, were still on fire a month later, smoking like pagan censors as they filled that silent, empty forest with a strangely evocative perfume.

Painfully clear is the fact that there is no way for firefighters to combat these superheated, firebreak-leaping fires—with or without a tornado in their midst. Water has little effect on a high-intensity wildfire, and fire retardant drops are about as effective as firebreaks. Among the structures burned near Redding was a fire station. There was a time not so long ago when a fire like this one, which forced the evacuation of forty thousand residents, destroyed more than fifteen hundred structures, and burned nearly four hundred square miles across two counties, might have been a monstrous anomaly, but now such fires are becoming the norm. Simultaneous with the Carr Fire was the Mendocino Complex Fire, the largest ever recorded in California. As of August 2018, seven of the most destructive wildfires in the state's already fire-prone history had ignited within the previous twelve months alone. Collectively, they caused more than one hundred deaths, destroyed twenty-five thousand homes, and nearly bankrupted PG&E, the state's largest power company, whose poorly maintained transmission lines were blamed for many of the fires. There is no respite in sight: since then, most of those record-setting fires have been knocked out of the top ten by even bigger fires. The colossal August Complex Fire, which burned an area larger than the state of Rhode Island in 2020, was more than twice the size of 2018's Mendocino Complex; it took four months to contain. According to Cal Fire, nine of California's twenty largest fires have occurred just since 2020. "The fire season used to run from May to October," a Cal Fire deputy chief named Jonathan Cox told me. "Over the last decade, it's changed to year-round—and also to twenty-four hours."

The impact on firefighters is exhausting and dangerous: when the Carr Fire first broke out, many local firefighters worked around the clock—for days, just as their counterparts in Fort McMurray had done. The changes in fire behavior have, in turn, changed the role firefighters play in these events. "Firefighters are never going to not engage," Deputy Chief Jonathan Cox told me, "but now firefighters are having to retreat sooner."

Redding offered a good example of what that looks like: "It shifted from a firefighting effort to a life-saving effort," Cheryl Buliavac, a Cal Fire spokesperson, told me. And that was even before the tornado formed.

"Fires are making their own behavior," Cox said. "The anomalies are becoming more frequent and more deadly." Even among anomalies, the Redding fire tornado was in a class of its own. "We've never seen anything like this," he said. " 'Extreme' is an understatement."

22

If all three realms are ruined—sea and land and sky—
Then we shall be confounded in old Chaos.
Save from the flames what's left, if anything can still
be saved.
Think of the Universe!

—Ovid, *Metamorphoses*

From 2016 onward, the fire seasons in both hemispheres have been relentless. In 2017, British Columbia, a huge coastal province bigger than Alberta, bigger than Chile, and more than twice the size of California, set a new global record. On July 7 alone, 142 separate wildfires ignited; by the end of the day, the province was in a state of emergency. A familiar combination of high heat, drought conditions, and wind caused many of those fires to grow rapidly into uncontrollable blazes. A month later, many of them were still burning. On August 12, four of the larger fires, along with one across the border in Washington State, erupted almost simultaneously into pyroCb thunderstorms, a phenomenon never observed before. David Peterson, a meteorologist at the U.S. Naval Research Laboratory in Monterey, California, speaking to the CBC, declared it "the most significant fire-driven thunderstorm event in history. Nothing else even comes close." Once in the stratosphere, the mass of particulate was swept into the jet stream, where it circled the globe for four months.

The only mercy shown British Columbia that summer was that

most of the fires ignited in sparsely inhabited areas. Even so, more than forty thousand people were displaced across the province, fire-fighting costs alone exceeded half a billion dollars, and almost five thousand square miles of forest burned. Though it was hundreds of miles from the biggest fires, the sky in Vancouver turned a burnt-orange color for weeks, and the air quality was rated some of the worst in the world. British Columbia's historic aerosol injection, which has come to be known as the "Pacific Northwest Event," was more than twice the size of any previously documented pyroCb.

But if twenty-first-century fire has taught us anything, it's that there is no top end. It wasn't long before the Pacific Northwest Event, "the mother of all pyroCbs," had company. In 2020, the—once again—record-breaking fire seasons in Australia, California, and Oregon generated similarly volcanic pyroCbs. Australia's, however, was—how many times can one say this?—unprecedented. Most readers will be familiar with the horrific fires that appeared to envelop that country in December and January of 2019–20, and with the shocking number of animals that perished. But high above Earth, something else was happening, too. An abstract describing it in the journal *Communications Earth & Environment* reads like a scientist's description of a cataclysm from the Old Testament:

> The Australian bushfires around the turn of the year 2020 generated an unprecedented perturbation of stratospheric composition . . . The resulting planetary-scale blocking of solar radiation by the smoke is [three times] larger than any previously documented wildfires and of the same order as the radiative forcing produced by moderate volcanic eruptions. A striking effect of the solar heating of an intense smoke patch was the generation of a self-maintained anticyclonic vortex measuring 1000 km. in diameter and featuring its own ozone hole. The highly stable vortex persisted in the stratosphere for over 13 weeks, travelled 66,000 km and lifted a confined bubble of smoke and moisture to 35 km altitude.

In other words, ferocious heat convection drove a climate-altering quantity of ash and particulate eight miles into the stratosphere, where it then formed an aerosol blob, six hundred miles wide and two miles thick. Because it contained so much water vapor and black carbon, it absorbed solar energy, which caused it to heat up and rise still further—en masse, like a black balloon the size of Texas—until it was more than twenty miles above the earth, twice as high as any previously known pyroCb injection. Once there, this half-million-cubic-mile pyrogenic carbon blimp drifted for more than three months around the Southern Hemisphere, covering forty thousand miles before finally dissipating.

In the 1990s, pyroCbs were a disturbing but exhilarating novelty wondered at, and discussed by, a small group of meteorologists. Now, they are not only a signature of major wildfires, they are actively growing in size and frequency—to the point that they are mimicking volcanoes, previously Earth's most rapid and powerful climate-changing agents. PyroCbs are now being observed all over the world in places they have never been reported before. As these events multiply, they are altering, in significant and measurable ways, the chemical composition of what atmospheric scientists refer to as the "stratospheric overworld."

This is the power of atmospheric CO_2. It expresses itself through heat retention, and its "vocabulary" appears to be growing, most obviously through variations in weather, fire, and related phenomena, but in other, less visible ways as well, most notably in the oceans. The oceans absorb approximately 30 percent of all emitted carbon dioxide. Over the course of the Petrocene Age, this global system, home to more than half of the world's species, has grown 30 percent more acidic, signaling the most rapid shift in ocean chemistry in the past 50 million years.

What the atmosphere and oceans are telling us is that carbon dioxide doesn't get the respect it deserves. Others have been saying this, too—for a long time. Roger Revelle said as much to a congressional subcommittee in 1956. Eunice Foote said it to the American Association for the Advancement of Science in 1856. Any climate scientist or environmental studies teacher will tell you every chance they get. Until very recently, most of them have been tuned out, brushed off, or appeased in ways that bear a strong resemblance to the experience of people reporting incidents of sexism or racism: "Where? *I* can't see it."

The conclusion arrived at by the *Communications* article's dozen authors is that a big enough pyroCb "eruption" could inject enough carbon into the stratosphere to alter the planet's climate, just as large volcanic eruptions have done in the past. Pollution and air quality aside, the carbon dioxide generated by events of "planetary scale"—like twenty-first-century wildfires—exceeds the annual CO_2 output of many states and countries. To put this in perspective, the CO_2 emitted by the Australian bushfires of 2019–20 more than compensated for the global reduction caused by the coronavirus pandemic.

It hardly needs to be said that more CO_2 leads to more heat retention, which leads to more fires, which leads to more pyroCbs . . . We are, right now, witnessing the early stages of a self-perpetuating and self-amplifying feedback loop, accompanied by myriad "cascade effects."[*] In human terms, this has been a long time coming, but in geologic terms it has taken place overnight—roughly seven human generations, or two life-spans. So limited are we by the brevity of our lives and, lately, by the kaleidoscopic swirl of technological advancement, further amplified by a twenty-four-hour news cycle, that it's hard to appreciate how far we've come (and gone) in such an extraordinarily short time.

I was born in the 1960s, but I personally knew people born in the 1870s and '80s, when the petroleum industry was in its infancy and Standard Oil was a start-up. The Civil War, waged by horse- and manpower, was a raw and recent memory then; Queen Victoria reigned over a global empire held together by sailing ships, and the climate visionaries Svante Arrhenius and Arvid Högbom were still in high school. In 1875, Chicago was still rebuilding after its great fire, the battle at Little Big Horn had not yet been won or lost, and boreal explorers were still fantasizing about how men might one day turn Alberta bitumen into money. Back then, 1.3 billion people walked the planet—literally, because there were no cars. Nor was there plastic, and the Keeling Curve of CO_2 had only just begun its relent-

---

[*] For example, when it gets too hot, helicopters can't fly and can't fight fires. When polar bears can't catch seals due to lack of sea ice, they raid seabird colonies and town dumps. When permafrost melts, Arctic roads and pipelines sag and rupture.

less upward bending. That world—the same one into which people whose hands I touched were born—is so close temporally (I looked into their eyes; I felt their breath), and yet it is so remote chemically, biologically, atmospherically, technologically, *anthropogenically* from the world we inhabit now, the world we are currently unmaking, the world our children are inheriting that resembles, less and less, the one that made us.

Our unprecedented success (and emissions) are due first to our mastery of fire, and second to our exploitation of fossil fuels in all their varied forms. In terms of its implications for life on Earth, our historically brief experiment with a fossil fuel–driven civilization is, in essence, a high-intensity carbon release project. Nature accomplishes the same thing with forest fires and volcanoes, but not nearly as efficiently, or as quickly, as we are doing now. Every year, this global industry releases ten gigatons of carbon in the form of coal, oil, and gas formerly sequestered in the planet's crust.* This is a rate roughly ten times faster than anything scientists can find in the geological record for the past 250 million years, and about one hundred times faster than natural systems were releasing it in more recent pre-industrial times. This is how Earth will remember us: thanks to fire and our appetite for its boundless energy, we have evolved into a geologic event that will be measurable a million years from now.

Viewed through this lens, twenty-first-century fire is not so much an aberration as a by-product of our principal accomplishment. Setting aside the ephemeral distractions of culture and civilization, modern humanity—*Homo flagrans*—will be remembered, above all, for building, and for *being*, the greatest combustion engine ever devised. In terms of heat, energy, and emissions, we are a supervolcano representing the largest, most rapid release of combustive energy, carbon dioxide, and methane since the Permian Age.

And that is saying something.

---

* A gigaton is a billion metric tons, roughly equivalent to three thousand Empire State Buildings.

The Permian Age commenced roughly 300 million years ago, and it started well enough. Jungles and forests of ferns, cycads, and conifers flourished across the landscape. Into this oxygen-rich environment evolved the first truly massive creatures, including sail-backed reptiles like dimetrodon, and ambiguous proto-mammals like the gorgonopsids, whose saber teeth were set in jaws that opened as wide as bear traps. Flying overhead were the largest insects the world has ever known. Meganeuropsis was a dragonfly-like creature with a thirty-inch wingspan whose four thrumming wings must have sounded like the rotors on a quadcopter drone. Down below, the Permian ocean, though poorly represented in the fossil record, supported fish and squid in sufficient abundance to feed whale-sized sharks, some with continually cycling teeth oriented like buzz saws (or bucketwheels). But what makes the Permian noteworthy, and relevant now, is not how it began, but how it ended.

During this period of 50 million years, Earth's landmass was a supercontinent, a compressed jigsaw puzzle now called Pangaea. There was ice at the poles then, too, and in its Northern Hemisphere, in what is currently Siberia, a series of massive volcanic eruptions began at the tail end of the Permian Age, about 250 million years ago. Without them, the Permian might have continued for another 50 million years. As the planet tore open, and magma jetted to the surface like an arterial wound, millions of square miles of the surrounding landscape was inundated by flowing lava, which hardened over time into basalt layers more than a mile thick, forming what geologists now call a "large igneous province." These planetary growing pains also drove magma horizontally through seams in the planet's crust, effectively fracking vast fossil fuel deposits laid down by even older seas and forests. The results were spectacular; according to geologists, some of these molten injections detonated gas explosions that left craters a half mile wide. This pressurized magma also found its way into ancient coal beds, where it smoldered and burned for hundreds of millennia.

While not as systematic or as broadly distributed as our fossil fuel emissions, this period of extended, volcano-driven combustion did the same thing we are doing today: it burned everything it came across as rapidly as possible. Then, as now, the inevitable result was an extraordinary amount of smoke, ash, water vapor, and carbon dioxide in the atmosphere—thousands upon thousands of gigatons, far more than existing plants, oceans, and chemical weathering could absorb and process. Thousands of gigatons may sound like a lot, but for most readers (and this writer), it is a meaningless measure, especially in the context of air, which doesn't appear to weigh anything, or the atmosphere, which appears infinite. Simply put, it's the carbon that gives carbon dioxide its weight, and carbon dioxide weighs about two ounces (fifty grams) per cubic foot (the Chevy Silverado emits about two pounds of CO_2 for every mile driven).

Numbers aside, we've all seen a polluted city, and we've all been in a smoky room, and we've all seen a surface coated with soot. We can imagine that it takes an extraordinary quantity of carbon dioxide to register as weight on a scale—in grams or ounces, let alone gigatons. And we all know that the smokier a room gets the smaller it feels and the harder it is to see, to breathe, to think, to live. Imagine this room filling with so much ash, methane, and CO_2 (from your barbecue, from your vehicle, and from the pig farm, the coal-fired power plant, and the oil refinery up the road) that virtually everything (plants, animals, birds, fish, and insects) overheated, suffocated, and died.

That's how the Permian ended.

Paleontologists call it the Permian-Triassic Extinction Event. Lee Kump, dean of the College of Earth and Mineral Sciences at Pennsylvania State University, calls it "the worst thing that's ever happened." It left behind a planet unrecognizable and virtually uninhabitable. In addition to an Alaska-sized slab of lifeless basalt in Siberia, the geologic record reveals Martian red rock and vast deposits of salt in Kansas. In between, in all directions, more than 70 percent of all terrestrial species went extinct. Then, as now, roughly 30 percent of all emitted carbon dioxide is absorbed by the oceans, where it is converted into carbonic acid. At the end of the Permian, with CO_2 levels in the thousands of parts per million and ocean temperatures above 100°F at the equator, those ancient seas became so acidic, and so hot, that more than 90 percent of all marine species died out. In recent times, the mysterious tooth whorls of the "buzz saw" shark, helicoprion, have turned up on riverbanks in western Australia and in phosphate mines in Idaho, but little else remains. There have been five major extinctions in Earth's history, but only the end-Permian has been called "the Great Dying."

It is to this terminal catastrophe—caused, not by meteorites, or by shifts in Earth's orbit, but by unrelenting combustion—that geoscientists are comparing our own Petrocene Age. Our fire-powered civilization is now in the early stages of replicating that "once-in-a-lifetime" extinction event. It is widely understood in the scientific community that a sixth major extinction is under way, and that it is wholly due to human activity. As confronting as this idea may be, it shouldn't come as a surprise: never in Earth's history has there been a disruption like us: *billions* of large, industrious primates whose evolving behavior is almost entirely dependent on the universal burning of hydrocarbons. Nor has Earth ever had to carry (at the same time, no less) *billions* of methane-emitting livestock the size of pigs and cattle.*†

There is a terrible symmetry in this. What we are allowing to happen now with carbon dioxide and methane is what cyanobacteria did with photosynthesized oxygen billions of years ago: gassing the planet to death.

Men have become the tools of their tools.

— Henry David Thoreau

In his delightful and illuminating book *The Botany of Desire*, Michael Pollan demonstrates how four local plants with pleasing and

---

* American bison, believed to be the most numerous large mammal ever to inhabit Earth, likely did not number above 40 or 50 million animals at their peak.
† Human flatulence alone generates about three-quarters of a billion liters of methane per day, or 30 million cubic feet—enough to meet the daily cooking and heating needs of 140,000 northern city dwellers.

useful qualities—potatoes, apples, tulips, and marijuana—harnessed human beings to propagate them globally, thereby changing the world. Fire is the ultimate expression of this "domestication" of desire. As Orkney Islanders and voyageurs did for the Hudson's Bay Company, and as Newfoundlanders have done for the bitumen industry, human beings are doing for fire: we are its willing servants, working for a pittance compared to the fabulous and world-changing "profits" (increased flammability and carbon dioxide) being derived from our labors. As far as Pollan's plants, or our fire, are concerned, humans are simply zombie hosts obediently disseminating their seeds, tubers, sparks, and gases around the globe. In the end, the geologic record will show that it is we who served fire, who enabled it to burn more broadly and brightly than it ever has before. Fire, thus far, has mastered us.

We are still a long way from Permian levels of carbon dioxide and temperature, but we are currently on pace to replicate the much more recent mid-Pliocene Warm Period, which ended 3 million years ago. With the seas and continents close to their current configuration, the mid-Pliocene offers a useful analog for our near future. At that time, our ancestors were still in Africa. Lucy (*Australopithecus afarensis*) was laying the groundwork for us in present-day Ethiopia, walking upright and experimenting with the crudest of stone tools. The Pliocene world was certainly habitable, but in a dramatically different way—not so much because of who lived in it, but because of the amount of atmospheric carbon dioxide. In Lucy's day, CO_2 levels were in the 400 ppm range, commensurate with ours right now, but average temperatures were 4–6°F warmer, the current prediction for the end of this century. With far less year-round ice, global sea levels were about eighty feet higher than they are today. Currently, almost half of the human population lives in coastal areas that were underwater when Lucy lived.

Just as the elders among us knew people who lived in the pre-Anthropocene climate, young people today are going to experience a version of this new Warm Period scenario, which has already begun. In this sense, the generations alive today represent a bridge between the lost world of a pre-industrialized atmosphere and a future defined ever more sharply by the rapid, increasingly violent discontinuities we are experiencing now. We may be a force of Nature, but we are not a mature one: like adolescents through all of time, petroleum burners want the power, but not the responsibility. In this way, we (the species) are not so different from a fire.

The immediate future—the next decade or so—is a kind of ultimate test: Are we, in our teeming, burning billions, capable of achieving some kind of equilibrium with the planet's carrying capacity, and with its ability to buffer methane and carbon dioxide? The long-lost creatures of the Late Permian had no choice, and no recourse, and the same is true for every species alive today—except us. The current moment is the greatest challenge humanity has faced since we (almost) mastered fire. This time, it is not fire we have to master, but ourselves. If we fail this test, there will be another one, and another after that, but each time the stakes will be higher and the price of failure steeper. On the bright side, life—in one form or another—has always won out against the unregulated, hyper-consuming impulses of fire and its most durable by-products, ash, methane, and carbon dioxide. That there will be life at the end of the Petrocene Age is a certainty, but whose, how much, and where is less clear.

⸺

In 1959, almost a decade before the Great Canadian Oil Sands plant was opened, Robert Dunlop, the president of the company that built it, was warned, explicitly, by one of the most influential scientists of his day, about the warming effects of carbon dioxide generated by fossil fuel. Two years before that, the oceanographer and climate scientist Roger Revelle had warned Representative Albert Thomas and members of his House subcommittee of an impending phase shift in the rain and climate patterns of the southwestern United States. Variations on these warnings have been repeated, in every medium and format, ever since. If the powers that be couldn't see it coming, it was not because they weren't warned. Lucretius might just roll his eyes and say, "What did you expect?"

Those early predictions, along with many others like them, have come to pass, and with devastating consequences that intensify by the year. In the summer of 2021, the U.S. Bureau of Reclamation, which manages water resources for 40 million people in the Colorado River Basin, issued its *first-ever* water shortage declaration. Recent photos of reservoirs across the Southwest and California show trees and lake bottoms that haven't been visible since they were submerged many decades ago. Farther north, on the Oregon-California border, the Bureau of Reclamation shut down the Klamath River Basin's extensive irrigation system for the first time in its 115-year history, impacting everything downstream from waterfowl and salmon habitat to agriculture and cattle ranches.*

"This isn't a 'drought,'" wrote the climate journalist Bob Berwyn in 2020, "because that implies recovery. This is aridification." Aridification precedes desertification.

This is what Revelle saw coming in the 1950s: a persistent new regime with no foreseeable endpoint. Based on tree ring analysis, it has been determined that the American West is currently in the most severe drought of the past 1,200 years. Rainfall cycles are naturally variable and difficult to predict, but CO_2, it has been said, is like steroids for the atmosphere: as fossil fuels have empowered us, so have our emissions turned a heavy hitter (the heat-retaining capacity of our atmosphere) into a record-breaking slugger. The implications for fire go without saying.

Evidence of a similar drying trend has been observed in Canada's boreal forest, also since the 1950s. In the meantime, scientists have determined that for every 1°C of warming (about 2°F) a 15 percent increase in precipitation is necessary to compensate for the increased evaporation. This is the exact opposite of what is happening in the boreal forest. Around Fort McMurray, average temperatures for the coolest months (October–April) have warmed by 3.4°C (about 6°F) over the past fifty years while precipitation has dropped by half. A similar phenomenon is being observed across Russia and Alaska, and it goes a long way toward explaining why fires in the circumboreal are burning earlier, faster, hotter, longer, and farther north: the boreal

---

* In December 2020, in response to worsening drought conditions, Wall Street began trading California water futures as a commodity, a first for a basic human right.

forest is no longer the same forest. This is not proprietary information; it is available to anyone interested in the boreal ecosystem, or in protecting the people who live and work there.

In the twenty-first century, not even tundra is immune. During the summer of 2007, Alaska saw what was then the largest tundra fire ever recorded. Ignited by lightning, the Anaktuvuk River fire burned a treeless area covering four hundred square miles, in spite of the permafrost that lay melting beneath it. With the scientist's characteristic understatement, Syndonia Bret-Harte, a coauthor of a *Nature* article on this landmark event, pointed out, "Fire has been largely absent from tundra for the past 11,000 or so years." Her colleague Michelle Mack dropped another bomb: "The amount of carbon released into the atmosphere from this fire is equivalent to the amount of carbon stored [annually] by the [pan-arctic] tundra biome. This was a boreal forest-sized fire."

A decade later, in 2017, Greenland experienced its first significant wildfire, a stunning development given that Greenland is treeless and frozen solid save for tenuous patches of tundra on the coastal margins of the ice cap. In 2016, Tasmania, on the opposite end of the planet, experienced its driest spring and hottest summer in recorded history, during which rain forests that have not burned in a thousand years caught fire. In 2015, fires in Indonesia, many of which were associated with land clearing for palm oil plantations, burned ten thousand square miles of forest. In 2012, wildfires burned more than eight thousand square miles of boreal forest in Siberia. In 2010, pan-Russian wildfires burned more than three thousand square miles of forest, impacting heavily populated areas around Moscow. According to the global reinsurer Munich RE, *56,000* deaths that summer were attributable to smoke and record-breaking temperatures. All of these fires may have ignited several years and thousands of miles apart, but the link between them is persistent heat and drought.

That, and peat.

Underneath the trees, grass, and shrubs that compose many forests from Alberta to Siberia, and from Indonesia to Tasmania, lies a layer of compressed, decayed plant material. Like tundra (and muskeg), peat functions like a sponge, and its natural state is to be wet, often underlaid by permafrost—in other words, unburnable. However, when dried sufficiently, peat burns very well indeed, and house-holds across northern Europe have heated with it for millennia. Prior to the twenty-first century, it took a lot of human intervention—digging, cutting, and drying—to make peat flammable, but now it is drying out and igniting in situ. Once lit, peat fires are extraordinarily difficult to put out. In Russia, following the worst fire season in the country's history (2021), smoldering peat was reported in the dead of winter at -76°F.

Like the Anaktuvuk River tundra fire, one of the principal rea-sons the Tasmanian, Canadian, Russian, and Indonesian fires burned as they did is because the peat beneath them has dried in place, trans-forming these historically soggy forest floors into vast beds of fire-ready biofuel. Trees are no longer necessary to sustain fire. Troy O'Connor, the owner of a commercial firefighting company in Red Deer, Alberta, who was on hand for the Fort McMurray Fire, described to me mus-keg bogs that have desiccated lately to depths of *eight feet*. Under these conditions, peat bogs (or tundra) can smolder indefinitely, like a coal seam fire. And, like coal dust, peat dust is explosive. The Indonesian peat fires of 2015 emitted nearly a billion tons of CO_2, comparable to the annual emissions of New York, Texas, and California combined. According to Mike Flannigan, research director of the Western Part-nership for Wildland Fire Science at University of Alberta, the boreal forests of Canada and Alaska contain thirty times as much peat as Indonesia.

With every degree of warming, there is a 12 percent increase in lightning[*] activity, a common cause of wildfires (and the only cause in the uninhabited Arctic and boreal regions). The drier and hotter peat bogs and forests become, the easier they are to ignite by light-ning and other means and, with milder winters, the earlier in the season they are able to do so. The drier the fuel and the hotter the air, the more explosive the fires, the more intensely they burn, the harder they are to extinguish, and the more likely they are to produce their own weather in the form of wind and pyrocumulus clouds, which can generate fire whirls, tornadoes, and more lightning, resulting in

[*] On a single June day in 2015, fifteen thousand lightning strikes were recorded in Alaska's interior.

yet more fires that will perpetuate themselves for as long as fuel and weather conditions allow. Over the past several decades, this feedback loop has been tightening, not only in the circumboreal forest but across virtually every major forest system from Norway to Chile. In southern British Columbia, in 2021, Troy O'Connor saw fire weather drought codes over 1,000 (the forest fuel equivalent of a fifty-inch rainfall)—a number that would have been simply inconceivable even a decade ago.

Truly sobering in consideration of our current circumstances is that, even 120 years ago, early climate scientists recognized and accounted for what was then referred to as "mutual reaction," now commonly known as positive feedback. Should the feedback loop of heating and drying continue to intensify as it has been, there is in our future a potentially winter-less scenario in which fire weather is the only weather, and "fire season" never ends. Australia and the American West are facing this reality already.

The influence of feedback is measurable with rain gauges, thermometers, and calendars, but its effects can be quantified in more oblique ways as well. In Kyoto, Japan, where cherry blossom season has been a national event celebrated for centuries, the date of peak blooming has grown steadily earlier over the past 150 years, tracking almost perfectly with Guy Callendar's and NASA's temperature graphs, and with Keeling's CO_2 curve. March 26, 2021, was the earliest date for peak blooming in 1,200 years of continuous record-keeping. Meanwhile, in the Arctic and subarctic, tree species have been migrating northward ever since the end of the last glaciation, a natural response to retreating ice and warming soils. In Alaska, this northward march has occurred at a rate of roughly half a mile per century. In the past few decades, however, the pace has quickened, with one result being that landscapes that have not known trees in more than 100,000 years, since before the last ice age, are now hosting young forests. These new forests, which may be only a decade old, are already burning down as landscapes that have never known fire become increasingly susceptible to combustion due to radical increases in temperature, evaporation, and lightning strikes. This heat-driven acceleration has been observed in other contexts as well. During the summer of 2019, a record-breaker in Alaska, where wildfires and smoke filled the state, hundreds of thousands of salmon died of heatstroke before they had an opportunity to spawn. At the same latitude, Greenland's ice cap is now melting at a rate ominously described as "nonlinear"—that is, out of phase with any known precedent, pattern, or cycle. Over the past three decades, the average temperature there has increased by 5°F, with summer temperatures soaring 40°F above normal. During this same period, the rate of melting has increased by a more than a third. "What seems clear now," wrote Jon Gertner, historian and author of *The Ice at the End of the World,* "is that Greenland is no longer changing in geological time. It is changing in human time."

This is a first.*

Starting around 1980, when atmospheric CO_2 hit 340 ppm for the first time in a million years, annual global temperatures began rising more steadily, with smaller deviations. Since 2000, every year has been trending warmer than the last, and, since 2010, these annual increases have been steadily greater. By almost any measure, anyone born after 1990 is finding themselves in a new geological era, navigating a world fundamentally different from the one Baby Boomers and Gen Xers inherited. The chances of anyone alive today experiencing a year as relatively cool as 1996 are effectively nil.

Cristi Proistosescu, a professor of climate dynamics at the University of Illinois, suggested a more forward-looking way to conceptualize this warming trend. After posting a graph of rising global temperatures, he tweeted, "Just wanna make sure everyone understands what we're looking at here: Don't think of it as the warmest month of August in the last century. Think of it as one of the coolest months of August in the *next* century."

Two decades into the twenty-first century, the Anthropocene Epoch is proving itself to be one of radical, hitherto unimaginable, planetary shifts. Pick almost any system or species—glaciers, oceans, birds, insects, fish, even seasons—and there is clear evidence of rapid

---

* In June 2021, the riverside hamlet of Lytton, British Columbia, broke the heat record for Canada three days in a row, topping out at 121°F. On the fourth day, a wind-driven wildfire burned the town to the ground in half an hour. Two people died. "I'm sixty," said Lytton's mayor, Jan Polderman, "and I thought climate change was a problem for the next generation. Now I'm mayor of a town that no longer exists."

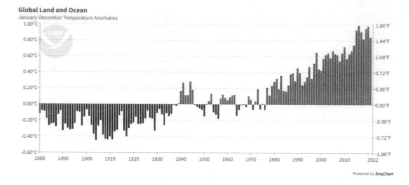

Global Land and Ocean
January-December Temperature Anomalies

change and intense stress. The heat-death of coral reefs, shellfish, and spawning salmon is calamitous and devastating, but for most people, it is remote—one more abstract tragedy that is painful to dwell on, and hard to identify with in a visceral way. But something analogous is happening in the world's forests, where many of us live. Like coral reefs, forests are "metropolitan" organisms: complex, sprawling, slow-growing armatures of life that support thousands of other species from virtually every genus, including ours. For a couple of generations now, we have been told that, in addition to producing much of the oxygen we breathe, forests absorb carbon dioxide and store it, thus helping to modulate Earth's climate. This is true, but not as true as it used to be. Roughly half the weight of a given tree is composed of stored carbon, which returns to the atmosphere quickly when that tree is burned, and more slowly if the tree rots away in the forest, or in the form of discarded paper, cardboard, or furniture in a landfill.

Since roughly 2000, an inversion has begun: the world's great terrestrial carbon sinks—the Amazon rain forest and the circumboreal forest, along with many other less famous forest systems around the world—have become net carbon *emitters*. In other words, what used to be a reliable source of carbon storage is now generating more CO_2 than it is sequestering. This grave reversal is one of the most pernicious developments of the Petrocene Age. As forests heat up and die—from disease, beetle infestation, fire, logging, land clearing, and drought, they skew the CO_2 balance even further. It is not that living, growing trees don't continue to absorb carbon, it is that they are no longer keeping pace with the emissions of their sick, dead, and burning neighbors.

Something similar is happening above the tree line in the tundra, and other regions where permafrost is melting. As these environments grow warmer (for the first time since the last glaciation), the frozen, inert organic matter beneath them thaws and then decays, releasing still more CO_2. Tundra may be the least charismatic of terrestrial ecosystems, but it possesses hidden depths: there is twice as much carbon dioxide bound up in frozen Arctic soils as is currently in our atmosphere. For the first time since the Ice Age, this process of thaw-and-release is currently under way across the Northern Hemisphere. Vast quantities of methane are trapped in these ice formations as well. In 2020, we saw the greatest annual increase in methane release since systematic measuring began in the 1980s.[*]

It is impossible to overstate the gravity of these developments or their implications for the future of Earth's climate, and for the beings who depend on it. For its entire brief history, human civilization has been able to count on the oceans, forests, tundra, and grasslands to offset emissions, whether from volcanic eruptions or our own fires. Then, around 1900, we crossed a threshold. While we have been burning through our vast and ancient fossil fuel inheritance with astonishing speed, we have also been borrowing against the planet's ability to absorb and process its by-products—carbon dioxide and methane, among many other gases and toxins. Despite decades of repeated, informed warnings, large, for-profit companies, together with their host (and sometimes captive) governments, have persisted in promoting and financing fossil fuels while downplaying the consequences. In so doing, the energy business, in conjunction with elected officials and the wealthiest 10 percent of the population, has mort-

---

[*] Anthropogenic methane emissions are underreported by 70 percent in the U.S. (and by as much as 100 percent in Canada—per Pembina Institute). According to the IEA, "Methane is responsible for around 30% of the rise in global temperatures . . . The energy sector accounts for around 40% of methane emissions from human activity . . . The methane leaked in 2021 could have provided 180 billion cubic meters for the market (~ all the gas used in Europe's power sector)."

gaged the atmosphere.* Now, the "bank"—our climate—is collecting on those overdue, overleveraged loans. As Wallace "Wally" Broecker, the pioneering climate scientist and early advocate for climate action, said back in 1998, "The climate system is an angry beast, and we are poking it with sticks."

One of the things so dismaying to people who study and advocate on behalf of our atmosphere is that greenhouse gases aren't like dust, or ash, or diesel exhaust. They don't go away or settle out after a few days or months; it's not like banning cars in Beijing for a couple of weeks and then hosting the Olympics. Once you hit, say, 400 ppm (which we did in 2015), that becomes the new baseline for lifetimes to come. In 2021, we reached 420 ppm, a 50 percent increase over pre-industrial CO_2 levels. If you increase *anything* important by 50 percent—house prices, blood pressure, rat population, rainfall—it will be very noticeable, and often bad.

Currently, we are on pace to re-create—in a couple of centuries—climatic conditions that previously took millions of years to bring about. Such a cataclysmic rate of change will outrun most species' ability to adapt, and a new equilibrium will be a long time coming. Whatever that new world looks like, it will be a lonelier place, inhabited by a relict fraction of today's biodiversity. The true impact of the Petrocene Age represents a legitimate Lucretius Problem: no one has been here before, or seen its consequences. No one can truly imagine what this means for life on Earth.

---

* As the Penn State climate scientist Michael Mann put it, "The vast proportion of historic greenhouse gases have been emitted as by-products of the choices and activities, not of the masses of ordinary people, but rather of the wealthy minority of the world's population . . . It's a very specific part of humanity that has created these problems."

23

And where two raging fires meet together,
they do consume the thing that feeds their fury.

—William Shakespeare, *The Taming of the Shrew*, Act II, Scene 1

Ever since Syncrude's bitumen upgrading plant came on line in 1978, it had never shut down—not once. Huge, complicated, expensive, and temperamental, it made more sense, and more money, to keep the sprawling complex running. Suncor's operations were even bigger, and Cenovus, Canadian Natural, Husky/BP, and Imperial/Esso were also huge, but the threat posed by the Fort McMurray Fire was bigger still. Starting on May 5, as the fire continued to trend north and east, mines, upgraders, and SAG-D operations began scaling back or shutting down altogether. Rolling smoke from the fire was choking workers and reducing visibility to mere yards, making conditions even more hazardous. A few miles north of the city, the fire was encroaching on Suncor's tailings ponds along with the roads and infrastructure that serviced them. Because of the fire's rapid growth and the heavy smoke, Syncrude shut down its Aurora mine, and its Mildred Lake plant. Suncor shut down its operations on Tar Island, and other companies followed.

In Canada, May 7 is National Wildfire Community Preparedness Day, and, in 2016, it fell on a Saturday. By then, a week into the fire, China's Nexen Long Lake facility, which the REOC had abandoned three days earlier, was completely surrounded by burned or burning forest. There, south of the city, the fire was also threatening U.S.-bound pipelines. This, combined with the site closures, cut the industry's daily production of dilbit, synthetic crude, and diesel by more than a million barrels a day—nearly half of their normal output. In monetary terms, this translated to about $60 million in losses every twenty-four hours. These vast plants, some of which cover square miles of land, are always occupied, day or night, winter or summer, and they are always busy. "It's super loud," explained Aron Harris, a member of Suncor's fire department. "Banging noises, flame, steam—you name it. It's just obnoxious."

And then, suddenly, it wasn't. By May 7, all the plants within a half hour of Fort McMurray were empty and shrouded in blinding smoke. With their thousands of workers evacuated, hundreds of trucks and bulldozers were idled while cranes and power shovels stood motionless on the horizon. The vessels this mechanical army served—insatiable crushers and rocket-sized boilers, hydrotreater reactors and vacuum towers—were all shut down.

"It was *off*," said Harris. "It was completely off."

Harris was part of a crew of twelve left in charge of Suncor's enormous Tar Island base plant, a site so big it has its own cloverleaf highway interchange, and its own bridge across the Athabasca River. Its tank farm alone is a half mile square. Harris spent his shifts with three of his mates in an ARFF truck, patrolling the sprawling facility, looking for spot fires in the smoke. "It was eerie," he told me. "The flare stacks were turned off. You're so used to loud noises down there, I said to Nick, 'You hear that?' He's like, 'Exactly. It's so quiet.'" Harris and his comrades were experiencing something most people only see in dystopian films: the world after humans, when the engines of civilization have stopped running. Like their municipal counterparts downtown, and Paul Ayearst with his son up in Beacon Hill, Harris and his crew were the only ones left. Most striking, perhaps, was how quickly the transition began. "You'd see deer down there," Harris said. "You could see coyotes."

It is almost impossible to imagine a scenario in which wild animals would willingly roam a bitumen plant, a truly toxic place reeking of men and petrochemistry, ablaze with fire and floodlights, and thunderously loud. Maybe these creatures sensed the void and were curious; maybe they were disoriented by the smoke; more likely, they were refugees, too, and Tar Island was one of the few places left that wasn't actively on fire. "*Weird* is the only way I can describe it," said Harris. "But I guess at this point, what's weird? Nothing was out of the ordinary anymore. Something caught fire? Well, of course, that's what it's supposed to do now."

On Sunday, May 8, strong winds from the west caused a significant eastward run of the wildfire and created a long and dangerous flank directly south of the oil sands operations. By midday on the 9th, when journalists and the premier were allowed their first limited access under escort, the wildfire had ballooned yet again to nearly half a million acres. The Fort McMurray Fire was barely a week old and it had already entered the record books: 90,000 people evacuated; 2,500 structures destroyed and another 500 damaged; nearly a thousand square miles of forest burned; an internationally significant mining and pipeline operation severely curtailed; overland transportation crippled; hundreds of firefighters and dozens of aircraft engaged with no end in sight. It was an unprecedented scenario, in both the annals of modern urban fire and peacetime resource extraction.

On May 9, Chief Darby Allen addressed the public on TV from Fire Hall 5. Standing behind him was Rachel Notley, the premier of Alberta. "I've never seen anything like this," Allen said. "I spoke to my colleagues from Forestry, and many of the fire conditions, and the way the fire behaved—no one's ever seen anything like this. This is rewriting the book—the way this thing happened, the way it traveled, the way it behaved. So, they're rewriting their formulas on how fires behave based on this fire."

Strictly speaking, this was not true. The formulas did not need rewriting—not if you were familiar with the Chisholm or Slave Lake fires; not if you had seen NOAA's Seasonal Fire Assessment, or Alberta Forestry's fire weather index, and not if you had studied how recent wildfires behaved under similarly explosive conditions in places like California and southern Australia. Twenty-first-century fire weather conditions have the same effect on landscapes all over the world. On that first summerlike weekend in May, an accurate and timely forecast had been made: a high pressure system in drought conditions accompanied by record-setting temperatures and rock-bottom humidity with a rising wind veering toward a thickly settled area in convoluted terrain. This combination equates to firestorm potential anywhere from Melbourne to Fairbanks. Whether it's eucalyptus in Australia, chaparral in California, or black spruce in the boreal, they play the same role in fueling wildfires, and the wind plays the same role in casting embers. Like us, fire is an omnivorous and adaptable consumer.

What was distinctive in 2016 was not even so much the scale (the Richardson Fire of 2011 was huge, and caused a half-billion dollars in damage to the bitumen industry), it was how people perceived it and responded to it. In other words, it was the Lucretius Problem: *"I've never seen anything like this."* It resembles the hubris that also seems to go hand in hand with large capital investments and petroleum, and it takes many forms. In 2015, ConocoPhillips CEO Ryan Lance told Bloomberg that it would be another fifty years before electric cars had a measurable impact on oil demand; five years later, few in the oil industry would make that claim. (As of 2023, Lance was still CEO.)

On Wednesday, May 11, by which time the fire had been burning in the city for *eight days,* Darby Allen went on live TV again. "We think we got this thing beat in McMurray," he said. The days continued hot and dry, the fire burning unabated in all directions around the city as it ebbed and flowed with the wind and sun. There was no rain in the forecast. Smoke from the fire was now visible on the Atlantic coast, more than 2,500 miles away. Reinforcements had been mobilizing from all over Canada, and by Monday, May 16, more than a thousand firefighters and firefighting personnel, along with 130 pieces of heavy equipment, scores of helicopters, and a dozen air tankers, were assigned to Fort McMurray, with more on the way. It was hard to imagine how fire weather conditions could get any worse than they were already, but they did.

On the 16th, the fire weather index hit an astonishing 42, handily surpassing the record-breaking conditions of May 3. In town, as efforts were being made to restore gas and electricity service, more houses ignited, for a variety of reasons. A suspected gas leak in a home in Dickinsfield caused an entire house to detonate with such force that it left a crater, drove two-by-fours through the walls of half a dozen neighboring homes, and blew out the windows in many more. Meanwhile, in the forest, the fire raged on, pushing ever closer to the mines, plants, and man camps. South of town, all available resources were rallied to defend Enbridge's Cheecham Terminal, a 2-million-barrel tank farm used to feed southbound pipelines, now in grave danger of being overrun and exploding.

The following day, May 17, Aron Harris was on patrol at Suncor when he saw a burst of flame on the horizon a few miles to the west of Tar Island. It was the Blacksands Executive Lodge; all 665 units of the deluxe new camp burned to the ground in less than thirty minutes. No one even tried to stop it. Several nearby camps with housing for more than two thousand people had fire burning right up to the fences. They were saved only by a fleet of ARFFs, pumpers, and water trucks forming a curtain of water around the burning perimeter. That same day, eight thousand occupants of these and other nearby camps had to be evacuated (again), not simply because of fire risk, but because of the smoke. Under normal conditions, the province of Alberta uses a 1–10 scale to rate air quality, 10 being the worst. That day, the air quality around Fort McMurray was 38. Without a clock, it was hard to tell what time of day or night it was. By then, smoke from this and other fires had cast a pall across the entire continent— south to the Gulf Coast of Texas, east as far as the Bahamas, and northward all the way to Labrador.

Blanketed in acrid smoke and stiflingly hot, the city and the surrounding forest went into a kind of crepuscular dormancy. Fire crews continued to search for hot spots and fight fire, a task they would be performing well into the fall. Police patrolled for looters, but mostly they found solitude and desolation. With the exception of the rec center, which was now being used as a staging area for firefighting operations, the electricity and gas were still off across much of the city, and its residents were still gone. With the sole exception of New Orleans after Hurricane Katrina, no modern North American city had been disinhabited for so long.

While deer and coyotes made tentative forays into the silenced bitumen plants, other creatures were exploring these new voids as well. In the city's twenty-five thousand or so surviving homes, apartments, restaurants, and grocery stores, a regime change was under way. It began, most often, in the kitchen. After a week or two in the unseasonal heat, perishables began to ripen and then to rot. Alberta is beef country, but hunting and fishing are popular here as well, and the abundance of meat, combined with high wages and a big-box mentality, meant that, in addition to refrigerators, meat freezers were common items in garages and basements. As the temperature rose, and all that meat decomposed, gases were generated, pressure built, and seals failed. By mid-May, many of the city's fridges and freezers stood in pools of clotting blood. Inside, the contents took on lives of their own. Dairy products that weren't cheese already were well on their way, and any leftover meals from May 3 were unrecognizable. Even through the heavy smoke, the rank and wayward odor of putrefying flesh was compelling, and flies caught wind of it. With the flies came maggots, which begat more flies. Warm and contained, with unlimited food and nothing to disturb them, breeding conditions were ideal. Outside, patrolling police, firefighters, and gas and electricity technicians saw nothing out of the ordinary as they made their rounds, but inside, any building with food in it was being colonized and transformed.

Insurance adjusters were some of the first to enter these putrid, teeming habitats, and one compared some of the sights he encountered to "CSI murder scenes." By then, many generations of flies had hatched, multiplied, and died; the growth was exponential. Another local adjuster opened a utility door in a client's house only to find the interior seething with mice. In the malls around town, the supermarkets had transformed into Olympic-sized petri dishes; floors and shelves were carpeted with dead flies while the cavernous spaces above buzzed and swirled with the living. Left to their own devices, the lobsters in the live tank at Save-On-Foods had turned on each other, but the mice never had it so good. Where there weren't vermin, there was rampant mold in a rainbow of colors. One adjuster marveled at the bread aisle, where every bag had blossomed into a psychedelic terrarium—save for one brand. After three weeks, the Wonder Bread looked as fresh as it had on May 3. Meanwhile, even in the tidiest homes, mold—from decaying food, water leaks, and neglect—

proliferated along with its attendant odors, which were further intensified by the penetrating stink of burned trees, cars, and houses that permeated everything. All told, about twenty thousand refrigerators and freezers were declared biohazards and had to be thrown away. Strapped shut and wheeled to the curb, they still stank, and this drew bears in from the surrounding woods. Nature would not be denied.

And neither would Wayne McGrath. Following his evacuation from Abasand on May 3, McGrath, the Suncor welder who fought the fire and lost, didn't go far. Unable to abandon his Harley Road Glide, he shuttled from one small community to the next in his pickup—staying with friends in Gregoire Lake, camping in Lac Labiche and then Wandering River. Like this, he hovered for a week, until he got word from a friend that his strategy had worked: because he had moved them out into the open at the last minute, both the Harley and the Cutlass had survived the fire. On May 10, he made his move: "I snuck back into town with a whole fleet of bulldozers," he told me. "There was red trucks—red mechanics' trucks—so, I got in between them, got waved through."

When he arrived in the city, McGrath was shocked by what he saw. "It was like Vietnam in there—fucking helicopters going everywhere. The fire was still going, right? Found my Harley downtown. Friend of mine put it on a float truck [along with his Cutlass sedan]. That was a good favor. There were cops fuckin' everywhere. Two cops came over and I said, 'Yeah, these are mine and what I'm going to do is take them.' Cop said, 'You can't leave with them.' 'What do you mean? They're mine.' 'Yeah, well, this is called a disaster area now. It's the property of the city.'

"I got more or less told I had to go because I didn't have an armband. What I did is, I made one out of tape, but the new ones were painted orange so I had to hustle back to Wandering River, which is two and a half hours away, and find an orange marker. There's a guy with me—he stayed down low in the seat and I just went like that with the armband and I got in again. I got to my Cutlass, said, 'If this car starts, bud, I'm gone.' He said, 'Well, I guess I might as well get out, too.' Cutlass started; buddy took the truck. The Cutlass ran out of gas [about nineteen miles] outside town. I had to leave it, flag my buddy down who was coming behind me. Back to Wandering River.

Had to go find gas cans, gas up, come back up, get the car. Had to go back for the bike. My Harley means a lot to me, right? Went back for the bike."

It is unlikely that any other unauthorized person came and went from Fort McMurray as many times as McGrath did (only three of numerous clandestine trips are described here). "I'm pretty cagey," he said. "Even my previous ex is like, 'I knew you'd go back for your bike. I knew you weren't leaving without it.'

"But the keys burned, right? The fobs. So, I called in a lot of favors over a few days. Ended up bumming a trailer in Wandering River, but I had nobody to come up and give me a hand. So, I met this guy in Wandering River who's hanging out at the gas station— French guy, alcoholic. I bought beer, didn't give a rat's ass. Got talking to him, told him what I was up to and said, 'You want to give me a hand?' Turns out he used to rob banks in Montreal when he was a kid. Gave me the whole story. He was eighteen, used to rob banks.

"Got up to town, pulled up right to my motorcycle with the trailer. Cops everywhere. I said to him, 'Don't look at anybody, just put the bike on the trailer, strap it down, and that's all there is to it.' So we're putting it on and strapping it down. I was just getting in the truck and this cop comes walking over. I had beer in the truck everywhere. Cop says, 'What year's your bike?' I started crying."

We are in a crowded bar and McGrath is choking up at the memory of this moment that was, for him, a fulcrum leveraging everything that followed. "Cop said, 'I knew it was your bike. I knew it was your bike a couple days ago when you were here, but I was working with another officer and they were here from all over Canada.' "

Mercy comes in many forms, and McGrath knew then that he was safe. "He couldn't say, 'Yeah, take it,' because he'd been with this other cop. I said to him, 'Here, I'll show you the registration.' He said, 'No, bud, I don't need to see it. I know it's your bike.' I *cried*, man. He was a biker cop—rolled up his sleeve: *full* of tattoos. He was from just outside Calgary—wherever they had that big flood. He lost everything, too."*

---

* Prior to the Fort McMurray Fire, the Calgary Flood of 2013, which displaced more than 100,000 people, was the costliest disaster in Canadian history.

After replacing his burned key fobs, McGrath rode his Harley to Newfoundland, a road and ferry journey of four thousand miles.

By the time the fire was three weeks old, it had burned over a million acres of forest, more than two thousand square miles. Still out of control and growing by the day, the fire had spread eastward into the neighboring province of Saskatchewan. With active fire now largely out of the city proper, and the smoke hazard somewhat less severe, technicians set about restoring gas and electricity to Fort McMurray's unburned neighborhoods. In an effort to keep down the prodigious amounts of ash and toxic dust, a pale gray shroud of adhesive "tackifier" was sprayed over the ruins.

Carol Christian, a writer and mother of a grown son, came up to Fort McMurray in 2007, intending to stay for six months. She took to the place and built a life that included writing for the local paper, *Fort McMurray Today*, and working the desk for a local politician. On May 3, Christian, like so many of her neighbors, only knew to evacuate when a friend called her. She rushed out of her Abasand townhouse with the barest of necessities, confident she would be returning in a day or two. As days stretched into weeks, Christian couch-surfed with family and friends, unsure when she would be allowed home again. The city was closed for a month. When residents were finally allowed back in early June, it was to a ghost town. Formerly vibrant, landscaped neighborhoods were now unrecognizable to the people who had once called them home. "The first time we were allowed up to Abasand," Christian recalled, "you still got that little bit of hope that maybe, just maybe. But no." She had seen the footage on the news, but she still wasn't prepared for the vision that greeted her. "That's a sight to behold," she said, "when you see your bathtub is in your furnace."

Christian's townhouse complex had been five stories tall. When she returned in June, there was nothing left but ash, bathtubs, and the shells of heavy appliances heaped together in the underground parking garage, which was now open to sky. "The people who were back at that time—myself included—people were walking around like zombies," said Christian. "They were shell-shocked."

The aftermath of a major wildfire has its own palette, one that reflects fire's omnivorous appetite by reducing urban, rural, and wild to a unified color scheme of total oxidation that ranges from bone white through taupe to charcoal gray and the glossiest raven black, the rest of the spectrum burned away. After one of these big fires, the only actual color left on the landscape is in the terragraphs of orange fire retardant, some of them hundreds of yards long, draped across the ridges and valleys like the Nazca Lines, or a work by Christo.

In town, at ground level, metal shower stalls are among the few things to survive. They stand alone in the ruins, a morbid joke now, while washer-dryer sets stare back like blank eyes in a roofless skull. The mottled frames of stove, air conditioner, freezer, and fridge are warped out of shape, or collapsed. In the driveways, and on the street, the scorched and hollow shells of cars look less like vehicles than like the exuviae of gigantic insects. Ash covers everything—the memories, the histories, smells, recipes, comforts, reduced now to the barest elements—carbon, stone, steel—all cloaked in smoke and suffused with the acrid reek of burning. Outside, for there is no longer an "inside," families stand on the sidewalk wondering where their houses went.

Those whose homes have burned are struck by how much is gone, and also by what remains: a carpet preserved by leaking water from a ruptured water pipe; books, ghost white with every page intact, until you touch them and they collapse in a cloud of ash. Home is our memory palace, and there is an existential cruelty in the razing of it. To burn them down by the hundreds and thousands, as wildfires are doing now across the western U.S. and Canada, is a brutal affront to the order we live by, and to the communities and habitats that give our lives meaning. Their loss shocks the heart like a sudden death. Left behind are juxtapositions so bizarre and disorienting that to describe them sounds like the mutterings of an insane person: garbage can puddle; melted guns on a platter; cars bleeding aluminum; piles of tire wire. Is this really where I lived, where I raised my children? Where did their beds go? Their bedrooms? The photos, the *evidence*? In their place, a void, the shadow of a burned tree where the kitchen table used to be, pools of once-familiar things gone molten, settled now into new forms, rigid and unrecognizable.

Fort McMurray selects for stoicism; it has to, with that industry and those winters. Many who lost homes shed their tears, sucked it up, and said, "It's just stuff." But it's more complicated. People were glad to be alive, but that didn't lessen the material bereavement of losing everything they owned, everything they had labored in those mines and plants for, year in and year out. As part of the reentry and recovery process, church groups like Samaritan's Purse offered to help sort through the ruins on the off chance something might have survived. Wearing protective gear, volunteers climbed into the basements of burned homes with shovels, rakes, and sieves. While the owners looked on, suggesting where to look, they sifted through the ashes. This is a level of compassion and care that few disaster survivors get, and it says a lot about the community and faith of Fort McMurray. Despite these generous efforts, however, very little was recovered; the fire's heat was simply too intense.

"Ashification" is another one of those words, like "spalling," that doesn't enter the conversation below 1,000°F or so, but that, in a word, is what happened. Many residents were perplexed by how a large, sturdy structure that had once housed them, their families, their cars, and all their possessions could be reduced to a pile of ash and scrap metal that would fit in the back of a pickup. Even forensic fire investigators were puzzled by the absence of normally fireproof objects like toilets and sinks. They weren't there, because they had vaporized.

Carol Christian had the sifting done at her place in Abasand. "I wanted to find *something*," she said. "I found a colander and some barbecue utensils. I thought, 'Well, I'll clean that up and I'll make a mobile'—kind of a symbol that you can still go forward. But when you stand there and think, *'This used to be my house . . .'* "

During those first days back, Christian went up to Abasand several times, in effect to confirm that this unthinkable calamity had actually occurred. "I was surprised I didn't see more people up there," she said, "but why would they come? There was *nothing*." And so, as one often does when visiting the grave of a loved one, she found herself alone. "It's like you're the last person alive on Earth," she told me. "I was waiting for bears—kept looking behind me just to make sure. It was weird. It was just—so—*weird*. When you knew what was there, what had been there, and that's what you kept picturing, but it's not."

Christian was describing a phantom feeling similar to what some amputees experience after losing a limb. "I'm thinking how many people here are in the same boat," she said. "They've emigrated from somewhere, brought their life with them, and now it's gone." Christian spoke with some authority on this subject: her family is from the Isle of Man, and this was the second home she'd lost to fire. "Your home," she wanted me to understand, "isn't just a building; it's your identity, a reflection of who you are when you gather with your family. It's your art gallery, museum, library—I lost a small fortune in books. Christmas is going be so hard because people are going to reach for things."

Not everyone was as eloquent or measured. "I hate that fucking fire," said a Newfoundlander named Pauline Vey. "It took everything."

Those who owned no property and had grown up far from Canada told some very different stories. While standing in line at the Tim Hortons downtown, a Sri Lankan security guard described to me his experience of the fire, and it didn't take long: "I grew up in a war zone," he said flatly, "so it makes no difference to me."

Carol Christian didn't see him, but Wayne McGrath was up in Abasand, too. McGrath rode his Harley all the way back from Newfoundland, returning to Fort McMurray in early June. Once there, he went through a process similar to Christian's—staring at the ruins, trying to assimilate an absence that was simply too big to take in. "I won't deny—for three weeks I was drunk," he told me. "A lot of people were. I hung out at my house for maybe ten days, drinking and crying and sitting and watching. Down on my hands and knees, I found a couple of things my dad gave me. Found a coin he got for serving with the mining company back in Labrador. Nobody seems to understand how much you lose. I have buddies that lost everything. They're sometimes the only ones I can talk to. Since I lost the house, though, I just slowed down and I'm sad. I haven't slept right since."

McGrath had nowhere to go, so he went back to work. "Went back to regular shift and I told them, 'Well, I don't have a house.' 'Well, you can stay in the camp, Wayne—for $125 a day.'" McGrath had eighteen years in with the company, and he took umbrage. "That's how much Suncor's changed," he said. "If this was five years ago, I could stay in that camp 'til my house was rebuilt, but now—new owners. I could lose my job for saying that."

McGrath had lost more than his house in the fire; something else fundamental was missing, too—something between faith and trust. "My foreman come to me and said, 'Wayne, you don't want to be here, do you?' Turning point for me was I had to go do this course one day, a driving course. I completely failed the exam. I didn't give two fucks, just more or less got up and walked out of the room. Foreman said, 'Wayne, what are you doing?' 'Just want to get out of here.' Went to the doctor and told him what happened and he seen I was a mess. He gave me a thing. He said, 'Take six weeks off, come back and see me.' So I did—jumped on my bike, rode across Canada again."

By late fall, McGrath had settled in a Suncor-owned rental in town, and things appeared to stabilize somewhat. A sympathetic agent was handling his insurance claim, and he was going to try to buy the lot next door so he could expand his workshop. For a while, this kept him going, but the claim and the rebuild progressed slowly. In the meantime, McGrath still found solace and escape in fast machines; he always pushed them hard, and, always, skill and luck had protected him. On December 15, 2018, he had plans to meet with his insurance agent, Sue McOrmond, for a progress report. "He texted me and said that he wanted to go sledding [snowmobiling]," McOrmond told me. "I just told him, 'No rivers,' as they didn't seem solid yet. The last text I got was a sad face saying, 'No rivers.' I never heard anything after that."

McGrath was reported missing two days later, and it took a helicopter to find him. He was spotted miles from town on a frozen river, which in winter are irresistible racetracks for many snowmobilers. He hadn't gone through the ice; he had crashed. Because there was nowhere to land the helicopter, McGrath's body couldn't be retrieved until the following day. By then, his body had been out in the elements for three nights and had frozen as hard as the ice.

Hope was hard to come by in June 2016, but it was there, in the kindness and compassion many people were shown, and also in the lawns and gardens. Deep beneath the ash and tackifier, the growing impulse prevailed. "The grass was *neon*," said one early returner. It was also knee-high. Tulips burst, phoenix-like, from the ashes to flaunt their vivid reds and yellows, as if to say, "We're still here. Life goes on."

That fall, Carol Christian, like just about everyone else, was still waiting for her insurance money to come through. Her townhouse was covered, but she was told the rebuild would take two years. In the meantime, she was sharing a two-bedroom rental with her newlywed son. It wasn't ideal. "I had eight of those reusable shopping bags," she told me, "and I was thinking, *This is the sum of my existence right now.* It's so sad, but I still had more than some people. I have one friend— all she got was her husband's ashes."

In different ways, everyone in Fort McMurray was having to integrate this traumatic assault on their lives, to psychologically metabolize the damage it inflicted, to somehow reckon all those voids where trees, buildings, and other landmarks had once stood, and where memories and meaning had once been made. Many of those voids, it turned out, were human beings. Of the roughly ninety thousand people who left on May 3, twenty thousand—nearly a quarter of the working population—did not return. The only other North American city to see this level of disaster-related attrition is New Orleans after Katrina.

One of the cruelest aftershocks for homeowners was the realization that just because their house was gone didn't mean the mortgage was. Likewise, many business owners who rented from remote landlords received no forgiveness on their rent for the time the city was closed down. Most people were insured, but the sheer magnitude of the event overwhelmed the system. Craig MacKay works for a local branch of ClaimsPro, and he summed it up this way: "On May second we had no claims; on May third we had fifty thousand claims." It wasn't just houses; it was cars, trucks, tools, firearms, and recreational vehicles. There was smoke damage, water damage, and residual heat damage. A lot of things melted: vinyl siding, rubber seals on doors and windows, outdoor wiring, toys, fences, garden furniture, and even the joints on water pipes. Whether it was due to fire, water, smoke, vermin, or mold, thousands of surviving houses were compromised in a myriad of ways, including shrapnel damage from the many things that exploded. Even months later, roofs and water pipes that had appeared unscathed began to leak, and electrical wiring shorted out for no discernible reason.

Because of the extraordinary volume of claims, insurance adjusters were flown into Fort McMurray from all over Canada and the U.S. It's easy to lose perspective on a place when you've lived there a long time, and these outsiders, most of them first-time visitors, arrived with fresh eyes. Vonda Pikes flew in from Texas in mid-May; Fort McMurray had not been on her radar before she got the call, but she ended up spending most of the summer there, processing claims out of a portable building. Pikes was surprised by how remote the place was, and also by how people treated her. She was fairly sure that she was the first African American many of her new colleagues had ever met. "I was born in Louisiana, in the woods," Pikes told me, "and I know some things about prejudice. All I worked with in Fort McMurray were white people. They didn't seem prejudiced; they were curious about my culture, and they asked about my experiences with racism and police brutality—how did we deal with the prejudice, and things like that."

It made an impression on Pikes because these were total strangers from another country who were under terrific stress from the thousands of claims they had to process, and yet they were curious about her life and concerned about her well-being. Pikes also met a number of people from the African diaspora, most of them in service and transportation jobs, and they worried about her, too. "Black people I met up there asked me, 'How do you live in Texas? I'd never want to live in the States.'"

Some of the cultural differences were funny, and they revealed the extraordinary breadth of the British colonial legacy—from the Arctic coast of Canada to the Gulf Coast of the United States. Pikes and her clients were all native English speakers, but "the people from Newfoundland didn't know what I was saying, and I didn't know what they were saying."

Even from twenty-five miles away, Pikes could smell the bitumen from the plants, and she encountered a lot of people with asthma; a number of them were policyholders who told her they needed new furniture because of smoke damage. "Smoke damage" is hard to quantify and would prove an ongoing source of conflict between adjusters and policyholders. "We wrote some big checks," she said, "and nothing was wrong with those things." Pikes has seen a lot in her career, but Fort McMurray managed to surprise her repeatedly. "A lot of people were in debt with an upside-down mortgage," she told me. "A lot of people had lost their jobs. Many of them were glad their house was burned down." But it was the real estate prices that really shocked her. "Those houses are not worth that much money," she said, unable to contain her derision. "You got a million-dollar home with vinyl siding on it. Those houses were built *cheap*—a $600,000 house in Fort McMurray is built like a $100,000 house in Texas. I'd *never* pay that much money."

But when it came to the people, it was different story. "I felt so good in Canada," she told me. "I would move to Canada. Everyone I met—they were so polite to me. I thought maybe they felt sorry for me or something."

Most owners whose homes had burned were presented with a choice: take a check that would be significantly less than replacement cost, or rebuild on the same site with full coverage. As high as property values were, they were still down from their peak in 2014 before the price of oil dropped, and the fire drove them down further. Those who still had jobs and means opted to replace their often out-of-date starter homes with bigger, more luxurious houses. Wine cabinets were popular that year, and so were heated floors, gas fireplaces, and jet tubs. Due to the remote location, and the fact that hundreds of houses were starting at once, prices for everything—from two-by-fours to concrete—skyrocketed. Houses that might have cost $200 a square foot to build before the fire jumped to $350, and then kept rising. The rebuild was a multiyear, multibillion-dollar enterprise that would be Fort McMurray's newest boom, and possibly its last. Builders and tradesmen flocked to Fort McMurray for what promised to be a bonanza.

But before anything could be rebuilt, the wreckage had to be cleared away, and in Fort McMurray, this process was an unusually thorough one—more in keeping with recovery from a war than from a fire. Because of the intense heat and the spalling, even concrete foundations were compromised and had to be removed. Formed of solid cement and laced with steel reinforcement rod, house foundations weigh about fifty tons apiece. Across the city, there were hundreds and hundreds of these bunker-like structures that had to be demolished, excavated, and carted away. Backhoes equipped with hydraulic hammer attachments were brought in to break them apart, but the fragments were still unwieldy. Besides being terrifically heavy, they were often connected by long strands of reinforcement rod. Somehow, these had to be cut apart and broken up, the recyclable steel separated from the useless concrete. There isn't much call for them in normal life, but there are machines designed for just this purpose.

Concrete pulverizers are what you get when you bolt the jaws of an allosaurus to the end of a backhoe boom, endow them with a thylacine gape filled with serried rows of hardened steel teeth, and empower them with a crushing force of two hundred tons per square inch. To watch these animatronic beasts at work, sometimes several together, as they gnaw at the ruins, reducing torso-sized chunks of concrete hung with entrails of reinforcement rod to bits and pieces one could move with a push broom, is to witness a world, and a scale, that has little to do with human beings or even life, and everything to do with forcing intransigent substances into something they were never meant to be, which is, in a nutshell, the primary function of Fort McMurray.

24

In our country the lie has become not just a moral
category but a pillar of the state.

—Alexandr Solzhenitsyn

The media called the tens of thousands of displaced residents from
Fort McMurray "evacuees," but now, in retrospect, the term "climate refugees" might also apply. In 2016, the latter term was considered provocative, even traitorous, especially in a prosperous,
bootstrappy resource town like Fort McMurray. This remains true in
Alberta's petroleum community, where the emphasis has been less on
reflection or evolution and more on reinforcing the status quo. But
there is no going back; since then, neither the jobs nor the population
have recovered.

There is an urban legend circulating in Fort McMurray concerning the twenty-three-year-old hauler driver who grossed $334,000 in
one year. Even if true—and it very well could be—it really is a legend
now, a stirring tale from another time. "De-manning" is an ugly neologism you will not hear in political speeches, nor in a sit-down interview with a CEO. It's a private, behind-the-scenes verb that comes
up in boardrooms and shareholder meetings. De-manning is how
you make money from a marginal product without needing actual
workers. Flying out of Fort McMurray in 2017, I sat next to an engineer from Liebherr, the German manufacturer of cranes and mining
equipment. "What brings you to Fort Mac?" I asked.

"We've been up at Site," she said, "field-testing autonomous hauler trucks for Suncor."

Most of Fort McMurray's inhabitants are competent, good-hearted, and optimistic, but they are also totally dependent on an eighteenth-century business model. Interviews with residents and officials alike reveal virtually no talk or reflection about doing things differently in the future. Despite thousands of layoffs, and billions of dollars in losses due to sagging oil prices and then the fire, tar sands developers are still bullish. Oil prices have rebounded, but for how long is an open question. The fires in town were not yet doused before vows to rebuild were being made at all levels, and efforts have been ongoing, in spite of many setbacks. This dogged loyalty to business as usual invites a comparison of this young city and its employers to an extraordinarily expensive and destructive self-perpetuation machine. Not so different from a boreal fire, it only knows how to do one thing. Narrow specialization and overdependence on a rigid status quo are lethal to species, industries, and civilizations alike. Fort McMurray is not a town that is likely to survive the impending energy transition, but the natural instinct is to fight for one's life, and for one's livelihood, and Fort Mac surely will. It's a long way back: the fire's $10 billion price tag is equivalent to two banner years of royalty revenue, numbers the province hasn't seen since the last boom (until Russia invaded Ukraine in February 2022 and oil prices briefly spiked).

Fort McMurray is not an isolated case, but just one expression of a confluence of forces emerging from a changing climate, a changing energy market, and a growing awareness of the ways these global forces influence each other. What is happening now with bitumen is already well under way with coal. In 2016, people who raised the question of climate change in the context of Fort McMurray, or its fire, were ignored, accused of exploiting a tragedy or, worse, kicking a man when he was down. The province's brief and contentious dalliance with a slightly more liberal government happened to overlap with the fire and ended abruptly afterward with a return to, and hardening of, the industry-friendly United Conservative Party, among whose devotees Donald Trump is considered an ally and, increasingly, a role model. Today, Alberta's relations with the

Liberal government are even more fractious than they were in the 1980s.

In 2019, three years after the fire, and still singularly focused on mining and selling bitumen, the Conservative government allocated $30 million of taxpayer money to a "War Room" whose purpose was to promote Alberta bitumen while investigating, smearing, and otherwise discrediting anyone who might question or criticize that objective. Perceived enemies targeted by war room–related publicity have received death threats and online attacks; others have been forced out of their jobs. (Around the world, many climate scientists, and even weather forecasters, have experienced similar abuse and derision.) Environmental groups are a particular focus, but Alberta's premier also vowed to publicly shame oil companies and banks that divested from bitumen. This strategy—to lash out at enormous, diversified corporations for whom Alberta's business is expendable—recalls a tantrum thrown by King Xerxes I of Persia in the fifth century BCE. After a storm washed away the bridges he had built to launch an invasion of Greece, the king punished the sea by having it publicly whipped.

In late 2019, Moody's, the global risk assessment firm, downgraded Alberta's credit rating to Aa2, its biggest drop in twenty years. It cited an "opinion of a structural weakness in the provincial economy that remains concentrated and dependent on non-renewable resources . . . and remains pressured by a lack of sufficient pipeline capacity to transport oil efficiently with no near-term expectation of a significant rebound in oil-related investments." In early 2020, Moody's downgraded TransCanada's doomed Keystone XL dilbit pipeline from "stable" to "negative." It was around this time that X-Site Energy, an oilfield services company out of Red Deer, began distributing pornographic decals depicting the teenage climate activist Greta Thunberg, emblazoned with their company logo. In April, the worst flood in fifty years inundated much of Fort McMurray's downtown, forcing the evacuation of thirteen thousand people and causing half a billion dollars in damage. In October, Moody's downgraded Alberta's credit rating again. With the continent crippled by COVID-19 and its attendant health restrictions, Alberta's energy minister was one of the few to find a silver lining: "Now is a great time to be building a pipeline," she said, "because you can't have protests of more than 15 people."*

As the Albertan petro-state remains fixated on bitumen, the ideological dissonance grows ever more acute. In 2019, Alberta energy experts and business executives surveyed by the industry journal, *Daily Oil Bulletin*, declared that government's chief priority should be "improving market access and implementing favourable regulatory regimes" for bitumen, oil, and gas. Only 4 percent said it should focus on diversification, and only 1 percent supported national greenhouse gas and carbon trading policies. *None* supported a national climate policy, or stronger ties with Indigenous communities (whose territories and traditional activities are disproportionately impacted by fossil fuel development and pipelines).† The mood among survey respondents was not only regressive but grim: fully two-thirds said they would not encourage young people to pursue a career in the petroleum industry, and more than three-quarters were pessimistic about the industry's future in Canada. It is worth noting that this survey was completed before the coronavirus pandemic, which saw the barrel price of synthetic crude drop through the floor.

A crane operator named Randy, who lives in a man camp and works for Syncrude, put it another way: "It's the quietest I've seen since I got here in ninety-nine," he told me. "There's no security now. I know management guys who are worrying about their jobs. I say fuck it; it's out of your control. The refinery's been hiccupping ever since they shut it down [due to the fire]. The coker's been down for two days. I just pull the levers, do what I'm told—it's easy as fuck." But the future looked as bleak from his crane cab as it did to the executives in the *Oil Bulletin* survey. "Oil is going down," he told me. "It's going down."

---

* Four days into Russia's invasion of Ukraine, Alberta premier Jason Kenney tweeted, "Now if Canada really wants to help defang Putin, then let's get some pipelines built!"
† In a 2020 University of Houston survey of the Texas Oil and Gas Association, 76 percent of respondents (the most for any category) ranked the election of Democrat Joseph Biden as "the greatest threat to company growth prospects," ahead of oversupply, weak demand, cyberthreats, coronavirus, and renewable energy, among others.

At the United Conservative Party's 2021 convention (five years after the costliest fire in the country's history and eight years after its costliest flood), 62 percent of the Alberta delegates voted against a proposal to include climate-related language in the party's policy book. Among the rejected phrases and concepts were "Climate change is real" and "Canadian businesses classified as highly polluting need to take more responsibility." Those who endorsed these additions, along with support for "innovation in green technologies," were rebuked by their peers.

David Mattson, a grizzly bear expert, retired from the U.S. Geological Survey, wrote the following in response to outdated policies on game management, but it applies equally well to energy and climate policies:

> The problem with despotic institutions is that they rarely adapt constructively to changing environments. Instead, the pattern is one of entrenchment against emerging threats at the enthusiastic behest of those who are most privileged by established arrangements. The result is an increasingly brittle institution destined for catastrophic failure, much like the Soviet Union at the end of the 1980s.

Ever since the downturn in 2014, due in part to an ongoing Saudi-Russian oil price war, investors had been cooling on the tar sands. By 2016, auction prices for bitumen leases had already dropped 80 percent from their peak a decade earlier. Until recently, these changes seemed purely market driven, but there were other forces at work. In late September 2015, seven months before the Fort McMurray Fire, the governor of the Bank of England, a Canadian economist from Alberta named Mark Carney, became the most prominent capitalist to name the elephant in the boardroom: "The challenges currently posed by climate change pale in significance compared with what might come," Carney told his audience at a Lloyd's of London dinner. "Once climate change becomes a defining issue for financial stability, it may already be too late." Investors, he warned, face "potentially huge" losses from climate change action that could make vast reserves of oil, coal, and gas "literally unburnable."

These were shocking words to be uttering in such company—as jarring and historic in their way as Edward Teller's to the API in 1959. In retrospect, Carney's warning rings like a death knell for the Petrocene Age.

The term "stranded assets" was new to investors' vocabulary in 2015, but it resonated, especially among those with heavy fossil fuel portfolios. In December 2016, fourteen months after Carney's prophetic speech, and six months after the Fort McMurray Fire, Norway's Statoil sold off its half-billion-dollar position in the tar sands. This was not significant by itself; investors move assets all the time. But then, just days later, Koch Industries abandoned plans for a SAG-D project west of Fort McMurray. A month after that, in January 2017, Exxon-Mobil and its subsidiary, Imperial Oil, announced a "write-down" on 2.8 million barrels of potential synthetic crude due to unfavorable economic conditions, effectively writing off a $20 billion investment on one of the region's biggest bitumen projects. "Write-down," like stranded assets, was unfamiliar language for people in the fossil fuel business, and it bode ill. A month later, Houston-based ConocoPhillips warned that it, too, might be forced into a multibillion-dollar write-down on its bitumen leases. In March, Royal Dutch Shell got out of Alberta altogether, selling its assets in the oil sands to Canadian Natural Resources for $13 billion. That same month, Houston-based Marathon Oil pulled out, too. In June of that year, Sweden's largest pension fund divested its bitumen holdings, as did the huge Dutch bank ING. In October, the French Bank BNP declared that it would no longer finance "pipelines that primarily carry oil and gas from shale and/or oil from tar sands," and will sever "business relations with companies that derive the majority of their revenue from these activities." In December, the World Bank announced that it would stop financing oil and gas exploration and extraction projects by 2019.[*] Two months later, in early 2018, Japan's Mocal Energy sold its position in the oil sands.

This was only the beginning of a mass exodus that is ongoing. But all those foreign companies weren't just walking away, they were selling their shares, leases, upgraders, and equipment to someone, and

---

[*] As of 2022, the World Bank has not followed through on this commitment.

that someone has been Canadian companies like Cenovus, Canadian Natural, and Suncor. Suncor, which purchased a controlling interest in Syncrude just days before the fire broke out, has now taken over operations of its smaller neighbor with the blessing of its Canadian and Chinese partners. The impact on workers is significant: with every sell-off, merger, and consolidation comes a new wave of job cuts and another blow to morale.* In the man camps north of Fort McMurray, entire wings have woken up in the morning to find pink slips under their doors. These events have ominous implications for Alberta, whose landlocked, subarctic motherlode of bitumen is exceptionally vulnerable to becoming a stranded asset. Even without factoring in long-term remediation costs for the destroyed and poisoned landscape, Canadian taxpayers could be left holding an awfully big bag. Anticipating this, a Calgary-based consortium persuaded former president Donald Trump to grant them a "presidential permit" in 2020, allowing them to build a two-thousand-mile, $20 billion rail line from the tar sands to Alaska, a tortuously roundabout way of moving tank cars of diluted bitumen to Pacific ports. Who will pay for this railroad, and who will buy this dilbit, remains to be seen.

Who will insure it may be the more pressing question.

Nobody tracks the incidence and costs of natural disasters as closely as insurance companies. In so doing, they provide us with some of the most objective and reliable data on the impacts of climate change and its accelerating severity. In September 2020, the *Insurance Journal* published highlights from a report commissioned by the U.S. Commodity Futures Trading Commission whose associates include financial giants like Citigroup, Goldman Sachs, Morgan Stanley, and Standard & Poor's. "Climate change," the report stated, "poses a major risk to the stability of the U.S. financial system and to its ability to sustain the American economy . . . Climate change is already impacting or is anticipated to impact nearly every facet of the economy, including infrastructure, agriculture, residential and commercial property, as well as human health and labor productivity."

---

* These shell games also make it harder to hold companies accountable for future environmental cleanup, reclamation, and climate-related lawsuits.

This awareness, combined with pressure from shareholders and activists, has led to a serious reconsideration of the risks associated with fossil fuels. In response, a growing number of the world's biggest insurance companies, including AXA, Swiss RE, Munich RE, Zurich Insurance, and the Hartford, have announced intentions to stop insuring and/or investing in ultrahigh-emissions fossil fuel projects like coal and bitumen. Talanx Group, an international consortium, refused to underwrite the Trans Mountain dilbit pipeline, a $20 billion project running from Alberta, through the Rocky Mountains, to a tidewater terminal in Vancouver.* "This type of project is not currently within Argo's risk appetite," explained another international insurer after refusing to renew its coverage. In September 2021, Chubb, the world's largest publicly traded property and casualty insurer, officially dropped Trans Mountain. The Australian Insurance giant Suncorp has gone a step further, announcing that it will no longer underwrite any oil and gas projects as of 2025. Lloyd's of London, Mark Carney's host when he made his grave climate predictions in 2015, has said it will stop insuring coal, bitumen, and Arctic oil by 2022, and all fossil fuel by 2030. Lloyd's represents 40 percent of the global insurance market. Time will tell if these pillars of capitalism-as-we-know-it follow through, but the implications are enormous. In most countries, you cannot operate a car without insurance. The same goes for mines, drilling rigs, refineries, and pipelines.

⸺⸺

In the meantime, the courts may settle this once and for all.

In April 2018, Boulder County, Colorado, sued both Exxon-Mobil and Suncor for damages and future mitigation costs due to climate change, and also for employing deceptive business practices.

---

* As of February 2022, the Trans Mountain Pipeline project, more than a year behind schedule and nearly $10 billion over budget, continues to be subsidized by the Canadian government and Canadian banks. The World Trade Organization defines a subsidy as "a 'financial contribution' by a government which provides a benefit including 'a potential transfer of funds or liabilities' (e.g., a loan guarantee)."

The plaintiffs claimed that the companies willfully disregarded the Stanford Research Institute's emissions study commissioned by the API back in 1967, as well as Exxon's subsequent climate research throughout the 1970s and '80s. This was just one of a burgeoning number of climate cases coming before the courts around the world. ExxonMobil, Standard Oil's most potent scion, has been singled out for criticism by numerous environmental groups and investors—and lawyers—not only because of its mounting debt and erratic profits, but also because of its cynical and duplicitous activities regarding climate research and disinformation. As of 2021, at least 1,500 similar lawsuits, often naming multiple oil companies, have been filed by other cities and states in the U.S., Canada, and elsewhere, including Annapolis, Baltimore, Minneapolis, and Washington, D.C., and by the states of New York, Massachusetts, Connecticut, and Delaware.

The most poignant and powerful cases are being brought by children. So far, most of these cases have been dismissed or delayed, most notably *Juliana v. United States*. The 2–1 verdict, handed down by the Federal Appeals Court, Ninth Circuit, in early 2020, allowed that, while the plaintiffs (twenty-one young people, aged twelve to twenty-three) "have made a compelling case that action is needed . . . Reluctantly, we conclude that such relief is beyond our constitutional power. Rather, the plaintiffs' impressive case for redress must be presented to the political branches of government." While this was discouraging in the moment, it was instructive and, ultimately, validating. The lone dissenter, Judge Josephine L. Staton, came as close to saying "WTF" as a judge can in federal court: "The government accepts as fact that the United States has reached a tipping point crying out for a concerted response," she wrote, "yet presses ahead toward calamity. It is as if an asteroid were barrelling toward Earth and the government decided to shut down our only defenses." *

Most climate scientists and wildfire analysts would agree.

Rather than discouraging future litigants, this and other losses have emboldened them. If history serves as an indicator, it is only a matter of time. The first lawsuits against tobacco companies met

---

* As of June 2021, the court had ordered the parties to discuss ways of coming to a resolution.

similar fates, and there are many parallels between these two industries, the harm they have knowingly perpetrated, and their defensive strategies.*

In January 2020, the same month as the *Juliana* decision, JPMorgan Chase, the world's largest financier to the fossil fuel industry, sounded yet another alarm to shareholders. In a report commissioned by the bank, the authors explained that a status quo approach to energy and investment "would likely push the earth to a place that we haven't seen for many millions of years." Climate change, they wrote, "reflects a global market failure in the sense that producers and consumers of CO_2 emissions do not pay for the climate damage that results . . . Although precise predictions are not possible, it is clear that planet Earth is on an unsustainable trajectory. Something will have to change at some point *if the human race is going to survive*" (italics mine).

Like Mark Carney's predictions at the Lloyd's of London dinner, these are startling words to be hearing from a bank (let alone a bank built on Rockefeller money that has loaned $75 billion to the fossil fuel industry since the Paris Agreement was signed in 2016). It should be noted that, despite their keen awareness and apparent concern, JPMorgan continues to finance fossil fuel projects. In fact, the world's sixty biggest banks have loaned $3.8 trillion to the industry between 2016 and 2021 alone.

There is a disconnect here that is reminiscent of the response to the Fort McMurray Fire just before it breached the city's defenses: while acknowledging openly that the fire was huge, out of control, and heading toward town in historic fire weather conditions, the leadership advised citizens to go about their business and to "have a plan." Fort McMurray is a servant to the bitumen industry in a remote corner of the subarctic, but JPMorgan is a major international bank managed by "the smartest guys in the room." It has the power

---

* A landmark finding in the Philippines has moved fossil fuel companies' climate liability into the realm of human rights, with broad implications for other cases.

to set agendas and influence policy across the globe, and yet bankers, it seems, are servants, too.

Currently, we live in a dangerously bifurcated reality where senior executives at forward-thinking, publicly traded global companies like Exxon, Shell, JPMorgan, and the Bank of England accept the science of anthropogenic carbon dioxide and the threat it poses, and still continue to—literally—pour gas on the flames. Like the Lucretius Problem, this behavioral dissonance reveals a glitch in human nature. It is a downside to our species' extraordinary adaptability that appears to be triggered by artificial abundance and exacerbated by the profit motive, which could be defined as "commercial entitlement" (or greed). Once an arbitrary standard has been set for production, profits, and their attendant gratification—something fire and fossil fuels have enabled us to do as never before—that expectation, however unsustainable, becomes the new baseline, and anyone attempting to revise it, or even question it, faces serious social and economic consequences.

Morals and ethics aside, modern humans are strangely susceptible to entitlement, even to the most miraculous and implausible of powers. If we need any more proof that we are merely animals, this trait provides it: cats don't question the food in the dish; they eat it happily and expect it to be there in the future, to the point that they may starve if the chain of continuity is broken. Birds at the feeder, pigs at the trough, hominids at the supermarket all develop similarly unexamined expectations. Given the difference in brain size between us, it is amazing that we share the same assumptions of magical abundance. How many of us have felt ill-used when our flight—in a machine we couldn't possibly build, operate, afford, or even fully explain—is delayed? Likewise, our irritation when a photo of our dessert fails to make the epic journey into space and overseas? In fairness, we also accept and expect the sun in the morning and the air we breathe.

Such a fluid "entitlement baseline" has serious implications for all species, and a poignant example can be found in the fabulously lucrative sea otter trade that thrived—and crashed—on the Pacific coast of North America between 1775 and 1830. Once it was discovered by Captain Cook that Chinese buyers in the port of Canton would pay extraordinary sums for the skins of North Pacific sea otters, the race was on. At its peak, the trade involved scores of ships from half a dozen countries across two oceans and three continents. But all things on Earth are finite, and sea otters are more finite than most; by 1800, three decades into the trade, they were growing harder to find. There was no discipline in this mercenary free-for-all, and no tomorrow, but that doesn't mean there was no awareness. The Indigenous hunters tasked with procuring the otters would have been the first to realize that, even as their own wealth and stature were growing beyond their wildest dreams, they were killing the golden goose.

By 1810, the otter hunters found themselves in a bind—the same bind that banks and oil companies find themselves in today: once the market has been created, they believe there is no choice but to participate. Any village or chief, or company or CEO, that doesn't will—they fear—become the losers in the ongoing race for membership, wealth, and status. Once astride a tiger like this, it appears suicidal to get off—even if staying on is sure to destroy you in the end. In 2021, after losing a historic climate lawsuit, Shell's CEO, Ben van Beurden, responded with what amounts to a cri de coeur for the status quo: "Imagine Shell decided to stop selling petrol and diesel today," he wrote on LinkedIn. "This would certainly cut Shell's carbon emissions. But it would not help the world one bit. Demand for fuel would not change. People would fill up their cars and delivery trucks at other service stations."*

Coastal sea otter hunters and Dene beaver trappers found themselves in the same quandary before those markets crashed, and so have many workers in the bitumen industry. The effects are already being felt: those eye-popping paychecks from 2010 are distant memories now, long since spent or mortgaged. Tens of thousands of jobs have been lost across the industry since 2014. In 2017, there were three thousand unemployed electricians in Alberta. In 2020, unemployment in Fort McMurray hit 11.2 percent, a record. As bitumen royalties plummeted (in spite of record production), Alberta's GDP dropped nearly 9 percent, leaving the province with a record deficit. In addition to having the highest debt-to-income ratio in the country, Albertans

---

* Lawyers for Alberta's bitumen industry used the same argument when they successfully overturned a new federal environmental assessment law in May 2022.

experienced a 19 percent mortgage deferral rate.* Meanwhile, in Calgary, office building vacancies jumped to 30 percent from near zero in just eighteen months. There hasn't been much relief: in June 2020, the French petroleum giant Total took a $7 billion write-down on two major projects in the tar sands. That same month, the American Keystone XL pipeline, designed to carry 800,000 barrels of dilbit per day into the U.S., was officially canceled. In mid-July 2021, a court upheld a local law in South Portland, Maine, banning a pipeline company owned by Suncor from exporting diluted bitumen for overseas shipment by tanker. A week earlier, the University of Calgary suspended its bachelor's program in petroleum engineering indefinitely, due to low enrollment.†

Just like those foreign sea otter traders, foreign oil companies are departing Alberta with no immediate plans to return. And it begs the question: Once a market has taken off, how do you "land" it without crashing? Or, in the case of petroleum, how do you transition?

No one in the sea otter trade was willing or able to answer this, and the inevitable result was collapse. By 1850, trading ships had stopped calling in those remote bays altogether, having moved on to seals, whales, timber, and mining. Meanwhile, the coastal tribes were left in a manufactured poverty they had never known before, afflicted by deadly imported viruses, and riven by irreparable social disruption. In the two centuries since, neither those communities nor the sea otters have recovered.

The Pacific coastal tribes' brief encounter with boomtown wealth was artificially fueled by a transient global trade whose template was a colonial one (which is to say, a capitalist one) modeled on the Hudson's Bay Company, and still dominant today: total exploitation in return for a conditional supply of trade goods, alcohol, and weapons. And yet its beneficiaries (on both sides) embraced these terms willingly—even eagerly. Not much has changed in Canada since then: the west coast fishing and logging industries have been offered

---

* A 2022 report by the CBC found that Albertans experience the highest rate of food insecurity in Canada.

† As of 2022, the number of U.S. petroleum engineering graduates had dropped 80 percent since peaking in 2017.

the same terms by their remote masters, and both those livelihoods are in states of protracted collapse, even as they wipe out the resource their existence depends on. In the case of petroleum, the collapse is one step removed: it isn't the supply that's crashing—far from it; it's our atmosphere's ability to absorb the market's "success."

Here, in the twenty-first century, executives and citizens alike are being confronted with a familiar scenario—only with petroleum instead of otter skins. Over the past century and a half, we have recalibrated our lives, culture, and economy to the point that we now feel both dependent on and entitled to the luxury of fire in waiting. In the developed world, it has become an unspoken expectation that one of government's jobs is to provide cheap and abundant energy, especially fossil fuels. It's hard for most people to imagine an alternative. (What if I have to change my behavior? What if it's unpleasant? What if it's somehow . . . less?) Shell's Ben van Beurden wants to keep selling oil and gas because people who are successful at what they do want to keep doing it.* Of course, sea otter hunters and traders felt the same way, and so do most bitumen workers. So do most of us.

Once again, how do you "land" a market safely once it's in flight, especially if the pilot refuses?

Maybe it takes a lawsuit.

---

* Van Beurden himself has acknowledged how the Russian invasion of Ukraine has led to banner profits (aka "war profiteering") for Shell, as it has for all the majors still active in Fort McMurray.

25

Maybe it wasn't as profitable as we thought it was,
because we haven't paid the bill.

—Kirk Bailey, former executive vice president, Suncor Energy

Virtue, then, is anything that moves you in the
direction of mastering delusions.

— George Saunders

If justice is served, May 2021 will be remembered as the beginning of
the end of the Petrocene Age.

On May 26, in a stunning conclusion to a historic lawsuit brought
by the Dutch chapter of Friends of the Earth (and seventeen thousand
co-plaintiffs), the Hague District Court ruled that Shell has a "duty
of care" to cut its emissions to 55 percent of 2019 levels by 2030. It was
the first time in history a court has imposed limits of this kind on a
private corporation. In its exhaustive ruling, the court systematically
dismantled and rejected a number of the greenwashing tropes and
hobby-horse arguments that defense lawyers, petroleum lobbyists, PR
hacks, and CEOs have relied on ever since the industry first came
under serious scrutiny in the 1980s. It was, all in all, an extraordinary
unveiling of the fossil fuel Wizard of Oz. That Shell will appeal is a
certainty, but there is no going back: #ShellKnew. Now the world
and its litigants know, too, and Big Oil's social license (which, until

now, has been closer to diplomatic immunity) is another step closer to being revoked.

Plaintiffs have approached the issue of holding big emitters to account from various angles—through tort cases seeking damages (*Boulder v. Suncor Exxon*), and through trust law focusing on a government's duty of care, that is, its responsibility to regulate damaging behavior and protect its citizens. On May 27, one day after the historic Shell ruling, a federal court judge made another kind of history in Sydney, Australia, ten thousand miles away. "This is the first time," said the seventeen-year-old plaintiff, Ava Princi, "a court of law anywhere in the world has recognized that a government minister has a duty of care to protect young people from the catastrophic harms of climate change."

The case, *Sharma v. Minister of Environment*, was a class-action suit brought by high school students from New South Wales who have grown up in the ash and horror of twenty-first-century Australian fire. The case, based on the so-called law of negligence, revolved around a permit for a coal mine expansion and the government's duty of care to not further endanger the future of young Australians by enabling the release of more heat-trapping CO_2. While the court did not block the mine expansion outright, it did the next best thing: it ordered the minister for environment not to cause personal injury to the plaintiffs and their peers when exercising her powers to approve (or reject) the mine proposal.* With the cataclysmic fires of 2020 vivid in his mind—fires in which more than 20 percent of the continent's forests burned and billions of animals perished—Judge Mordecai Bromberg issued a blistering statement from the bench:

> It is difficult to characterise in a single phrase the devastation that the plausible evidence presented in this proceeding forecasts for the Children. As Australian adults know their country, Australia will be lost and the World as we know it gone as well. The physical environment will be harsher, far more extreme and devastatingly brutal when angry. As for the human experience—quality of life, opportunities to partake in Nature's

---

* Australia continues to approve domestic coal projects.

treasures, the capacity to grow and prosper—all will be greatly diminished. Lives will be cut short. Trauma will be far more common and good health harder to hold and maintain. *None of this will be the fault of Nature itself. It will largely be inflicted by the inaction of this generation of adults, in what might fairly be described as the greatest inter-generational injustice ever inflicted by one generation of humans upon the next.* (Italics mine.)[*]

This is language that the late Roger Revelle, Gilbert Plass, and Wally Broecker would have used, too, had they known their urgent warnings would go unheeded for so long. And this is what James Hansen and a growing chorus of concerned scientists around the world have been saying for decades. Because, knowing what they know—what *we* know—it is the only rational response to any major fossil fuel project anywhere on Earth. Dumping the number one waste product of the Petrocene Age into the atmosphere with no restraint has, in one century, become, not only the ultimate act of colonization, but the ultimate tragedy of the commons. Its impact, the skewing of virtually every planetary system for centuries to come, will be our most enduring legacy. Given that extinctions will be an inevitable by-product of this ongoing disruption, it is a legacy that will go on humankind's "permanent record." For generations, individuals, companies, and governments who knew better capitalized on carbon dioxide's quiet invisibility, and their own ability to manipulate the law and public opinion. Meanwhile, the twenty-first century has been telling us, again and again, through every means of expression available, that CO_2 cannot and will not be ignored.

The hour is late, but this message is being heard in ever more resounding ways. In January 2022, a federal judge in Washington, D.C., invalidated 80 million acres (125,000 square miles) of drilling leases in the Gulf of Mexico because the U.S. Interior Department failed to adequately consider the climate impacts of downstream greenhouse gas emissions. This is another first. It suggests the readiness of the law to demand the true and full accounting that the petro-

---

[*] This landmark ruling was overturned by an Australian court in March 2022.

leum industry has fought to avoid, ever since the API disbanded its CO_2 and Climate Task Force back in 1984.

The years following the Fort McMurray Fire have been full of horror and paradox, and also shocking revelation. After *InsideClimate*'s damning exposé, other exposés followed, and it has since become much harder for Big Oil to plead ignorance, or to dissemble its role in manipulating governments and misleading the public regarding what it did and did not know about the impacts of industrial CO_2. Realizing there was nowhere to hide, Shell's CEO, Ben van Beurden, one of the most powerful oilmen in the world, came right out and said it: "Yeah, we knew," he told a *Time* reporter in 2019. "Everybody knew. And somehow we all ignored it."[*]

Ben van Beurden was only twelve in 1970, but by then, any executive with a connection to the American Petroleum Institute—whether they were in energy, manufacturing, public utilities, banking, insurance, or government—knew that atmospheric CO_2 was increasing and why. By 1980 (when van Beurden was twenty-two), the science was effectively confirmed, and there was no excuse not to act. And "act" they did: for the next forty years, the API, and its affiliates, actively engaged in what the futurist Alex Steffen calls "predatory delay." Predatory delay, as Steffen defines it, is "the deliberate slowing of change to prolong a profitable but unsustainable *status quo* whose costs will be paid by others."

"The evidence that had been gathered in the late '70s and early '80s was already unequivocal," Edward Garvey, a geochemist who worked on Exxon's early climate research, told *The New York Times*. "We had a significant window, but we squandered the opportunity." The consequences of this first-degree negligence speak for them-

---

[*] In September 2022, van Beurden told the *Financial Times*, "For me, . . . it is really important that you do the right thing so that in retirement, . . . you can look your grandchildren in the eye and say, 'This is what I did to make the world a better place for you.'"

selves. Its impacts are being felt ever more keenly now, not only in the most vulnerable regions of the world, but also in wealthy urban centers and in the salubrious beach and mountain towns where their inhabitants seek refuge. Like Ben van Beurden, even the faceless investors and customers he condescends to are realizing that there's nowhere to hide. Some of them saw this coming. For years now, the corporate offices and shareholder meetings of major petroleum producers have been targeted by protesters, and their efforts are taking a toll across the globe. In July 2019, the head of OPEC, which accounts for about 40 percent of the world's crude oil production, called climate activists "the greatest threat to our industry." In recent years, the movement has grown increasingly sophisticated.

A fossil fuel divestment campaign that began at tiny Swarthmore College* in 2010 and that was joined by the international climate action group 350.org in 2012 has been growing exponentially. Over the past decade, hundreds of university and church endowments, along with city, state, and union pension plans (including the University of California's $80 billion endowment), have pledged to divest their portfolios of fossil fuel stocks. In 2016, the Rockefeller Family Fund, built on the broad back of Standard Oil, announced its own divestment from fossil fuels, in its way a harbinger of the end of the Petrocene Age. As of mid-2021, actual fossil fuel divestments, combined with pledges, totaled nearly $40 trillion and counting.

In 2019, Shell acknowledged that divestment has had a "material adverse effect" on its bottom line. BlackRock, the world's largest asset manager, put it more bluntly: "Removing fossil fuel securities from New York State's quarter-trillion-dollar pension fund (one of the country's biggest) will have no negative impact on the rate of return federal employees get on their retirement savings." This is a head-spinning reversal. It is also a fund manager's way of saying the emperor has no clothes. In 2007 and 2008, with oil over $100 a barrel and the bitumen industry booming, Shell and ExxonMobil were two of the most profitable companies in the history of the world. ExxonMobil, already the world's biggest publicly traded company, was posting profits in excess of $40 billion a year, and Shell wasn't far behind.

---

* Swarthmore has yet to divest.

For generations, Big Oil has been considered a bulletproof blue-chip investment, beloved by pensions and pensioners alike, and enabled by politicians and governments the world over. As the self-designated keeper of the flame, Big Oil held the burning world in thrall.

In less than a decade, that picture changed: Big Oil is still big, but it is carrying debt loads unimaginable in the 2000s and, with the low-hanging fruit long gone, the cost of exploration and recovery is only growing. Meanwhile, dividends, a big attraction for investors, are harder to afford when, like ExxonMobil, you have been carrying $50 billion in debt and your stock price is volatile, encouraging buybacks and layoffs to shore it up (the oil price spike supercharged by Russia's invasion of Ukraine is unlikely to sustain itself over the long term). The biggest companies, including Exxon, Shell, BP, Chevron, and Total, have all effectively mortgaged themselves in order to maintain their dividend payouts—with mixed results. In 2021, Home Depot was worth more than ExxonMobil, and NextEra, a wind and solar company most people have never heard of, had a greater market capitalization. These comparisons would have been inconceivable in 2010, but the changes have been in the works for some time. In 1980, the oil and gas sector represented 28 percent of the S&P 500 index's value; in 2019, it represented less than 5 percent, and it is likely to keep shrinking. ExxonMobil, which was the most valuable company in the world in 2013, was removed from the Dow Jones Industrial Average index in 2021. In 2016, such a proposition would have gotten you laughed off the squash court.

Meanwhile, the industry's champions are turning on them, using ever bolder language. In 2021, Aviva, one of the United Kingdom's largest asset managers, promised to deploy its "ultimate sanction"—liquidating its oil, gas, and mining positions—in order to force action on climate. "We have an obligation to clients and society to not fund something we believe is catastrophic to the world and capital markets." For someone born in the 1960s, or even the 1990s, reading statements like this in the *Financial Times*, with Margaret Thatcher's party in power, is a brand-new experience.

Truly unnerving to anyone heavily invested in oil and gas is the recent news from the International Energy Agency, a longtime ally of the industry. In their "World Energy Outlook" for 2020, the IEA announced that solar power is now the "cheapest electricity in history." The threat is serious enough that Moody's has declared renewable energy a "significant business and credit risk"—to the petroleum industry. Governments are paying attention; in June 2021, the European Parliament passed a historic law that made Europe's aspirational emissions targets legally binding. What a Dutch court imposed on Shell, the entire European Union, less than a month later, imposed on itself: by 2030 the European Union must reduce its emissions by 55 percent from 1990 levels with net emissions cut to zero by 2050.

Canada, by contrast, is the standout laggard among the G-7 countries. While every other member country has been reducing its CO_2 emissions steadily since at least 2010, Canada's emissions have been rising. In 2021, they were 25 percent over 1990 levels, due largely to the bitumen industry whose emissions have more than doubled since 2005. If this trend is to be reversed, something miraculous will need to happen in Alberta. In a statement to the *Toronto Star* in October 2021 (ahead of the UN Climate Change Conference meetings in Glasgow), "the Canadian Association of Petroleum Producers said it wants oil and gas companies to sell more fossil fuels in the coming years, and that the industry will require 'substantial collaboration' from the federal government to make sure emissions go down while that happens."

You can almost hear Roger Revelle and Gilbert Plass choking on their seltzer water.*

⸺

Fire has no heart, no soul, and no concern for the damage it does, or who it harms. Its focus is solely on sustaining itself and spreading as broadly as possible, wherever possible. In this way, fire resembles the unspoken priorities of most commercial industries, corporate boards and shareholders, and, more broadly, the colonial impulse. It has taken decades, but the dissembling, distracting, gaslighting, bribery, bullying, and outright lying perpetrated by the fossil fuel

---

* According to the 2023 Climate Change Performance Index, Canada ranks sixth from the bottom, next to Russia.

industry—before and after disbanding the API's CO_2 and Climate Task Force—is being exposed in ever-harsher light.

Recently, those beams have penetrated the inner sanctum of shareholder meetings, forcing an internal reckoning that is long overdue. For years, progressive shareholders have attempted to mitigate—to humanize, if you will—the damage that fossil fuel companies are profiting from. Here, too, May 2021 was a watershed. By a nearly two-thirds vote, Chevron shareholders demanded cuts to the tailpipe emissions of the company's products—a first. On the same day, Exxon shareholders replaced a quarter of the company's conservative twelve-member board with new members determined to address climate change and the need to shift the company toward renewable energy and a net zero future. Another first.[*]

Earlier that same month, the International Energy Agency, the global adviser and advocate for the fossil fuel industry, issued a landmark "flagship report" entitled "Net Zero by 2050," a document as significant in its way as the creation of the API's CO_2 and Climate Task Force back in 1980. Historically, the IEA's role has been to advise governments and companies in their efforts to anticipate and meet future energy needs around the world. This is the first time such an influential agent of industry has said, in essence: *Stop. No more coal plants. No new gas, oil, or bitumen projects.* For the record, this is what climate scientists have been saying, softly and then loudly, since Gilbert Plass correctly anticipated our current situation two generations ago.

"Achieving net-zero emissions by 2050," the IEA's report states in a bold pull quote, "will require nothing short of the complete transformation of the global energy system." It goes on to say in unequivocal terms that, if the future is to be livable, it will be powered by carbon-neutral energy.

---

[*] In November 2022, a similar climate-focused incursion took place on the board of energy giant AGL, Australia's biggest carbon emitter.

26

it is a fire which consumes me
but I am the fire.

—Jorge Luis Borges, "A New Refutation of Time"

B y mid-June 2016, halfway through the hottest year in recorded history (prior to 2020), the Fort McMurray Fire had impacted 2,300 square miles of forest—an area larger than Delaware—and it was still burning, with four and a half months of fire season to go. For the hundreds of firefighters now on scene, fire had become a kind of natural state, and, as with any long-lasting war, it had found its own rhythm. Bulldozer operators "walked" their massive machines through trackless forest for weeks, scraping out firebreaks as they rode herd on the fire—all the way into Saskatchewan. Meanwhile, in Fort McMurray, and throughout the surrounding woods, hotspots revived themselves, sometimes days later, no matter how much water you dumped on them.

Fire 009—the Horse River Fire, now the Fort McMurray Fire—would not be declared "extinguished" until August 2, *2017*, fifteen months after a listener called in a new fire to Chris Vandenbreekel at Mix 103. People familiar with disasters of this kind were astonished by the outcome. Acting Captain Mark Stephenson, who saw his home burn in Abasand, summed up the general feeling: "With the amount of loss we had in this town, I'm still shocked that there weren't bodies turning up in basements." Glenn McGillivray, managing director at the Institute for Catastrophic Loss Reduction in Toronto, attributed this to "sheer grace." "We could have been looking at a staggering loss on the financial side," he told the *Globe and Mail*, "and how we didn't lose hundreds of people, I don't know."

The halal butcher Ali Jomha had an answer. "That's true that the fire took place and it was a disaster," he told me, "but you had the feeling that God didn't like to harm anybody. When close to 100,000 flee the town without even one single person getting burned, or injured, or killed—this itself shows you."

But the fire injured people in other ways, and the toll it took—on first responders, on evacuees, on the city, the industry, the forest, the rivers, and the atmosphere—is still being assessed. Many of those who stayed and fought for Fort McMurray suffer from persistent respiratory problems, and are resigned to the fact that prolonged exposure to the fire's unrelentingly toxic air has shortened their lives. There are more obvious and easily quantifiable measures of the fire's impact: the most homes lost in a Canadian wildfire; the costliest disaster in Canadian history. And then there are the thousands of small and secret costs that continue to accrue, among them the father choking back tears as he describes his child's response to the national emergency broadcast signal. This ear-piercing alert sounded repeatedly over the radio as residents made their slow-motion escape on May 3. Months later, in their new home far from Fort McMurray, "We had the radio on," he told me, "my thirteen-year-old daughter sitting in the living room, not paying attention. The emergency broadcast system goes off and she drops to the floor in the fetal position, crying hysterically. I mean it was *instant*. Talk about Pavlovian response. I think the term for it now is 'post-traumatic stress.'"

For many thousands of people who endured and survived May 3, 2016, the sound of helicopters or sirens, the smell of smoke or the sight of fire, the experience of heavy traffic—all kinds of things trigger involuntary responses. Some of these delayed reactions are visible to others, but many are not. John Knox, the program director for Country 93.3, was well tuned to the mood of his city. "After the fire," he said, "I kept being asked how the community was, and my answer was always the same: 'Imagine a city—thousands of people—all living in everyday harmony, each and every one with some aspect of PTSD.'"

Preemptive evacuation is now becoming a standard policy in regions with high fire risk, and it has been deployed to good effect during many recent fires. It goes without saying that the practice saves lives, but it also saves untold thousands from unnecessary psychological trauma, especially children. For those who stayed behind and fought, those wounds would be harder to avoid. Wayne McGrath wasn't the only one who got drunk and stayed drunk. More than one firefighter described to me being on his knees, doubled over in the shower—the only place he could safely lose it without frightening his wife and children, who were managing terrors of their own. In Fort McMurray, after the fire, there was a marked uptick in church attendance, and in requests for mental health services.

Throughout the spring of 2016, survivors' stories relayed on TV, through news sites, and on social media were viewed by most who saw them as glimpses of a frightening and aberrant event outside normal time, in a place safely far away. In fact, those articles, posts, and photos were messages from the near future, and from very close to home. Just since 2016, extraordinary changes have taken place across the entire climate file—from the atmosphere, to energy, forestry, finance, and insurance, but especially to fire itself. In June 2016, Tennessee, Florida, Texas, Oklahoma, Kansas, Nebraska, California, Oregon, Colorado, New Mexico, Montana, Alaska, British Columbia, Australia, New Zealand, Chile, Brazil, Paraguay, Greenland, Sweden, Ireland, Portugal, Spain, France, England, Greece, Sardinia, Lebanon, Turkey, Algeria, South Africa, Siberia—among many other places—had yet to ignite in the alarming and historic ways they soon would. Wherever they live, most people are aware of recent changes in weather and climate, and now fire is a growing part of that awareness. Since 2016, people all over the world who had never heard of the WUI have discovered they are living in it.

Planet Earth may be a sphere whose surface is mostly water, but scattered across it is arrayed an infinity of fire triangles, large and small—fuel, weather, topography—in endless combination. From the steep Victorian neighborhoods of San Francisco to the shingled seaside villages of Cape Breton to the pool-dotted suburbs of Canberra to the logging camps of the Russian Far East—wherever trees, grasses, and human settlement are found together, and even where they're not. Since 2016, even places we never considered capable of burning, like the Arctic, the Amazon, the moors of Britain, or that we thought somehow inviolable, like Gatlinburg, Tennessee, Boulder, Colorado, Notre-Dame Cathedral, or the National History Museum of Brazil, have caught on fire and burned.

Until Notre-Dame caught fire in April 2019, few people knew that the cathedral's three-hundred-foot-long roof was supported by a "forest" of 1,300 oak trees—another kind of wildland-urban interface, dating back eight hundred years. Ours has long been a blended world; there are forests in cities everywhere from Berlin to Rio de Janeiro, and there are cathedrals in forests. Between August and December 2020, a ferociously hot and stubborn fire burned through California's Sequoia National Park, killing thousands of these majestic giants. Sequoias can live for millennia, in part because they have evolved to withstand fire—even freak thousand-year events. But twenty-first-century fire is something else again. Hotter and drier now, the atmosphere has been tilted in fire's favor.

In addition to Paul Ayearst's and Carol Christian's, thousands of individual and family histories were lost in Fort McMurray, and most of us would classify these as personal tragedies. But how do we classify the loss of a national museum when it contained 20 million artifacts collected over the course of two and a half centuries from a country as biologically and culturally complex as Brazil? Along with countless plant and animal specimens, Brazil's National History Museum in Rio de Janeiro, which was gutted by fire in 2018, housed human artifacts, sacred objects, and photos and voice recordings, most of them irreplaceable because the people, the cultures, the languages, and the worldviews that conceived them are now extinct. Museums, like libraries and cathedrals, are communal memory palaces, not just for us but for those who came before and for those yet to come. Built with great care, often out of stone, with the future—and fire—in mind, they project safety and stability by design. Lately, they seem to offer neither.

In April 2021, a wildfire driven by high temperatures and strong winds swept across Table Mountain, into the WUI, and onto the campus of the University of Cape Town, where it gutted most of the two-hundred-year-old Jagger Library. Inside that pillared stone building was housed the university's Institute for Communities and Wildlife in Africa, and its Plant Conservation Unit, dedicated to studying the long-term history of climate change in South Africa. Irreplaceable records and entire careers of research were destroyed.

Cathedrals, museums, and libraries are all common sanctuaries where we can safely learn, reflect, explore, and simply be. Sanctuaries, almost by definition, are not supposed to burn. Nor are rain forests, marshes, or sequoia groves. The human and natural world, our collective sanctuary and memory palace, has entered a new precarity.

In May 2021, Kent Porter, a veteran news photographer from California, summed up the risk posed by twenty-first-century fire in a single tweet: "The WUI is everywhere now."

EPILOGUE

How we get there is where we go.

n August 2018, shortly after the Redding fire tornado tore through Willie and Larry Hartman's neighborhood, I noticed some things easy to overlook in that heartbreaking wreckage. The Hartmans' kindness, optimism, and determination to keep living there stood out, but there was something else. Despite the fact that there was no grass or topsoil, and most of the surviving trees had been completely denuded of their leaves, there were signs of emergent life. Less than a month after one of the most violent fire events in the history of the continent, new shoots had burst through the scorched hardpan, nourished by the still-vital roots of those flayed and blackened trees. So out of place were these tender greenlings in that wasteland that they could have been mistaken for plastic plants. But they were real: alive and undeterred. A hundred yards away, next to the Hartmans' house foundation, where every organic thing had burned and only stones and twisted metal remained, I saw three amaryllis flowers pushing through the ash.

This is revirescence. It is in the nature of Nature to do this at every opportunity. It might sound funny, even trite, but Earth's capacity for revirescence is without parallel in the known universe. Again and again it has occurred, without fail, for the past three billion years, no matter how violent or broad the disruption.

That we have become such a disruption grows ever harder to ignore. The consequences of burning millions of years of accumulated fossil energy in a period of decades will be ongoing and dramatic. The effects will influence everything that matters in more ways than we can imagine for the foreseeable future and probably far longer. In light of this, it is almost unbearable to consider that our reckoning with industrial CO_2 is only in its infancy, and that future generations will bear this burden far more heavily than we do now. The willful and ongoing failure to act on climate science is unforgivable; recrimination is justified, but none will be sufficient. In this case, at the planetary level, there is no justice; the punishment will be shared by all, but most severely by the young, the innocent, and the as-yet unborn.

In the meantime, life will persist, and so will we.

Three years on, the charred and broken trees around the Hartmans' property have sprung back to life. Revirescence after an assault so brutal seems like a kind of forgiveness, or even deliverance, not unlike Fort McMurray's miraculous escape on May 3rd. In spite of otherworldly fires, those trees and people were given another chance. Meanwhile, the forest around Fort Mac is slowly reviving, too, and the leveled neighborhoods are filling in with new homes, trees, and flowers. The Hartmans have rebuilt their tornado-blasted house, and so have their neighbors.

This miraculous process of renewal and continuity was of great interest to Hildegard of Bingen. Hildegard was a twelfth-century Benedictine nun who was also a polymath. In addition to being an abbess, a scholar, a composer, and a true visionary (vivid paintings of her visions survive to this day), she wrote extensively about medicine, theology, and the natural world. Six hundred years before there were natural philosophers like Joseph Priestley, Ferdinand de Saussure, and Benjamin Franklin, there was Hildegard and her fascination with life itself—specifically, the mysterious vitality that compels things to *grow*. Hildegard didn't call it "the life force." Instead, she used a term more specific to what she saw in her native Rhine Valley: *viriditas*. In Latin, Hildegard's written language, *viriditas* means "greenness," but in the context of Hildegard's writing, it implies something closer to "greening energy," the innate impulse in all living things to be healthy, whole, and regenerative. Oxygen, carbon dioxide, and photosynthesis would have been unknown to her, but Hildegard's intuition was strong. She understood that, as far as humans are concerned, "greening energy" is what makes life not only possible, but also wonderful, holy, and renewable.

Viriditas, as Hildegard understood it, is Earth's standard operating procedure; it is also our planet's response to catastrophe. *Viriditas* is amaryllis driving through the ash after fire and ruin; it is tulips blooming in Abasand when all hope appears lost.

Human beings planted those flowers with no idea what was to come. This—devoting our energy and creativity to regeneration and renewal rather than combustion and consumption—is what Nature is modeling for us and inviting us to do. *Homo sapiens* got us to the Petrocene Era. *Homo flagrans* is who we have become. *Homo viriditas* can guide us forward—and, possibly, back.

ACKNOWLEDGMENTS

This book has been a collaborative project; one way or another, the words and hands of hundreds of people are present in its pages. Gratitude goes first and foremost to the citizens and firefighters who took the time to speak with me about wildfire and its impacts: Your courage, candor, and insight were crucial to my understanding of the power and hazards of twenty-first-century fire.

Among the many kind individuals who contributed to this book in tangible and intangible ways are Daniel Aarons and Val Rendell, Angie Abdou, Daniela Aiello, Dominic Ali, Paul Ayearst, Kirk Bailey, David Beers, Garry Berteig, John Blanchette, Dick Bon, Michael Christopher Brown, Cheryl Buliavac and the media team at Cal Fire, Steve and Carrie Bustillos, Warren Cariou, David Carruthers, Cedric from Nova Scotia, David Chapman, Carol Christian, Paul Colangelo, Jamie Coutts, Ryan Coutts, Jonathan Cox, Evan Crawford, Walt Cressler, Joanna Croston, Paul Curtis, Roseann Davidson, Geoff Dembicki, Pastor Doug Doyle, David Dubuc, Pat Duggan, Trevor Dumba, Jill Edwards, Saul Elbein, Emma Elliott, Timmy Joe Elzinga, Larry Farough, Douglas Fox, Dusty Gyves, Elee Kraljii Gardiner, Ian Gill, Matthew Heikel, Vishavpreet Hundal Gill, John Gillis, Denise Golden, Jay Goupil, Brad Grainger, Bruce Grierson, Jude Groves, Chad Gunrow, Scott Hamilton, Aron Harris, Nate Harskamp, Larry, Willie and Christel Hartman, Steven Higgs, Chris Hubscher, Joyce Hunt, Tim Hussin, Ali Jomha, Raman and Amrit Kaler, John Kidder, Pastor Rick Kirschner, John Knox, Dr. Sonya Leverkus, Shandra and Corey Linder, Ronnie Lukan, Craig MacKay, Harvey Marchand, Chris McGillivray, the late Wayne McGrath, Sue McOrmond, Rob Moore, Steve Morrison, Andrew Nikiforuk, Troy O'Connor, Troy Palmer, Jack "Torchy" Peden, Vonda Pikes, Ryan Pitchers, Kent Porter, Chad Price, Cristi Proistosescu, Jim Rankin, Jessica Reed, Trevor Rivers, Louis Rondeau, Denis Roy, Rikia Saddie, Lavalle St. Germain, Danielle Schwanke, Carol Shaben,

Joan Shaben, Mathieu Simard, Professor David Smith, Don Smith, Tom Smith, Paul Spring, Randy and Lynne Stefanizyn, Mark Stephenson, Dale Thomas, John Topolinski, Chris Turner, Cordy Tymstra, Dale and Pauline Vey, Lucas Welsh, Nettie Wild, Lou Wilde, Park Williams, Sampson Woldemichael, Wayne Woodford, Andy Wright, and Stephen Wright.

I was not on the ground in Fort McMurray during the fire, but many brave journalists were, among them Reid Fiest, Brianna Karstens-Smith, Vince McDermott, Laurent Pirot, Briar Stewart, Chris Vandenbreekel, and Marion Warnica: Your diligent reporting has enriched this book immeasurably.

Twitter has been a blessing and a curse as long as I've been on it. For this project, it has also been an ally, colleague, informant, fact checker, and research assistant, as well as science and history teacher. I am deeply grateful to all the journalists, scientists, policy wonks, historians, archivists, activists, economists, and photographers I've had the good fortune to follow. You answered questions I never thought to ask, provided sources I didn't know existed, and made this a better book.

My research would have been much more difficult without the steadily growing database documenting who knew what and when regarding the causes and impacts of anthropogenic CO_2. My starting place was Spencer Weart's comprehensive timeline and history, "The Discovery of Global Warming," which can be found on the American Institute of Physics website, history.aip.org/climate/. For corporate and government documents, these sites are treasure troves: Inside Climate News (inside climatenews.org), Climate Files (climatefiles.com), Smoke and Fumes (smoke andfumes.org), Desmog (desmog.org), Koch Docs (kochdocs.org), and the Climate Investigations Center (climateinvestigations.org). Meanwhile, fossilfuellobbyists.org tracks the petroleum industry's ongoing efforts to expand markets, influence policy, and sway public opinion.

In the all-too-real world of wildfire, I am deeply indebted to the wisdom and generosity of these elder statesmen: Marty Alexander, Rick Arthur, Mike Fromm, Peter Murphy, Dennis Quintilio, Brian Stocks, and Cliff White. Special thanks to Professor Mike Flannigan for writing so lucidly about fire behavior, and for answering so many questions; to Professor Jen Beverly for introducing me to the Lucretius Problem; to Vyto Babrauskas for making the Hamburg connection, and to Brad Johnson (@climatebrad) for guiding me to Roger Revelle and the Congressional Record of 1956. Thanks also to Professor Michael Griffin for his translation of relevant lines from Lucretius's epic poem, *De Rarum Natura*. Steven Johnson's *The Invention of Air* was a wonderful way to learn about early efforts to understand our atmosphere. For anyone seeking ways of understanding wildfire on Earth, and across history, the works of Stephen J. Pyne are indispensable. For those seeking a comprehensive history of the bitumen industry in Alberta, *Local Push—Global Pull* by Joyce Hunt is an invaluable resource. Thanks to Linnea Bell, Lynne Daina, Lindsay Griffin, Melanie Kage, Rosemary Pryce-Digby, Tulene Steiestol, and the Provincial Archives of Alberta for their assistance with research. I am particularly grateful to Sheila Gazlay for her tireless help with interview transcriptions, and to my parents, siblings, and in-laws for their steadfast enthusiasm.

For their generous support of my work, very special thanks are due to Beatrice Monti della Conte and all at the Santa Maddalena Foundation, and to the Paul D. Fleck Fellowship and the Banff Centre for the Arts. Louise Dennys, Andrew Miller, and Martha Kanya-Forstner are a truly formidable team of editors and advocates who believed in this project from the beginning and saw it through to a book I can be proud of. Tiara Sharma kept it all from flying apart. As always, my deepest gratitude to Stuart Krichevsky and all at SKLA.

To Nora, Roan, and Isla, who shared a home with me during this long and arduous process: My heart would be empty without you, and so would the pages of this book.

NOTES

PROLOGUE

1   "In this great chain of causes": Alexander von Humboldt and Aimé Bonpland, *Essay on the Geography of Plants* (Chicago: University of Chicago Press, 2009), p. 79.

4   "You go to a place": Dean Bennett, "Alberta Premier Notley to Greet Evacuees; Recalls Early Days of Wildfire," *Canadian Press*, May 31, 2016.

5   "No one's ever seen anything like this": *Global News*, May 9, 2016.

CHAPTER 1

8   "The foliage is so thick": "World's Largest Beaver Dam Explored by Rob Mark," *CBC News*, September 19, 2014.

8   Girdling the Northern Hemisphere: Hanna Corona, "World Boreal Forests— Largest Biome Taiga," *Boreal Forest*, August 30, 2022.

10   The fire generated: Tymstra, *The Chinchaga Firestorm*, p. 67; Harry Wexler, "The Great Smoke Pall—September 24–30, 1950," *Weatherwise* 3 (1950).

10   Rising forty thousand feet: Tymstra, p. 70.

10   lavender suns and blue moons: Ibid., p. 72.

10   Prior to the Chinchaga Fire: Ibid., p. 66.

10   Carl Sagan was sufficiently: Ibid., p. 90.

10   The *Outlook* includes maps: See cpo.noaa.gov/.

11   "We're just a colony": Rick Kirschner interview, February 6, 2017.

11   around 4 million barrels: Canada Energy Regulator, "Crude Oil Export Summary," neb-one.gc.ca/. Canada accounts for 40 percent of U.S. daily imports. Despite its lingering prominence in the public mind, Saudi Arabia now accounts for less than 10 percent of U.S. petroleum imports. See nrcan.gc.ca/.

11   Of this vast quantity: "Crude Oil Export Summary."

11   In 2016, two years past: "Median Total Income of Households in 2015" ($195,656.00 CAD) and "Census Profile, 2016 Census, Fort McMurray, Alberta," Statistics Canada, www12.statcan.gc.ca/.

12   As one insider put it: Name withheld.

13   "It happens every year": Shandra and Corey Linder interview, November 4,

2016. (All subsequent quotations from Shandra Linder derive from this interview.)

CHAPTER 2

15   "It is natural": Clark, *The Bituminous Sands of Alberta*, p. 32.

16   "TRUDEAU WANTS EVERY DROP": Bruce Grierson, personal communication; see also "Lougheed Retaliates Against Trudeau for NEP," *CBC News*, November 2, 1980.

16   Those two men may have: "Oil and Gas Liabilities Management," alberta.ca/.

16   If all of Alberta's pipelines: "Pipeline Performance in Alberta," Alberta Energy Regulator.

17   "A typical oil sands deposit": "Oil Sands Geology and the Properties of Bitumen," *Oil Sands Magazine*, updated February 28, 2020.

19   In an effort to articulate: Louis Rondeau interview, February 5, 2017.

19   The tailings ponds alone: "Total Area of the Oil Sands Tailings Ponds Over Time," osip.alberta.ca.

20   these skyscraping, fire-breathing gnomons: They were finally surpassed in 2018 by two high-rise office buildings in Edmonton and Calgary.

21   "Nature's Supreme Gift to Industry": International Bitumen Company (IBC).

21   "Magic Sand-Pile" campaign: *Fortune,* September 1947, p. 172.

22   By 1930, the United States: "Map of Oil Trunk Pipe Lines," *The Oil and Gas Journal,* August 30, 1928.

22   In language that sounds: See crudemonitor.ca/.

22   Had the industry discovered it: "McMurray Formation," Wikipedia, en.wikipedia.org/.

22   more than a trillion: Ian M. Head et al., "Biological Activity in the Deep Subsurface and the Origin of Heavy Oil," *Nature*, November 20, 2003.

23   Situated at the lower limits: "Exploring the Limits of Life in the Deep Biosphere," Deep Carbon Observatory, July 23, 2015, deepcarbon.net/.

23   "The oil sands and heavy-oil belts": Stephen R. Larter and Ian M. Head, "Oil Sands and Heavy Oil: Origin and Exploitation," *Elements* 10 (August 2014): 277–84.

23   When Larter tried to calculate: Ibid.

24   To be capable of ignition: Tristin Hopper, "Can Oil Sands Catch Fire? No, If They Could They Would Have Burned Centuries Ago," *National Post*, May 6, 2016.

24   These diluents are usually: Malone Mullin, "What We Know—and Don't Know—About Diluted Bitumen," *Globe and Mail*, January 30, 2018.

24   The million-gallon (24,000-barrel) spill: Garrett Ellison, "New Price Tag for Kalamazoo River Oil Spill Cleanup: Enbridge Says $1.21 Billion," *Grand Rapids Press*, November 5, 2014.

24   The market value of the spilled dilbit: Assuming a 2010 price of @ $50.00/bbl, inflationdata.com/.

24   In both SAG-D and surface mining: Rachel Nuwer, "Oil Sands Mining Uses Up Almost as Much Energy as It Produces," *Inside Climate News*, February 19,

2013. Shell's bitumen operation in Fort McMurray requires roughly six times the energy input to produce the same amount of energy as its conventional oil and gas holdings elsewhere, see cbc.ca/news/.

25   the energy equivalent of 350,000 barrels of oil: Based on the following calculation: one barrel of oil contains the energy equivalent of 5,800 cubic feet of natural gas.

25   Canada is the fourth-largest producer: "Oil Sands Production Using Nearly One-Third of Canada's Natural Gas," *BNN Bloomberg*, April 19, 2017.

25   Even after it has been separated: Rhys Baker, "Making Crude Oil Useful: Fractional Distillation and Cracking," *Owlcation*, April 4, 2022.

25   One way to measure: Nafeez Ahmed, "Will Covid-19 End the Age of Big Oil?," *Le Monde Diplomatique*, April 24, 2020.

25   Alberta has taken: Geoff Dembicki, "This Energy Analyst Says the Oilsands Are 'Done,'" *Tyee*, May 11, 2020.

26   A rough estimate: Chris Varcoe, "When the Oil Sands Hit Pay Dirt," *Calgary Herald*, September 26, 2017.

CHAPTER 3

27   "Whereas many incredible miracles": Diodorus of Sicily, *Bibliotheca Historica*, II.12, via Forbes, *Bitumen and Petroleum in Antiquity*, p. 21.

28   a "Great River": York Factory Post journals, pdf from microfilm reel 1M154, folio 43, Archives of Manitoba (manitoba.ca).

29   In return for actionable intelligence: Newman, *Company of Adventurers*, p. 87. See also "Royal Charter of the Hudson's Bay Company," Hudson's Bay Company History Foundation, hbcheritage.ca/.

29   "a fist in the wilderness": Tracy Sherlock, "The Elusive Mr. Pond: A Geopolitical Visionary," interview with author Barry Gough, *Vancouver Sun*, November 7, 2014.

31   Some particularly arduous: Morse, *Fur Trade Canoe Routes of Canada*, p. 3.

31   still cost forty beaver skins: Heming, *The Drama of the Forests*, pp. 19–20. (Today, a dressed beaver pelt starts at around 125 Canadian dollars.)

32   "A state in the guise": William Dalrymple, "Before the East India Company," *Lapham's Quarterly*, September 27, 2019; see also newleftreview.org/. The Hudson's Bay Company, though now under new ownership and much changed, is still in business, making it the oldest continuously operating company in the English-speaking world.

32   "Since [1840], the dividends": M'Lean, *Notes of a Twenty-Five Years' Service in the Hudson's Bay Territory*, vol. 2, p. 262, via Murphy, *History of Forest and Prairie Fire Control Policy in Alberta*, p. 44.

33   Employees in Alberta's: Chris Fournier et al., "Canadians Are Feeling the Debt Burn," *Bloomberg*, March 26, 2019; see also Chris Fournier et al., "Rising Insolvency Readings Raise Red Flags in Canada. Sort of," *Bloomberg*, November 18, 2019.

33   In 2019, Canadian household debt: Fournier et al., "Canadians Are Feeling the Debt Burn."

34   One of those intrepids: "Oil Sands: Sidney Ells," Alberta Culture and Tourism, history.alberta.ca/.

34   "a dozen primitive log cabins": Ells, *Reflections of the Development of the Athabasca Oil Sands*, p. 15.

35   "Scow tracking south": Ibid., p. 12.

36   first productive New World oil wells: Dug by a carriage maker named James Miller Williams, near Oil Springs village in Lambton County.

36   "It penetrates, purifies, soothes, and heals": Gale, *Rock Oil*, pp. 42ff.

39   he and his pioneering refinery: Mitch Waxman, "North American Kerosene Gas Light Company," *Brownstoner*, July 14, 2014.

39   "A Parisian, by the name of Lenoir": *Scientific American*, January 28, 1860, p. 1 (via Gale, p. 46).

41   There, in the summer of 1897: Thomas Court, "A Search for Oil," *Alberta Historical Review* 21, no. 2 (1973).

41   enough to heat every house and building: Based on two hundred cubic feet per household, per day.

41   it burned, intermittently, for twenty-one years: A dramatic example of a gas well on fire can be seen in video footage presented at a Calgary meeting of the Petroleum History Society, on November 27, 2013. The film, entitled "Killing the King Christian D-18 well, Arctic Islands, 1970-71," can be found on Vimeo .com.

42   2019 International Monetary Fund report: David Coady et al., "Global Fossil Fuel Subsidies Remain Large: An Update Based on Country-Level Estimates," International Monetary Fund working paper, May 2, 2019, p. 35, imf.org/.

42   Canadian taxpayers contributed: Erin Gray et al., "Canada's Fossil Fuel Subsidies Amount to $1,650 per Canadian. It's Got to Stop," *Narwhal*, October 3, 2019.

42   "Calgary never saw": Henry, *History and Romance of the Petroleum Industry*, p. ix.

43   In that single day: *Winnipeg Free Press*, May 16, 1914, p. 1; see also David Bly, "May 14, 1914—Dingman Discovery No. 1 Blows in Turner Valley," *Calgary Herald*, May 15, 2012.

CHAPTER 4

44   Pew called the quarter-billion-dollar: "The Oil Sands Story (1960s, 1970s, & 1980s)," suncor.com/.

44   While Sunoco put up: Tracy Johnson, "The Oilsands at 50: Will They Still Be Producing in 100 Years?," *CBC News*, September 29, 2017.

45   "His conversations": Bob McClements, Petroleum History Society, Petroleum Industry Oral History Projects, p. 5, petroleumhistory.ca/.

45   Still engaged in the Lord's Work: GCOS Annual Report, 1967, p. 4.

47   Many men shared this dream: "Oil Sands: Setting the Stage—Ernest Manning," Alberta Culture and Tourism, Alberta's Energy Resources Heritage, history.alberta.ca/.

47   "This is a historic day": "The Great Canadian Oil Sands Project in 1967—Suncor Energy," YouTube, posted on June 16, 2010.

47   "We're gathered here": Johnson, "The Oilsands at 50."

47   "a red-letter day": "The Oilsands: A Political Timeline," *Edmonton Journal*, December 18, 2013; January 7, 2014.

48   *"Prepare ye the way of the Lord":* Isaiah 40:3–5, King James Version.

48   When it came on line: Peter Mackenzie-Brown, "Bitumen: The People, Performance and Passions Behind Alberta's Oil Sands," October 2, 2019, languagein stinct.blogspot.com/.

49   "the scar on the side": Nate Harskamp interview, January 20, 2017.

50   The last time bitumen: M. G. Lay, *Ways of the World* (New Brunswick, N.J.: Rutgers University Press, 1992), p. 50.

50   Despite the twin chasms: "Babylonia," *Plumbing & Mechanical Magazine* (July 1989).

50   Gates of Ishtar: Forbes, *Bitumen and Petroleum in Antiquity*, p. 51.

50   "I . . . laid its foundation": Koldeway, *Excavations at Babylon*, p. 12.

50   identified eighty first languages: "Fort McMurray, Alberta, Canada—Knowledge of Official Languages, Language Spoken Most Often at Home and at Work," City-Data.com.

51   Canada's fourth straight year: See eia.gov/.

51   When the price of oil drops: The Investopedia Team (Charles Potters, Timothy Li), "How & Why Oil Impacts the Canadian Dollar," updated March 5, 2022, investopedia.com/.

51   Harper was in London: Met Office, "Past Weather Events," July 2006, metoffice .gov.uk/.

51   "Canada's emergence": Paul Wells and Tamsin McMahon, "How Ottawa Runs on Oil," *Maclean's*, March 23, 2012.

52   "This is not a normal world": Louis Rondeau interview, February 5, 2017.

52   "We've kind of focused": Lucas Welsh interview, January 23, 2017. (All subsequent quotations derive from this interview.)

53   Between January 2021: Amanda Stephenson, "Suncor's Safety Record in the Spotlight as Activist Investor Calls for Change," *Canadian Press*, May 2, 2022; and Gabriela Panza-Beltrandi, "Contractor, 26, Killed at Suncor Mine near Fort McMurray," *CBC News*, July 7, 2022. Since 2014, there have been at least twelve deaths at Suncor sites, more than all of its oil sands rivals combined.

53   COVID-19: Wallis Snowden, "Oilsands Workers Inside Alberta's Largest COVID-19 Outbreak Fear for Their Safety," *CBC News*, May 12, 2021.

53   alcoholism and suicide: Alberta Health, "Community Profiles: Health Data and Summary," version 2, March 2015.

54   In 2018, a Fort McMurray: Liam Nixon, "Record Narcotics Bust at Southern Alberta Border Crossing," *Global News*, March 22, 2018.

54   people like Jake McManus: Not his real name.

55   When *Global News* reported: Emily Mertz, "1 Employee Injured in Explosion and Fire at Syncrude Upgrader North of Fort Mcmurray," *Global News*, March 14, 2017.

56   "The front end": Stephen Wright, personal communication, March 1, 2018.

56   "Most of them are head-ons": Name withheld.

56   "light pillars": "Icy Temperatures Bring 'Alien' Light Pillars to Alberta Night Sky," *CBC News*, February 12, 2019.

57   ". . . *from the arched roof*": John Milton, *Paradise Lost*, l. 726–30.

CHAPTER 5

58   "Our intent entered the world": Wendell Berry, *Horses* (Monterey, KY: Larkspur Press, 1975).

59   "reaction zone": Kit Chapman, "The Complexity of Fire," *Chemistry World*, July 20, 2020.

59   Volcanic eruptions start fires: M. Onifade and B. Genc, "Spontaneous Combustion of Coals and Coal-Shales," *International Journal of Mining Science and Technology* 28, no. 6 (November 2018): 933–40.

60   a South African cave: Kenneth Miller, "Archaeologists Find Earliest Evidence of Humans Cooking with Fire," *Discover*, December 16, 2013.

61   At the most basic level: Michelle Nijhuis, "Three Billion People Cook Over Open Fires—with Deadly Consequences," *National Geographic*, August 14, 2017.

61   Already, we are into: Fred Pearce, "Collateral Damage: The Environmental Cost of the Ukraine War," *Yale Environment 360*, August 29, 2022.

61   quarter billion trucks, buses, and vans: "Car Production," Worldometer, worldometers.info/.

61   200 million motorcycles: "Motorcycling," Wikipedia, en.wikipedia.org/wiki/Motorcycling.

61   25,000 passenger and cargo jets: Hugh Morris, "How Many Planes Are There in the World Right Now?" *Telegraph*, August 2017.

61   more than 50,000 cargo ships: "Shipping Facts," International Chamber of Shipping, ies-shipping.org/.

62   Whether it is a teaspoon: Less than 10 percent of global petroleum production is used as feedstock for plastics, fertilizers, lubricants, and so on. See "Oil and Petroleum Products Explained," U.S. Energy Information Administration, eia.gov/energyexplained/.

62   one hundred tons of marine biomass: Smil, *Oil*, p. 66.

63   According to another energy historian: Maxine Joselow, "Daniel Yergin on Peak Oil, Pandemic, Rom-Coms," *GreenWire*, *E&E News*, October 30, 2020, politicopro.com/.

63   100 million barrels: Amanda Cooper and Christopher Johnson, "Now Near 100 Million bpd, When Will Oil Demand Peak?," *Reuters*, September 20, 2018. A standard barrel holds 42 gallons, or 159 liters.

63   another 40 million barrels: "Transport Volume of Crude Oil in Global Seaborne Trade from 2010 to 2020," *Statista*, statista.com/.

63   More than a third of global shipping: "Oil Tankers in Canadian Waters," Clear Seas, clearseas.org/.

64   Long after other components: Natalie Wolchover, "How Does Oil Form?,"
      LiveScience, March 2, 2011.

65   4 billion years ago: Nicola K. S. Davis, "World's Oldest Fossils Found in Can-
      ada, Say Scientists," *Guardian*, March 1, 2017.

65   Abundant and mercurial: David Biello, "The Origin of Oxygen in Earth's
      Atmosphere," *Scientific American*, August 19, 2009.

66   Without it, Earth: Nick Lane, "The Rollercoaster Ride to an Oxygen-Rich
      World," *New Scientist* 205, no. 2746 (February 2010).

68   Then, like a smothered fire: "US Woman Dies in Iron Lung After Power Fail-
      ure," *NBC News*, May 28, 2008.

CHAPTER 6

73   "I balanced between destiny and dread": Seamus Heaney, *The Spirit Level* (New
      York: Farrar, Straus and Giroux, 1996), p. 55.

74   By 10:00 a.m. on Monday the 2nd: "A Review of the 2016 Horse River
      Wildfire," prepared for the Forestry Division, Alberta Agriculture and For-
      estry, by MNP (Myers Norris Penny), June 2017, pp. 5, 61. (Hereafter "MNP
      Report.")

74   At 5:30 p.m. on the 2nd: "Media Briefing: Wood Buffalo Forest Fire Update—
      May 2, 5:30 p.m.," YouTube, posted on May 2, 2016.

74   During her tenure: Frits Pannekoek and Erin James-Abra, "Fort McMurray,"
      Canadian Encyclopedia, updated March 11, 2019.

74   Some of the Cats: Chris Hubscher interview, February 6, 2017.

78   "A lot of ash has fallen": Cordy Tymstra interview, November 1, 2016.

78   In the ensuing days: "Final Report from the Flat Top Complex Wildfire Review
      Committee," p. v, May 2012, open.alberta.ca/.

78   "Metal melted, concrete spalled": Jamie Coutts via email, February 20, 2017.

78   "It actually incinerated": Ronnie Lukan interview (with Jamie and Ryan Coutts),
      February 4, 2017.

79   "We're all sitting in Slave Lake": Ryan Coutts interview, July 27, 2017. (All quo-
      tations derive from this interview unless otherwise indicated.)

81   among the highest-paid firefighters: Brad Grainger interview, May 4, 2017.

81   Of those 185 firefighters: Ibid.

82   "We want them to know": Jamie Coutts interview, July 31, 2017. (All quotations
      derive from this interview unless otherwise indicated.)

83   Later that evening: MNP Report, Exhibit A-8, p. 63.

CHAPTER 7

85   "Fire synthesizes its surroundings": Stephen J. Pyne, "Pyromancy: Reading Sto-
      ries in the Flames," *Conservation Biology* 18, no. 4 (2004): 875.

86   "Here we are on another day": "Media Briefing: Wood Buffalo Forest Fire
      Update, May 3, 11 a.m.," YouTube, posted on May 3, 2016.

88   This wasn't just a little bit: "Average Maximum Temperature—Station 'A':

14.8°C," Past Weather and Climate, Government of Canada. See also "Almanac Averages and Extremes for Fort McMurray, May 3, 2016" and "Normals for Ft Mac, Station 'A':13.5°C/56°F," climate.weather.gc.ca/.

88   Fifteen percent humidity: "Almanac: Historical Information: Death Valley, CA," myforecast.co/.

89   Even on its first day: Paul Spring interview, November 7, 2016.

89   Schmitte had seen: David Staples, "Firestorm: How a Wisp of Smoke Grew into a Raging Inferno," Edmonton Journal, April 30, 2017.

89   Depending on who's telling: Dale Thomas and Dean Macdonald, personal communication, Kelowna Fire Conference, October 24, 2016.

89   topping out at forty-five thousand feet: "Final Documentation Report Chisholm Fire (LWF-063)," Sustainable Resource Development, updated 2001, open. alberta.ca/.

91   By the time it intensifies: Peter Attiwill, "Anatomy of a Bushfire," ABC News (Australia), 2009.

91   at its height: "Final Documentation Report Chisholm Fire (LWF-063)," p. 48.

91   "I told the guys": Troy O'Connor, personal communication, Kelowna Fire Conference, October 24, 2016.

91   "I have to be careful": Dale Thomas interview, Kelowna Fire Conference, October 24, 2016.

91   "The fire behaviour": "Final Documentation Report Chisholm Fire (LWF-063)."

91   At its peak intensity: Jim Sergent et al., "3 Startling Facts About California's Camp Fire," USA Today, updated November 21, 2018: "The [Camp Fire] grew rapidly near midday Nov. 8, consuming 10,000 acres in about 90 minutes . . . It charred more than 70,000 acres in a day's time."

92   a funnel cloud was observed: Michael D. Fromm and René Servranckx, "Transport of Forest Fire Smoke Above the Tropopause by Supercell Convection," Geophysical Research Letters, May 31, 2003.

92   The energy released: Daniel Rosenfeld et al., "The Chisholm Firestorm: Observed Microstructure, Precipitation and Lightning Activity of a Pyro-Cumulonimbus," Atmospheric Chemistry and Physics (February 2007).

92   "Chisholm was the forerunner": Dennis Quintilio, personal communication, June 2016.

93   "It was almost as if": John Knox interview, May 13, 2018. (All subsequent quotations derive from this interview.)

94   "I don't think it was": Chris Vandenbreekel interview, May 19, 2018. (All subsequent quotations derive from this interview.)

94   The last time a city: "Area of Fire—Wholesale District—Toronto, Canada," map, toronto.ca/wp-content/uploads/2017/09/8f34-goads_with_image_numbers.jpg.

96   Meanwhile, the temperature was rising: (12:30-1:00 PM—"A" Station). ["CS" station calls this change between 11 and noon/32.3 C. at 3 PM]. See also "Fort McMurray 'CS' Station, Hourly Data Report for May 03, 2016," Past Weather and Climate, Government of Canada.

96   "North," in Alberta's case: "Northern Arctic Ecozone," Ecological Framework of Canada, ecozones.ca/, arctic.uoguelph.ca/.

CHAPTER 8

97   "The greatest shortcoming": Albert Allen Bartlett, "Arithmetic, Population and Energy: Sustainability 101," lecture given at the University of Colorado at Boulder, February 26, 2005.

97   "We saw stories": Reid Fiest interview, May 24, 2018. (All subsequent quotations derive from this interview.)

98   In 1995, the Mariana Lake Fire: Alberta Soils Science Workshop, *Alberta Soils Tour Guide Book*, 2017, p. 26.

98   more than half a billion dollars: "Firefighters Concentrate on Richardson Fire," *Fort McMurray Today*, June 10, 2011.

98   "Fire-spawned stands": David Pitt-Brooke, *Crossing Home Ground* (Madeira Park: Harbour Publishing, 2017), p. 62.

99   itself a record-breaking year: MNP Report, p. 12.

99   So dry was the forest: Association of Alberta Forest Management Professionals, aafmp.ca/.

100   Over the past twenty years: MNP Report, p. 11.

102   "It's been a tough slog": *Fort McMurray Matters*, Mix 103.7, May 3, 2016.

103   "Maybe it'll get": Chris Vandenbreekel interview, May 19, 2018. (All subsequent quotations derive from this interview.)

106   those in the business of wildfire: For an example, see "Fort McMurray Wildfire Remains Out of Control After City Evacuated," *CBC News*, May 3, 2016.

106   "There is no way a man": Lynne Stefanizyn interview, January 22, 2017.

111   "When you listen to wireless": Herman Hesse, *Steppenwolf* (New York: Frederick Ungar, 1957), p. 301.

111   "The problem with exponential growth": Aubrey Clayton, "To Beat COVID-19, Think Like a Fighter Pilot," *Nautilus*, March 18, 2020.

CHAPTER 9

112   "Macbeth shall never vanquished be": William Shakespeare, *Macbeth*. Act IV, Scene 1, 105–7; Folger Library edition. https://shakespeare.folger.edu

112   It is one thing: 82°F, on May 3, 1945.

112   "The most important failure": "The 9/11 Commission Report, Executive Summary, General Findings," p. 9, govinfo.gov/.

113   While Bernie Schmitte: MNP Report, p. 41.

113   "We're streaming the press conference": Jamie Coutts interview (with Ryan Coutts and Ronnie Lukan), February 4, 2017.

115   Hall 5 is a brand-new: Marshall-Lee Construction, Completed Projects, Fire Hall #5, marshall-lee.ca/.

115   "It's just a hell of a lot": Evan Crawford interview, November 7, 2016. (All subsequent quotations derive from this interview.)

116   "We were looking down": Mark Stephenson interview, February 7, 2017. (All subsequent quotations derive from this interview.)

117   By 1:00 p.m.: MNP Report, chronology chart, p. 65.

118 prices tumbling to $16: Oil price charts, January 21, 2016, oilprice.com/.

118 new projects were being shelved: Yadullah Hussain, "How High Break-Even Costs Are Challenging New Oilsands Projects," *Financial Post*, January 22, 2015. See also Peter Tertzakian, "This Crude War Is About a Lot More Than Oil Prices and Market Share," *Calgary Herald*, March 9, 2020.

118 Syncrude, one of the smaller: Elise Stolte, "Syncrude Bison Left Behind as Fort McMurray Fires Force Further Oilsands Shutdown," *Edmonton Journal*, May 7, 2016.

120 "All of a sudden": Evan Crawford interview.

120 "When I got in the shower": "A Look Back at How the Fort McMurray Fire Unfolded," *Canadian Press*, April 27, 2017.

120 "The teachers asked us": Ryan Pitchers interview, January 29, 2017. (All subsequent quotations derive from this interview.)

122 "Had we but world enough": Andrew Marvell, "To His Coy Mistress," Poetry Foundation.

122 With the low-level jet: Scott Gilmore, "The Horror of Forest Fires Is Roaring Back," *MacLean's*, May 5, 2016.

CHAPTER 10

123 "Which kid you going to pick?": Ali Jomha interview, February 10, 2017. (All subsequent quotations derive from this interview.)

123 The dominant forest type: D. J. Downing and W. W. Pettapiece, eds., "Natural Regions and Subregions of Alberta," Natural Regions Committee, 2006, p. 149.

123 Radiant heat: Martin E. Alexander, "The Power of Crown Fires in Conifer Forests," a presentation prepared for the Canadian Forest Service, Northern Forestry Centre, April 10, 2009; revised February 17, 2020, p. 1.

123 By the time the fire: David Staples, "Alberta Battles 'The Beast,' a Fire That Creates Its Own Weather and Causes Green Trees to Explode," *Edmonton Journal*, May 7, 2016.

124 "Dad drug us up here": Paul Ayearst interview, May 1, 2017. (All subsequent quotations derive from this interview.)

125 It wasn't until 2:05: Regional Municipality of Wood Buffalo (RMWB) Twitter feed.

126 "The sky was red": Emma Elliott interview, January 22, 2017.

130 "I have to save my wife and children": Dave Dubuc interview, January 25, 2017.

131 "If you're driving in the car": Ali Jomha interview.

CHAPTER 11

133 "All halls! All halls!": Asher, *Inside the Inferno*, p. 55.

134 "She was at the door": Troy Palmer interview, April 30, 2017. (All subsequent quotations derive from this interview.)

137 "literally like lava": Jude Groves interview, April 24, 2017.

138 By then, the four-story Super 8: Struzik, *Firestorm*, p. 43.

140   more like lacrosse: Stewart Culin, *Games of the North American Indians* (New York: Dover, 1975), pp. 562ff. See also Joseph Boyden, *The Orenda* (Toronto: Hamish Hamilton, 2013), pp. 175ff.

140   Today, more than a third: Elizabeth Shogren, "What Fire Researchers Learned from California's Blazes," *High Country News*, December 11, 2017.

141   "Watching houses": Stephen J. Pyne, "Welcome to the Pyrocene," *Slate*, May 16, 2016.

143   "We could feel it": Hawley et al., *Into the Fire*, p. 9.

143   Fire trucks, invisible in the smoke: Ibid., p. 7.

CHAPTER 12

144   "We don't have a forest fire problem": Bill Gabbert, "The Home Ignition Problem," *Wildfire Today*, July 11, 2019.

145   Underwriters Laboratories conducted: "How Closing Your Bedroom Door Could Save Your Life in a Fire," *CTV Vancouver*, October 24, 2017. See also "New vs. Old Room Fire Final UL," YouTube, posted on December 17, 2010.

151   "the explosions were constant": John Topolinsky interview, November 5, 2016.

151   "only without the incoming": Pat Duggan interview, February 8, 2017. (All subsequent quotations derive from this interview.)

154   More alarming still: April 28, "Daily Data Report for April 2016," Fort McMurray A, Alberta, Past Weather and Climate, Government of Canada, climate.weather.gc.ca/.

155   In some cases, over-the-radio orders: Lucas Welsh interview.

CHAPTER 13

158   "I do my best to do my damnedest": Tim Hus (with Corb Lund), "Hurtin' Albertan," *Huskies and Husqvarnas*, album, 2006.

159   "There's four boys": Wayne McGrath interview, May 4, 2017. (All subsequent quotations derive from this interview.)

162   "So, folks, here's the video": "Labrador Man Wayne McGrath Captured a Frightening Scene as Homes Burned Around Him in Fort McMurray," *CBC Newfoundland and Labrador*, May 4, 2016.

164   The fury of the fire: This YouTube video of an acetylene tank fire in Texas gives an indication of the hazards firefighters faced throughout Fort McMurray: "Oxyacetilene Canister Plant in Texas Blowing Up," YouTube, see at 1:00. Posted May 20, 2010.

165   Taken from a security camera: "Fort McMurray Man Watches His Home Burn on Security Cam," *CBC News*, May 6, 2016.

167   "You're on your own": David Staples, "Firestorm Part Three: 'You're Out of Time,'" *Edmonton Sun*, May 3, 2017.

167   Running to each one: Paul Curtis interview, Fort McMurray Waterworks, June 14, 2019.

168   They could even drain: Ibid.

170   A lieutenant named Damian Asher: Asher, *Inside the Inferno*, p. 99.

CHAPTER 14

171   "We look at the present": Marshall McLuhan, *The Medium Is the Massage* (New York: Random House, 1967).

172   "So there was hail": Exodus 9:24 (King James Version).

172   There was none like it: On September 22, 2005, at least 2.5 million evacuated coastal Texas and Louisiana due to the approach of Hurricane Rita. The evacuation of Manhattan on September 11, 2001, included the largest sea evacuation in recorded history, with more than 500,000 being evacuated in nine hours by hundreds of boats. On October 8–9, 1871, about 100,000 people lost their homes during the Great Chicago Fire. Presumably, all of those people would have evacuated, along with others whose homes were at risk, but that fire burned for approximately thirty-six hours, with residents fleeing ahead of it as it moved north and east.

173   A photo taken: Via Jude Groves.

173   In the Northern Hemisphere: Robinson Meyer, "The Simple Reason That Humans Can't Control Wildfires," *Atlantic*, November 13, 2018.

173   Because of their size: Ed Struzik, "Fire-Induced Storms: A New Danger from the Rise in Wildfires," *Yale Environment 360*, January 24, 2019.

174   wildfire-generated pyroCbs: Michael Fromm et al., "The Untold Story of Pyrocumulonimbus," *Bulletin of the American Meteorological Society*, September 2010, pp. 1193-1209.

174   The plume it generated: Mike Fromm, personal communication, March 4, 2020.

174   While ember-generated fires: Andrew Dowdy et al., "Pyrocumulonimbus Lightning and Fire Ignition on Black Saturday in Southeast Australia," *Journal of Geophysical Research: Atmospheres* 122, no. 14 (July 27, 2017): 7342–735.

175   "All I can hear right now": Vince McDermott, @Vincemcdermott, Twitter, May 3, 2016, 2:15 p.m.

175   Jill Edwards, the business manager: Jill Edwards interview, February 6, 2017.

175   Radiant heat from the fire: Sarah Boon, "Northern Alberta Wildfires," *Science Borealis*, May 4, 2016.

176   This was true: Name withheld.

177   In 2012, he was hired: David Staples, "Firestorm Part Three: 'You're Out of Time,'" *Edmonton Sun*, May 3, 2017.

178   "At times," Staples wrote: Ibid.

178   "We were starting": Episode transcript, *CBC Radio: The Current*, July 27, 2016.

178   "Rather than learning": MNP Report, p. 41.

178   The RCMP found out: Struzik, *Firestorm*, p. 42.

179   *"Yes, and so any river is huge"*: Translation by Michael Griffin from the unrevised 1924 Loeb edition by W. H. D. Rouse, VI.674–77.

179   "The fool believes": Nassim Taleb, *Antifragile* (New York: Random House, 2012), p. 46.

179   "I didn't quite believe it yet": Reid Fiest interview.

179   "I thought, 'What the hell is going on here?'": Evan Crawford interview.

180   Given the speed and mobility: David Staples, *Edmonton Journal*.

180   "This gives the most people": Marion Warnica, "Battling the Beast: The Untold Story of the Fight to Save Fort McMurray," *CBC News*, July 27, 2016.

181   "I looked down at Waterways": John Knox interview.

182   While fires burned: "Apocolypse & Ash—Couple Narrowly Escape Fort McMurray Wildfire," YouTube, posted on May 10, 2016, see at 4:40.

182   By 7:00 p.m.: Julia Parrish, " 'Sad Day': Tens of Thousands Evacuated from Fort McMurray Due to Wildfire," *CTV Edmonton*, updated May 4, 2016.

CHAPTER 15

184   "When thou walkest through the fire": Isaiah 43:2 (King James Version).

184   "Basically, the fire behavior": "RMWB Wildfire Press Briefing, May 3, 2016, 10:00 p.m." (audio only), YouTube, posted on May 4, 2016.

185   "All I thought": Tina LeDrew Sager quoted in Eleanor Hall, "Canadian Wildfire: Weather Change Respite to Exhausted Fire Crews," *World Today*, ABC News, May 9, 2016.

185   The closest rain: See timeanddate.com/.

187   Taking their places: "Final Update 39: 2016 Wood Buffalo Wildfires (June 10 at 4:30 p.m.)," May 8, 2016, alberta.ca/.

189   Among more plausible belongings: Caitlin Hanson, "Fort McMurray Fire: What People Brought with Them," *CBC News*, May 6, 2016.

190   "The people's demeanor": Henry Lansdell, *Through Siberia*, vol. I (1879), p. 267.

190   "During the confusion": *New York Times*, June 16, 1886, p. 1.

CHAPTER 16

193   "It is a possibility": Jana G. Pruden, "A Week in Hell: How Fort McMurray Burned," *Globe and Mail*, May 6, 2016.

193   "This is a nasty, dirty fire": "Regional Fire Chief Calls Fort McMurray Wildfire 'Nasty' and 'Dirty,' " *Global News*, May 4, 2016.

194   "I really believed": Marion Warnica, "Battling the Beast: The Untold Story of the Fight to Save Fort McMurray," *CBC News*, July 27, 2016.

195   "This fire," Allen said: "Regional Fire Chief Calls Fort McMurray Wildfire 'Nasty' and 'Dirty,' " YouTube, posted on May 4, 2016.

196   The Great Chicago Fire: "The Great Chicago Fire and the Web of Memory: Inside the Burning City," greatchicagofire.org/.

196   A combination of drought conditions: Japan has endured more urban fires than any other country. In addition to many other disastrous fires over the past three centuries, a series of firestorms following the Great Kanto earthquake of 1923 killed well over 100,000 people in Tokyo and surrounding areas. In 1945, American firebombing raids targeted sixty-seven Japanese cities to devastating effect. These lethal raids on predominantly wooden cities preceded the nuclear attacks on Hiroshima and Nagasaki, which ignited additional firestorms. General Curtis LeMay, who oversaw the Japanese bombing campaign, said later that, had

the U.S. lost the war, he would have been prosecuted as a war criminal. See "Firebombing Japan," *The Pop History Dig*, pophistorydig.com/.

197   Twenty years ago: "It's Going to Be a Long, Hot Summer in Toronto, Environment Canada Says," *CBC News*, July 8, 2020. See also "Hourly Data Report for July 2000," Past Weather and Climate, Government of Canada, climate .weather.gc.ca/.

197   in 2020, it counted fourteen: "July 2020 Weather in Toronto," graph, www .timeanddate.com/.

197   During the same month: "Excessive Heat Warning Continues Throughout Arizona; Valley Ties High Temperature Record for August 3," *Fox 10 Phoenix*, August 4, 2020.

197   the new location: Janet French, "Fort McMurray Wildfire: A Small Fire Turns into the Beast," *Edmonton Journal*, May 13, 2016.

197   Miles overhead: MNP Report, p. 69.

198   Even with the addition: "Fort McMurray Wildfire Update #3—May 4, 2016 at 3:15 pm," YouTube, posted on May 4, 2016.

199   "When we actually come through": Paul Spring interview, November 7, 2016.

201   Thirty-six hours in: "Fort McMurray Fire—Hwy 63 Prairie Creek. Taken 1:46 am May 5, 2016," YouTube, posted on May 7, 2016.

201   it never gets tired: See night burning in Calf Canyon/Hermits Peak Fire, Las Vegas, New Mexico, in May 2022 (biggest in the state's history), "Wildfires in New Mexico Continue to Grow," *NPR*, May 10, 2022.

CHAPTER 17

204   If you see me running: Via Evan Crawford, November 7, 2016.

205   "the LORD rained upon Sodom": Genesis 19:24–25 (King James Version).

205   Working in conjunction: Douglas Fox, "Inside the Firestorm," *High Country News*, April 3, 2017.

206   "Fire severity in German structures": Standard Oil Development Company, "Design and Construction of Typical German and Japanese Test Structures at Dugway Proving Grounds, Utah," Standard Oil Development Project 30601, declassified government document, p. 6.

206   "The bombs," wrote one survivor: Miller, p. 126.

206   High above the city: Ibid., p. 39.

207   According to an official: Ibid., p. 143.

207   Twenty thousand civilians: Greig Watson, "Operation Gomorrah: Firestorm Created 'Germany's Nagasaki,'" *BBC News*, August 2, 2018.

207   Sixteen thousand apartment buildings: "Royal Air Force Bomber Command 60th Anniversary," Campaign Diary: 27/28 July 1943," National Archives, nationalarchives.gov.uk/.

207   In time, the bombing: "Battle of Hamburg," National Archives, national archives.gov.uk/.

209   The firestorm engineers: Standard Oil Development Company, "Design and Construction of Typical German and Japanese Test Structures at Dugway Proving Grounds, Utah," p. 9.

213   "If we leave that F-150": Ronnie Lukan interview (with Jamie and Ryan Coutts), February 4, 2017.

215   "Tearing into somebody's house": Jim Rankin interview (with Chris Hubscher), February 6, 2017. (All subsequent quotations from Rankin and Hubscher derive from this interview.)

215   "Every mountain and hill shall be made low": Isaiah 40:4 (King James Version).

216   "to crush and drown": Troy Palmer interview.

221   "It was on fire": Hawley et al., *Into the Fire*, p. 51.

221   At that moment in Prairie Creek: "Fort McMurray Fire—Hwy 63 Prairie Creek, Taken 1:46 AM May 5, 2016," YouTube, posted on May 7, 2016.

221   "They aren't just gonna": Hawley et al., p. 51.

221   "the glutton element": Seamus Heaney, trans., *Beowulf* (New York: Farrar, Straus and Giroux, 2000), l. 1124.

222   "The beast is still up": Video message posted online, May 5, 10:00 p.m., Bill Chappell, "'The Beast Is Still Up': Alberta Wildfires Rage; Evacuees Told to Wait It Out," *NPR*, May 6, 2016.

224   "The dark figure streaming with fire": J. R. R. Tolkien, *The Fellowship of the Ring* (New York: Ballantine, 1970), pp. 428–29.

PART THREE: RECKONING

225   "I gave them fire": Aeschylus, *Prometheus Bound*, trans. Joe Agee (New York: NYRB Classics), pp. 20, 31.

CHAPTER 18

227   "Dissonance": William Carlos Williams, *Paterson* (New York: Penguin, 1983), p. 176.

227   The first time the word "atmosphere": *Oxford English Dictionary*.

227   That there were elements: John W. Severinghaus, "Fire-air and Dephlogistication: Revisionisms of Oxygen's Discovery," *Advances in Experimental Medicine and Biology* (2003).

228   "Once any quantity of air": Ibid., p. 181.

228   "It is evident, however": Ibid.

228   "He had very little knowledge": Johnson, *The Invention of Air*, p. 73.

229   Priestley continued: Ibid., pp. 166ff.

229   "This observation": *Philosophical Transactions of the Royal Society* 62 (1772).

229   "exciting the attentions": Johnson, pp. 60ff.

229   "the vegetable creation": "From Benjamin Franklin to Joseph Priestley, [July 1772]: Extract," *Founders Online*, National Archives, founders.archives.gov (via Johnson).

232   Once in the atmosphere: Alan Buis, "The Atmosphere: Getting a Handle on Carbon Dioxide," NASA—Global Climate Change, October 9, 2019, climate.nasa.gov/.

232   Meanwhile, methane: Per EPA, epa.gov/.

232   A by-product of fracking: Phil McKenna, "To Counter Global Warming, Focus Far More on Methane, a New Study Recommends," *Inside Climate News*, February 9, 2022. See also Akshat Rathi, "The Case Against Methane Emissions Keeps Getting Stronger," *Bloomberg*, February 15, 2022; and "Methane Emissions from the Energy Sector Are 70% Higher Than Official Figures," International Energy Agency report, February 23, 2022: "The methane leaked in 2021 could have provided 180 billion cu. m. for the market (~all the gas used in Europe's power sector)."

236   "It is difficult to know": Timothy Casey, "Text of E. Burgess' 1837 Translation of Fourier (1824)," burgess1837.geologist-1011.mobi/, p. 11.

236   On a sunny July day: Horace-Bénédict de Saussure, "The Cause of the Cold That Reigns on the Mountains," *Travels in the Alps*, trans. Alastair B. McDonald, vol. 2, chap. 35. June, 2017, unpublished.

236   "The more dense the air": Ibid., pp. 16–17 (quote edited by author).

237   "a great mathematical poem": James R. Fleming, *Joseph Fourier's Theory of Terrestrial Temperature* (Oxford, 1998).

CHAPTER 19

238   "Some say the world will end in fire": Robert Frost, *New Hampshire* (New York: Henry Holt, 1923).

238   "An atmosphere of that gas": Eunice Foote, "Circumstances Affecting the Heat of the Sun's Rays," *American Journal of Science and Arts*, ser. 2, vol. 22, art. 31 (November 1856): 383.

239   It was on that late summer day: The male colleague was Joseph Henry, a pioneer in electromagnetism and the first secretary of the Smithsonian Institution.

239   "The receiver containing the gas": Foote, p. 382.

239   Étienne Lenoir built: Tabea Tietz, "Étienne Lenoir and the Internal Combustion Engine," *SciHi Blog*, January 24, 2021.

239   "When the heat is absorbed": John Tyndall, "On the Transmission of Heat of Different Qualities Through Gases of Different Kinds," *Notices of the Proceedings at the Meetings of the Members of the Royal Institution of Great Britain*, vol. 3 (London, 1862), p. 158. See also Roland Jackson, "Who Discovered the Greenhouse Effect?," rigb.org/explore-science/explore/blog/who-discovered-greenhouse-effect.

240   As similar as Tyndall's findings: There is a case to be made for questioning Tyndall's priority. In 1856, Eunice Foote's paper was published in the same issue of *The American Journal of Science and Arts* in which an article by Tyndall (on an unrelated subject) also appeared. Over the next year, Foote's experiment was written up in *Scientific American*, the *New-York Daily Tribune*, and several European journals, including the *Edinburgh New Philosophical Journal* and *German Advances of Physics in Year 1856*, an annual compendium of significant discoveries. Though based in England, Tyndall had trained in Germany, was fluent in German, and maintained close ties with that country's scientific community. Given this and his keen interest in the subject, combined with the timing of

his own discoveries and his noted eagerness to claim priority for them, it is reasonable to question how and why he had no awareness of Foote's revelatory work.

240   It would take another forty years: Svante Arrhenius, "On the Influence of Carbonic Acid in the Air upon the Temperature of the Ground," *London, Edinburgh, and Dublin Philosophical Magazine and Journal of Science* (April 1896): 237–76. (In 1895, Arrhenius presented a paper to the Stockholm Physical Society.) Arvid Högbom quotes (from 1894) below begin on p. 269; impact of coal on climate, p. 270.

240   After making tens of thousands: Julia Uppenbrink, "Arrhenius and Global Warming," *Science* 272 (May 24, 1996): 1122.

240   Arrhenius determined: Arrhenius, "On the Influence of Carbonic Acid," "Table VII: Variation of Temperature Caused by a Given Variation of Carbonic Acid," p. 266.

240   "very lively discussions": Ibid., p. 267.

240   the impacts of carbon dioxide: Henning Rodhe et al., "Svante Arrhenius and the Greenhouse Effect," *Ambio* 26, no. 1 (February 1997).

241   "This quantity of [CO_2]": Arvid G. Högbom, "On the Probability of Global Changes in the Level of Atmospheric CO2," *Svensk kemisk Tidskrift* (1894), quoted in Arrhenius, "On the Influence of Carbonic Acid," p. 270ff, p. 269. (Translation by Patrick Lockerby.)

241   At the turn of the nineteenth century: Smil, *Oil*, p. 161.

241   "The atmosphere may act": Nils Ekholm, "On the Variations of the Climate of the Geological and Historical Past and Their Causes," *Quarterly Journal of the Royal Meteorological Society* (January 1901): 19.

242   "[Lowell] takes into account": *London, Edinburgh, and Dublin Philosophical Magazine and Journal of Science*, ser. 6, vol. 14, no. 84 (1907): 749.

242   After describing the "hot-house": Arrhenius, *Worlds in the Making: The Evolution of the Universe* (New York, 1908), p. 51.

242   "The enormous combustion of coal": Ibid., p. 61.

242   To most people: Ibid., p. 63.

242   In the time since Arrhenius's: Industrial fossil fuel emissions increased from 1.5 billion tons in 1895 to 36 billion tons in 2018. See Hannah Ritchie and Max Roser, "Annual Total CO2 Emissions," ourworldindata.org/.

243   He had doubts: Frank Wicks, "The Engineer Who Discovered Global Warming," *American Society of Mechanical Engineers*, April 29, 2020.

243   "Few of those familiar": G. S. Callendar, "The Artificial Production of Carbon Dioxide and Its Influence on Climate," *Quarterly Journal of the Royal Meteorological Society* 64 (1938): 223–40.

243   Callendar made his calculations: Ibid., p. 233.

244   "The course of world temperatures": Ibid., p. 236.

245   Was it 2,000 years ago: Paul Stephenson, "Ancient Roman Pollution," *Lapham's Quarterly*, February 23, 2022.

246   In his lifetime: James R. Fleming, "What Role Did G. S. Callendar Play in Reviving the CO2 Theory of Global Climate Change?," Presidential Sympo-

sium on the History of the Atmospheric Sciences: People, Discoveries, and Technologies. Hans Seuss and Roger Revelle referred to the *"Callendar effect,"* defining it as "climatic change brought about by anthropogenic increases in the concentration of atmospheric carbon dioxide, primarily through the processes of combustion."

246 "Releases of carbon dioxide": *Washington Post*, May 5, 1953, p. 5.

246 "In the hungry fires of industry": "Science: Invisible Blanket," *Time*, May 25, 1953.

246 In June 1953, *Life*: Lincoln Barnett, "The World We Live In—Part IV: The Canopy of Air," *Life*, June 8, 1953.

246 "It is not usually appreciated": Gilbert Plass, "Carbon Dioxide and the Climate," *American Scientist* 44, no. 3 (July 1956): 302–16, 305.

247 "Today man by his own activities": Ibid., pp. 305–6.

247 "In the last fifty years": Ibid., p. 310.

247 *"There can be no doubt"*: Ibid., p. 312.

247 In his article: Smil, *Oil*, p. 161.

248 on that balmy March morning: "Arlington, VA, Weather History," Wunderground, wunderground.com/.

248 "Human beings during": "Hearings Before Subcommittees of the Committee on Appropriations on Second Supplemental Appropriation Bill," House of Representatives, 84th Congress, Second Session, 1956 (426), pp. 467ff.

249 "Didn't I read": Ibid., p. 473.

249 "Glacier studies": Ibid., p. 502.

249 "The reason [for this new warming cycle]": Ibid., p. 474.

250 "I think the best way": "Hearings Before Subcommittees of the Committee on Appropriations," House of Representatives, 85th Congress, First Session, 1957, pp. 104ff.

252 The 20 percent increase: Atmospheric carbon dioxide concentrations in 1957 were 315 parts per million. They reached 378 ppm, Revelle's cautioned 20 percent increase, in 2004.

252 has increased eightfold: "Indicators of Climate Change in California—Wildfires," Cal Fire publication, p. 185.

252 Meanwhile, the drought: Kasha Patel and Lauren Tierney, "These Maps Illustrate the Seriousness of the Western Drought," *Washington Post*, June 16, 2022.

252 As for fire tornadoes: Andy Park and Alex McDonald, "Former ADF Official Says Increasing Climate-Related Weather Events Could Overwhelm Defence Force," *ABC News*, April 19, 2012.

253 Given where things stand: Philip Shabecoff, "Global Warming Has Begun, Expert Tells Senate," *New York Times*, June 24, 1988.

CHAPTER 20

254 "Tomorrow came and went": Cormac McCarthy, *The Road* (New York: Alfred A. Knopf, 2006), p. 28.

255 (talk about colonization): Doreen Stabinsky, @doreenstabinsky, "Net zero is the great escape of developed countries. We refuse carbon colonialism, Diego Pacheco (Bolivia) on behalf of the Like Minded Developing Countries at #cop26," Twitter, November 13, 2021.

255 *The Unchained Goddess*: "The Unchained Goddess 1958—Bell Science Hour," YouTube, posted on August 8, 2015.

255 "man may be unwittingly changing": Ibid., at 50:51.

256 Columbia's monumental Low Library: Benjamin Franta, "On Its 100th Birthday in 1959, Edward Teller Warned the Oil Industry About Global Warming," *Guardian*, January 1, 2018.

256 "Many petroleum products": Allen Nevins et al., "Energy and Man: A Symposium," Trustees of Columbia University, 1960, p. 25.

257 Teller was a champion: Ibid., pp. 67–68.

258 "When the temperature does rise": Ibid., p. 70.

259 Humble Oil, a subsidiary of Esso/Exxon: February 2, 1962, pp. 88-89.

259 That same year: *Evolution of Canada's Oil and Gas Industry*, Canadian Centre for Energy Information, 2004, p. 37.

260 "At a single stroke": Tristin Hopper, "Nuke the Oilsands: Alberta's Narrowly Cancelled Plan to Drill for Oil with Atomic Weapons," *National Post*, August 2, 2016.

260 strontium 90, a known "bone seeker": "Alberta Technical Committee Report to the Minister of Mines and Minerals and the Oil and Gas Conservation Board," Alberta Government Publications, 1959.

260 Much to the relief: *Petroleum History Society Archives* 16, no. 4 (June 2005).

261 This sophisticated instrument: "A Breathing Planet, Off Balance," NASA, November 12, 2015, nasa.gov/.

261 A 12 percent increase: D. Luthi et al., "Graphic: The Relentless Rise of Carbon Dioxide," NASA, Global Climate Change, climate.nasa.gov/.

262 Today, there are monitoring: "Bloomberg Carbon Clock," bloomberg.com/.

263 It was called: Wallace S. Broecker et al., "Restoring the Quality of Our Environment," President's Science Advisory Committee Report on Atmospheric Carbon Dioxide, 1965, climatefiles.com/.

263 "Man is unwittingly": Ibid., pp. 126–27.

263 "There is evidence": James R. Garvey, "Air Pollution and the Coal Industry," *Mining Congress Journal* (August 1966).

264 The following year: E. Robinson and R. C. Robbins, "Sources, Abundance, and Fate of Gaseous Atmospheric Pollutants," Stanford Research Institute, February 1968, smokeandfumes.org/documents/document16.

264 "when the abundant pollutants": Ibid., p. 110.

265 There was wiggle room: Geoff Dembicki, "Has Suncor Seen the Climate Crisis Coming for 61 Years?" *Tyee*, July 21, 2020.

266 "Present thinking holds": J. F. Black, "The Greenhouse Effect," summary, Exxon Research and Engineering Company, July 1977, p. 2.

266 "foresee and prevent": John W. Zillman, "A History of Climate Activities," World Meteorological Organization, 2009, public.wmo.int/en/.

266   It was a hopeful beginning: Neela Banerjee, "Exxon's Oil Industry Peers Knew About Climate Dangers in the 1970s, Too," *Inside Climate News*, December 22, 2015.

266   Exxon was a leader: Neela Banerjee et al., "Exxon Believed Deep Dive into Climate Research Would Protect Its Business," *Inside Climate News*, September 17, 2015.

266   "STUDY FINDS WARMING TREND": *New York Times*, August 22, 1981, p. A1.

266   In 1982, Exxon released graphs: Exxon Research and Engineering Company memo from M. B. Glaser, Manager, Environmental Affairs Program: "CO2 'Greenhouse' Effect," November 12, 1982; Figure 9, p. 28 (via *Inside Climate News*).

267   "Exxon knew": David Hasemyer, "2015: The Year We Found Out #ExxonKnew," *Inside Climate News*, December 30, 2015.

267   "There is concern": M. B. Glaser, "CO2 'Greenhouse' Effect," internal memo, Exxon Research and Engineering Company, November 12, 1982, p. 5, climate files.com/.

267   In October of that year: E. E. David, "Inventing the Future: Energy and the CO2 'Greenhouse' Effect," speech, Exxon Research and Engineering Company, October 26, 1982, climatefiles.com/.

268   The bibliography goes on: Banerjee, "Exxon's Oil Industry Peers Knew." See also Shannon Hall, "Exxon Knew About Climate Change Almost 40 Years Ago," *Scientific American*, October 26, 2015.

268   "This is no wad of cash": Douglas Martin, "The Singular Power of a Giant Called Exxon," *New York Times*, May 9, 1982.

268   In October 1983: Walter Sullivan, "Study Finds Warming Trend That Could Raise Sea Levels," *New York Times*, August 22, 1981.

269   "Swiss RE and Munich RE": Don Smith, via Tom Smith, personal communication, July 8–15, 2021; 1985 joint UNEP/ICSU/World Meteorological Organization Conference in Villach, Austria. At the time, Don had been nominated for deputy secretary-general of the WMO.

269   This vexing quandary: "Crude Oil Prices—70 Year Historical Chart," Macrotrends, macrotrends.net/.

270   "Government is not": Inaugural Address, January 20, 1981, reaganfoundation .org/.

270   One Watt: Bill Prochnau, "The Watt Controversy," *Washington Post*, June 30, 1981.

271   For the first time: Philip Shabecoff, "Global Warming Has Begun, Expert Tells Senate," *New York Times*, June 24, 1988, A-1.

271   They called themselves: This was the brainchild of the publicist E. Bruce Harrison. Via Amy Westervelt, *Drilled* podcast, Season 3, Episode 8: "Meet the Harrisons."

271   The GCC's role: "Global Climate Coalition," desmog.com/. See also GCC Disinformation campaign: Kurt Davies, Climate Investigation Center, "Once Again the US Has Failed to Take Sweeping Climate Action. Here's Why," *NPR*, October 27, 2021.

271   Adopting the tobacco industry's: Amy Westervelt, *Drilled* podcast, Season 3, Episode 1: "The Mad Men of Climate Denial."

271   Lobbyists and receptive politicians: Amy Westervelt, *Drilled* podcast, Season 1, Episode 4: "Exploiting Scientists' Kryptonite: Certainty." See also María Paula Rubiano A., "How Economists Helped Big Oil Obstruct Climate Action for Decades," *Mother Jones*, October 11, 2021.

271   Exxon and the Koch family: Scott Neuman and Jeffrey Pierre, "How Decades of Disinformation About Fossil Fuels Halted U.S. Climate Policy," *NPR*, October 27, 2021. See also Jane Mayer, "'Kochland' Examines the Koch Brothers' Early, Crucial Role in Climate-Change Denial," *New Yorker*, August 13, 2019; Elliott Negin, "Will This Case Finally Bring Down ExxonMobil's Culture of Climate Deception?," *EcoWatch*, November 5, 2018; and Robert J. Brulle et al, "Corporate Promotion and Climate Change: An Analysis of Key Variables Affecting Advertising Spending by Major Oil Corporations, 1986-2015," *Climatic Change*, March 2020.

272   It's hard work: Matthew C. Nisbet and Teresa Myers, "Twenty Years of Public Opinion About Global Warming," *Public Opinion Quarterly* 71, no. 3 (fall 2007): 444–70.

272   Despite knowing full well: Dawn Stover, "Shaming: I Care About Climate Change, So Why Am I Driving an SUV?," *Bulletin of the Atomic Scientists*, November 18, 2019.

272   atmospheric CO_2 has increased: NOAA, Global Monitoring Laboratory (May 2021), gml.noaa.gov/ccgg/trends/.

272   According to a summary analysis: "SUVs Are Worse for the Climate Than You Ever Imagined," wired.com/.

272   "The number of trucks and SUVS": "Average Canadian Vehicle Size Rises 25%. Automakers Double Down on Trucks, SUVs," theenergymix.com/.

273   One of the few things: Fuel economy, 2020 Ford F150 Pickup, U.S. Department of Energy, fueleconomy.gov/. Fuel economy, 1988 Ford F150 Pickup, U.S. Department of Energy, fueleconomy.gov/.

273   Despite the abundance: "Ford F Series—US Sales Figures," fordauthority .com/.

273   If they got one thing wrong: Robert Lee Hotz, "World's Ice Is Melting Faster Than Ever, Climate Scientists Say," *Wall Street Journal*, January 25, 2021.

CHAPTER 21

276   "It will sometimes burst": Herman Melville, *Moby-Dick* (New York: W. W. Norton, 2002), p. 379.

276   Prior to that sweltering: John Thistleton, "Researchers Confirm First 'Fire Tornado' During 2003 Bushfires," *Sydney Morning Herald*, November 19, 2012.

276   Looking northward: "Mount Arawang Summit," alltrails.com/.

277   "Holy *shit* . . . Ho-ly *mackerel*": Tom Bates, "Fire Tornado Video," ACT Emergency Services Agency, esa.act.gov.au/cbr-be-emergency-ready/bushfires/fire -tornado-video.

277 The Canberra fire tornado: Richard H. D. McRae, "An Australian Pyro-tornadogenesis Event," *Natural Hazards*, October 12, 2012.

278 While you can have: Rick McRae and Jason Sharples, "Turn and Burn: The Strange World of Fire Tornadoes," *Conversation*, December 17, 2012.

278 It is fair to say: "7 of the Most Destructive Wildfires in Australian History," *Interesting Engineering*, January 24, 2020.

278 "There are no weather records": "Flashback: Black Saturday," *ABC News*, You-Tube, posted on February 6, 2010.

278 The ambient temperature: "Melbourne—Highest Temperature for Each Year," *Current Results*, currentresults.com/.

279 While none of them: Kennedy Warne, "60 Hours on Burning Kangaroo Island," *National Geographic*, January 22, 2020.

279 So otherworldly: Cameron Stewart, "The Australian 'Black Saturday' Bushfires of 2009," *Encyclopaedia Britannica*, britannica.com/.

279 The new, more dire classification: Ibid.

279 "For your survival": "Fire Danger Ratings," New South Wales Rural Fire Service, rfs.nsw.gov.au/.

280 Flying into Redding: In August 2018, I traveled to Redding on assignment for *The Guardian* with the photographer Tim Hussin.

280 Poking through the murk: Ian Livingston et al., "Mount Shasta Is Nearly Snow-less, a Rare Event That Is Helping Melt the Mountain's Glaciers," *Washington Post*, September 15, 2021.

281 "It was unreal": Name withheld.

282 "The lawn chair's in the house": Willie Hartman interview, August 28, 2018.

282 There is video: "Aerial Footage of Massive Fire Tornado That Killed California Firefighter," YouTube, posted by *The Oregonian* on August 17, 2018.

284 "A bomb," he said: Larry Hartman interview, August 28, 2018.

284 "It made a roaring sound": Christel Hartman interview, August 28, 2018.

284 "like it had been through": Dusty Gyves interview, August 27, 2018. (All subsequent quotations derive from this interview.)

285 "You could see the plume": Steve and Carrie Bustillos interview, August 28, 2018. (All subsequent Bustillos quotations derive from this interview.)

286 This might have been true: Cal Fire, "Carr Incident Green Sheet," July 26, 2018, p. 2.

286 "I'm zero for six": Kate Baker interview, August 29, 2018.

287 "Sometimes, that channel": Robinson Meyers, "The Simple Reason That Humans Can't Control Wildfires," *Atlantic*, November 13, 2018.

291 "peak gas temperatures": Bill Gabbert, "Report Concludes Fire Tornado with 136+ Mph Winds Contributed to a Fatality on Carr Fire," *Wildfire Today*, August 20, 2018.

291 "Well-constructed houses levelled": "Enhanced Fujita Scale," Wikipedia.

292 "this is a rare": David W. Goens, "NOAA Technical Memorandum NWS WR-129: Fire Whirls," National Weather Service Office, Missoula, Montana, May 1978.

292 "Natural fire never did this": Dusty Gyves interview.

292   With the exception: The only local comparable is a gigantic fire whirl, which occurred in thick forest forty miles east of Redding during the 2014 Eiler Fire. While it broke and uprooted mature trees, it was not documented as thoroughly as the Redding event.

293   Collectively, they caused: Since then, nearly two dozen PG&E executives and board members have been named in a civil action lawsuit brought by eighty thousand fire victims. See *PBS Newshour*, February 24, 2021, at 7:15.

293   "The fire season used to run": Jonathan Cox interview, September 10, 2018.

294   "It shifted from a firefighting effort": Cheryl Buliavac interview, August 26, 2018.

CHAPTER 22

295   "If all three realms are ruined": Ovid, *Metamorphoses*, trans. Allen Mandelbaum (New York: Everyman's Library, 2013), Book II, p. 51.

295   On July 7 alone: Lien Yeung et al., " 'We Are in This for the Long Haul': No Relief in Sight as B.C. Wildfires Rage," *CBC News*, July 8, 2017.

295   "the most significant": Bethany Lindsay, "B.C. Wildfires Triggered Mega Thunderstorm with Volcano-Like Effects," *CBC News*, April 26, 2018.

296   Even so, more than forty thousand: Michelle Ghoussoub, "Meet the 30-Year-Old Who Steered B.C. Through the Worst Wildfire Season on Record," *CBC News*, October 1, 2017.

296   British Columbia's historic aerosol injection: Mike Fromm, personal communication, March 2, 2021.

296   "the mother of all pyroCbs": Lindsay, "B.C. Wildfires."

296   "The Australian bushfires": Sergei Khaykin et al., " 'The 2019/20 Australian Wildfires Generated a Persistent Smoke-Charged Vortex Rising Up to 35 Km Altitude," *Communications Earth & Environment* 1, no. 22 (2020).

297   PyroCbs are now being observed: Ed Struzik, "Fire-Induced Storms: A New Danger from the Rise in Wildfires," *Yale Environment 360*, January 24, 2019.

297   The oceans absorb: Nicholas Gruber et al., "The Oceanic Sink for Anthropogenic CO2 from 1994 to 2007," *Science*, March 15, 2019.

297   Over the course: Jennifer Bennett, "Ocean Acidification," Smithsonian ocean portal, ocean.si.edu/.

298   To put this in perspective: "Wildfires Had a Bigger Climate Impact Than the Pandemic in 2020," *Yale Environment 360*, August 3, 2021.

299   In terms of its implications: "The ship is not built to outlast the world, but to incinerate it. A workhorse of the global market, its own gargantuan carbon emissions unrecorded in any nation's ledger, it accelerates the rate at which oil is burned to make power and plastic to make commodities to make money." Mark Bould, "Dulltopia," *Boston Review*, January 22, 2018.

299   Every year, this global industry: "Global Carbon Emissions," Co2-Earth, co2 .earth/.

299   This is a rate: Bärbel Hönisch et al., "The Geological Record of Ocean Acidification," *Science* 335, no. 6072 (March 2, 2012): 1058–63.

301   Simply put: Sherry Listgarten, "What Is a 'Ton' of Carbon Dioxide Any-way?," *Almanac*, December 1, 2019; see also "How Can Carbon Emissions be Weighed?," niwa.co.nz/.

301   "the worst thing": Peter Brannen, "Burning Fossil Fuels Almost Ended All Life on Earth," *Atlantic*, July 11, 2017.

303   With far less year-round ice: Maureen E. Raymo et al., "Departures from Eustasy in Pliocene Sea-Level Records," *Nature Geoscience*, April 17, 2011.

304   If we fail this test: Christina Goldbaum and Zia ur-Rehman, " 'Very Dire': Dev-astated by Floods, Pakistan Faces Looming Food Crisis," *New York Times*, Sep-tember 11, 2022. See also Lisa Cox, " 'Unprecedented' Globally: More Than 20% of Australia's Forests Burnt in Bushfires," *Guardian*, February 24, 2020.

305   In the summer of 2021: Josh Lederman, @JoshNBCNews, "Feds formally issue first-ever water shortage declaration for Lake Mead & Colorado River, trigger-ing water cuts to AZ, NV & Mexico starting in January," Twitter, August 16, 2021.

305   Farther north: Gillian Flaccus, "Water Crisis 'Couldn't Be Worse' on Oregon-California Border," *AP*, May 24, 2021.

305   "This isn't a 'drought' ": Bob Berwyn, @bberwyn, Twitter, March 2, 2020. See also Bob Berwyn, "New Study Projects Severe Water Shortages in the Colorado River Basin," *Inside Climate News*, February 20, 2020; and Allison Chinchar, "The US 'Megadrought' Sets Another Stunning Record," *CNN*, January 13, 2022.

305   Based on tree ring analysis: Margaret Osborne, "The Western U.S. Is Experienc-ing the Worst Drought in More Than 1,200 Years," *Smithsonian*, February 17, 2022.

305   In the meantime, scientists: Bob Berwyn, "Global Warming to Spur More Fires in Alaska, in Turn Causing More Warming," *Inside Climate News*, May 16, 2016.

305   Around Fort McMurray: Jason Markusoff et al., "Fort McMurray: The Great Escape," *Maclean's*, May 12, 2016. "Since the 1960s, per-decade average tempera-tures around the city for the seven-month period between October and April have risen a stunning 3.4°C. During the same period, Environment Canada records show, precipitation levels have plummeted from a total of 161 mm in the seven months between October and April to just 80, turning the densely forested area around the city into a giant tinder box."

306   "Fire has been largely absent": Michelle E. Mack et al., "Largest Recorded Tun-dra Fire Yields Scientific Surprises," *Science Daily*, July 27, 2011.

306   "The amount of carbon": Ibid.

306   A decade later: Brian Kahn, "Wildfire Burns Across (Formerly) Icy Greenland," *Scientific American*, August 8, 2017.

306   In 2016, Tasmania: Karl Mathiesen, "World Heritage Forests Burn as Global Tragedy Unfolds in Tasmania," *Guardian*, January 27, 2016.

306   In 2015, fires in Indonesia: Ann Jeannette Glauber, "Seeing the Impact of Forest Fires in South Sumatra: A View from the Field," *World Bank Blogs*, February 19, 2016.

306   In 2012, wildfires burned: Roman Vorobyov, "Siberia in Flames," *Russia Beyond*, August 3, 2012.

306 In 2010, pan-Russian wildfires: Andrew E. Kramer, "Past Errors to Blame for Russia's Peat Fires," *New York Times*, August 12, 2010.

306 According to the global reinsurer Munich RE: "Overall Picture of Natural Catastrophes in 2010—Very Severe Earthquakes and Many Severe Weather Events—Major Catastrophes Dominate the List of Losses," Munich RE, March 1, 2011.

307 In Russia, following the worst: Anna Liesowska, "Zombie Fires Burn at -60C Outside Oymyakon, the World's Coldest Permanently Inhabited Place," *Siberian Times*, December 2, 2021.

307 And, like coal dust: Bill Gabbert, "Explosive Peat Moss," *Wildfire Today*, July 15, 2019.

307 The Indonesian peat fires: Beh Lih Yi, "Southeast Asian Fires Emitted Most Carbon Since 1997—Scientists," *Thomson Reuters Foundation*, June 28, 2016.

307 comparable to the annual emissions: "Environment—Energy-Related CO2 Emission Data Tables," U.S. Energy Information Administration, 2019, eia.gov/.

307 a 12 percent increase: Ed Struzik, "1950 Monster Fire Burned Its Way into History," *Edmonton Journal*, May 22, 2011.

308 "mutual reaction": Julius Hann, *Handbook of Climatology*, translation of *Handbuch der Klimatologie*, 2nd ed. (New York and London: Macmillan, 1903), p. 389.

308 Australia and the American West: "Tundra Is Ablaze in Magadan Region in Out-of-Season Wildfire, Complicated by Wind and Zero Snow," *Siberian Times*, November 4, 2021. See also Fred Pearce, "Why 'Carbon-Cycle Feedbacks' Could Drive Temperatures Even Higher," *Yale Environment 360*, April 28, 2020.

308 March 26, 2021: Jason Samenow, "Japan's Kyoto Cherry Blossoms Peak on Earliest Date in 1,200 Years, a Sign of Climate Change," *Washington Post*, March 29, 2021.

308 landscapes that have not known trees: Bianca Fréchette et al, "Vegetation and Climate of the Last Interglacial on Baffin Island, Arctic Canada," *Palaeogeography, Palaeoclimatology, Palaeoecology* 236 (2006) 91–106

308 These new forests: Ben Rawlence, " 'The Treeline Is Out of Control': How the Climate Crisis Is Turning the Arctic Green," *The Guardian*, January 20, 2022.

309 "What seems clear now": Jon Gertner, "In Greenland's Melting Ice, a Warning on Hard Climate Choices," *Yale Environment 360*, June 27, 2019.

309 hit 340 ppm for the first time: "Earth in the Future—Carbon Dioxide Through Time," Penn State/NASA, e-education.psu.edu/.

309 The chances of anyone alive today: NOAA National Centers for Environmental Information, Climate at a Glance: Global Time Series, published October 2022, retrieved on November 11, 2022, from https://www.ncei.noaa.gov/.

309 "Just wanna make sure": Cristi Proistosescu, @cristiproist, Twitter, September 10, 2020.

309 "I'm sixty": Via Will Cole-Hamilton, @w_colehamilton.

310 Since roughly 2000: Brendan Montague, "Brazilian Amazon 'Releasing Carbon,' " *The Ecologist*, April 30, 2021. See also Barry Saxifrage, "One of Canada's Biggest Carbon Sinks Is Circling the Drain," *The National Observer*, May 7, 2021.

310   fire, logging, land clearing: Antonio José Paz Cardona, "Settlers Invading, Deforesting Colombian National Parks 'at an Unstoppable Speed,'" trans. Theo Bradford, Mongabay.com, May 19, 2021.

311   Tundra may be the least: Yang Chen et al., "Future Increases in Arctic Lightning and Fire Risk for Permafrost Carbon," *Nature Climate Change*, April 5, 2021. See also Steven Mufson, "Scientists Expected Thawing Wetlands in Siberia's Permafrost. What They Found Is 'Much More Dangerous,'" *Washington Post*, August 2, 2021.

311   In 2020, we saw: Chelsea Harvey, "Heat-Trapping Methane Surged in 2020," *Scientific American*, April 9, 2021. See also Hannah Osborne, "Giant, 90ft Deep Craters Are Appearing on the Arctic Seafloor," *Newsweek*, March 14, 2022.

311   Despite decades: James Gustave Speth, "They Knew: How the U.S. Government Helped Cause the Climate Crisis," *Yale Environment 360*, September 15, 2021.

312   Now, the "bank": "Mortgaging the atmosphere" via Lisa Song and James Temple, "Is California's Carbon Offset Program Actually Helping the Environment?" *High Country News*, May 11, 2021.

312   "The climate system": Andy Rowell, "RIP Wally Broecker, the 'Grandfather of Climate Science,'" *Oil Change International*, February 20, 2019. See also William K. Stevens, "Scientist at Work: Wallace S. Broecker; Iconoclastic Guru of the Climate Debate," *New York Times*, March 17, 1998.

312   As the Penn State climate scientist: Upstream podcast, December 28, 2021. See also Jason Moore's writings on the "Capitalocene."

CHAPTER 23

313   "And where two raging fires meet together": William Shakespeare, *The Taming of the Shrew*. Act II, Scene 1, 139–140; Folger Library, shakespeare.folger.edu/.

314   In monetary terms: Otiena Ellwand, "Fort McMurray Wildfire Destroys Work Camp, Encroaches on Oil and Gas Facilities," *Edmonton Sun*, May 18, 2016. "The Conference Board of Canada on Tuesday estimated average oilsands output would fall by 1.2 million barrels of oil a day for two weeks, translating into $985 million in lost gross domestic product."

314   "It's super loud": Aron Harris interview, May 1, 2017. (All subsequent quotations derive from this interview.)

315   "I've never seen anything": Darby Allen, "'No One's Ever Seen Anything Like This': Fire Chief on Fort Mcmurray Wildfire," *Global News*, May 9, 2016.

315   The formulas did not need rewriting: Mike Flannigan, personal communication, July 8, 2021.

316   In 2015, ConocoPhillips CEO Ryan Lance: Tom Randall and Hayley Warren, "Peak Oil Is Suddenly Upon Us," *Bloomberg*, November 30, 2020.

316   "We think we got this thing beat": "Fort McMurray Fire Chief Speaks to Residents," *Global News*, May 10, 2016, posted to YouTube, May 11, 2016.

316   On the 16th: MNP Report, p. 59.

316   A suspected gas leak: Julia Parrish, "Two Explosions in Fort McMurray Cause Damage to a Number of Homes," *CTV News*, May 17, 2016.

317   South of town: Marion Warnica, "Hazardous Smoke and Hot Spots Slow Re-
      entry Plans for Fort McMurray," *CBC News*, May 16, 2016.

317   Blacksands Executive Lodge: Otiena Ellwand, "Fort McMurray Wildfire
      Destroys Work Camp, Encroaches on Oil and Gas Facilities," *Edmonton Sun*,
      May 18, 2016.

317   That day, the air quality: Otiena Ellwand, "Fort McMurray Air Quality Health
      Index Has Risen to Extreme Levels," *Calgary Sun*, May 16, 2016.

317   By then, smoke: Bill Gabbert, "Wildfire Smoke from Canada Affects Much of
      the United States," *Wildfire Today*, May 8, 2016.

318   "CSI murder scenes": Via Craig MacKay, January 27, 2017.

318   the lobsters in the live tank: Ibid.

319   All told, about twenty thousand: Paige Parson, "Thousands of Refrigerators
      Emptied and Crushed as Fort McMurray Landfill Deals with What's Left of
      Destroyed Homes," *Edmonton Journal*, June 19, 2016.

324   "I hate that fucking fire": Pauline Vey interview, May 3, 2017.

324   "I grew up in a war zone": Name withheld.

325   "He texted me": Sue McOrmond via email, June 6, 2019.

326   "The grass was *neon*": Randy Stefanizyn interview, January 27, 2017.

326   "On May second we had no claims": Craig MacKay interview, January 27, 2017.

327   "I was born in Louisiana": Vonda Pikes interview, April 9, 2017.

328   "A lot of people were in debt": Geoffrey Morgan, " 'Mental Degradation': A Fresh
      Wave of Layoffs Is Pushing Albertans to the Edge—and in Danger of Losing
      Their Homes," *Edmonton Journal*, October 26, 2020.

329   Concrete pulverizers: "ShearForce Fixed Demolition Pulverizers for Excavators,"
      shearforce.ca/.

CHAPTER 24

330   "In our country": From a 1974 interview, in his memoir, *The Oak and the Calf*
      (New York: Harper & Row, 1980).

331   What is happening: Ketan Joshi, "The End of Coal Is Coming Sooner Than
      You Think," *Foreign Policy*, August 13, 2021.

332   a "War Room": Geoff Dembicki, "Alberta Inquiry Paid $28K for a Report
      Smearing Hundreds of Climate Journalists," *Vice*, January 25, 2021. See also Ian
      Austen, "Alberta Took on Environmental Groups, but Only Proved They Did
      Nothing Wrong," *New York Times*, October 22, 2021.

332   Perceived enemies: Nicholas Kusnetz, "In Attacks on Environmental Advocates
      in Canada, a Disturbing Echo of Extremist Politics in the US," *Inside Climate
      News*, February 24, 2021.

332   Around the world: "UK Heatwave: Weather Forecasters Report Unprecedented
      Trolling," *BBC*, July 29, 2022.

332   Environmental groups: Drew Anderson, "Alberta's Energy 'War Room' Launches
      in Calgary," *CBC News*, December 11, 2019.

332   companies and banks: Christopher Flavelle, "Global Financial Giants Swear Off
      Funding an Especially Dirty Fuel," *New York Times*, February 13, 2020.

332   "opinion of a structural weakness": Sarah Rieger, "Moody's Downgrades Alber-

ta's Credit Rating, Citing Continued Dependence on Oil," *CBC News*, December 4, 2019.

332 In early 2020, Moody's downgraded: "Rating Action: Moody's Changes Trans-Canada's and Most Subsidiaries Outlooks to Negative from Stable, Affirms Ratings," Moody's Investors Service, March 31, 2020.

332 It was around this time: Karen Bartko, "Alberta Energy Company Under Fire for Image Appearing to Depict Greta Thunberg," *Global News*, February 29, 2020.

332 In October, Moody's downgraded: Emma L. Graney, @EmmaLGraney, Twitter, October 2, 2020.

332 "Now is a great time": "Alberta Minister Says It's a 'Great Time' to Build a Pipeline Because COVID-19 Restrictions Limit Protests Against Them," *Canadian Press*, May 25, 2020.

333 "improving market access": "Oil & Gas Survey Report," *Daily Oil Bulletin*, 2019, p. 18.

333 The mood among: Ibid., pp. 4, 5.

333 A crane operator named Randy: Not his real name.

334 Those who endorsed: John Paul Tasker, "Conservative Delegates Reject Adding 'Climate Change Is Real' to the Policy Book," *CBC News*, March 20, 2021.

334 "The problem with despotic": David Mattson, "The Cult of Hunting and Its Timely Demise," *Grizzly Times*, April 19, 2018.

334 Ever since the downturn: Ernest Scheyder and Nia Williams, "Innovators Toil to Revive Canada Oil Sands as Majors Exit," *Reuters*, June 18, 2017.

334 By 2016, auction prices: Dan Healing, "Companies Abandon Nearly One Million Hectares of Alberta Oilsands Exploration Leases," *Canadian Press*, July 28, 2017.

334 "The challenges currently posed": "Mark Carney Warns Investors of 'Potentially Huge' Losses from Climate Change Risks," *Bloomberg*, September 30, 2015.

335 In December 2016: "Statoil Sells Oilsands Assets to Athabasca Oil in Deal Worth Up to $832 Million," *Canadian Press*, December 15, 2016.

335 But then: Patrick DeRochie, "Seven Oil Multinationals That Are Pulling Out of Canada's Tar Sands," *Environmental Defence*, March 14, 2017.

335 A month after that: "Exxon to Leave Up to 3.6 Billion Barrels of Tar Sands/Oil Sands in the Ground," *Energy Mix*, February 22, 2017.

335 In March, Royal Dutch Shell: "Royal Dutch Shell Signs Deals to Sell Oilsands Assets," *CBC News*, March 9, 2017.

335 In October, the French Bank BNP: "BNP Paribas Takes Further Measures to Accelerate Its Support of the Energy Transition," press release, October 11, 2017. See also "French Bank La Banque Postale Quits Oil & Gas, Sets International Precedent," press release, October 14, 2021, reclaimfinance.org/; and "Canada's 7th Largest Bank @Blaurentienne Will Stop New Financing of Coal, Oil and Gas to 'Differentiate' Itself from Fossil Banks," December 2021, via @reclaim finance.

335 Two months later, in early 2018: "Suncor Buys Out Mocal Energy's 5% stake in Syncrude to Increase Oilsands Ownership," *Global News*, February 27, 2018.

335 As of 2022, the World Bank: Nick Cunningham, "World Bank Continues Financing Fossil Fuels Despite Climate Crisis," *DeSmog*, October 6, 2022.

336 a controlling interest: "Suncor Takes Control of Syncrude in $937M Deal for Additional Five Per Cent Stake," *Financial Post*, April 27, 2016.

336 taken over operations: Sarah Rieger, "Suncor to Assume Operation of Syncrude by End of Next Year," *CBC News*, November 23, 2020.

336 Anticipating this, a Calgary-based consortium: Sarah Rieger, "Trump Issues Presidential Permit Authorizing $22B Railway Between Alaska and Alberta," *CBC News*, September 30, 2020.

336 "Climate change," the report stated: Don Jergler, "Report Urges Urgent Action from Financial Regulators to Address Climate Change," *Insurance Journal*, September 10, 2020.

337 the world's biggest insurance companies: "Munich Re Toughens the Tone on the Oil Sands," *Reclaim Finance*, May 12, 2020.

337 Zurich Insurance: "Sustainability—Exclusion Policies," Zurich Insurance Group, zurich.com/.

337 the Hartford: Stephen Singer, "Hartford Financial to Speed Exit from Tar Sands Investments to Year End" (originally published in the *Institute for Energy Economics and Financial Analysis* on November 10, 2021).

337 Talanx Group: Paul Lucas, "Talanx Group Dropping Support for Trans Mountain Pipeline," *Insurance Business Canada*, June 30, 2020.

337 "This type of project": "Insurance Provider for Trans Mountain Pipeline Says It Won't Renew Policy," *Global News*, June 3, 2021. See also David Thurton, "Finance Canada Defends $10 Billion Loan Guarantee for Trans Mountain," *CBC News*, May 11, 2022; and "Royal Bank of Canada, TD, Scotia, CIBC, BMO, National Bank Front $10 Billion to Finance Financially Risky Trans Mountain Pipeline, Analysis Reveals—Unprecedented Loan Guaranteed by Public Money, Deal Inked Three Weeks Before Being Made Public," *Stand. Earth*, May 31, 2022.

337 In September 2021: Lyle Adriano, "Chubb Exits from Covering Tar Sands Projects," *Insurance Business Canada*, September 15, 2021. See also "Swiss Re Leads Insurance Industry's Exodus from Oil and Gas," press release, March 17, 2022, global.insure-our-future.com/.

337 The Australian Insurance giant: Graham Readfearn, "Insurance Giant Suncorp to End Coverage and Finance for Oil and Gas Industry," *Guardian*, August 21, 2020.

337 Lloyd's of London: Julia Kollewe, "Lloyd's Market to Quit Fossil Fuel Insurance by 2030," *Guardian*, December 17, 2020.

337 The same goes for mines: Corbin Hiar, "Coal, Oil Sands Companies Feel Growing Insurance Squeeze," *E&E News*, September 20, 2021. "There is always the concern in the back of insurers' minds that climate change may be the next asbestosis . . . Senior executives mentioned the fact that their kids at dinner table start talking to them about, you know, what is your business doing about climate change, and why are you not changing things?" she said. Those questions have caused leaders Surminski has spoken with to rethink the impacts of their work and to feel more personally accountable for their company's climate-related decision making. "I'm surprised how often I hear that from senior executives," she said.

338 The plaintiffs claimed: "Boulder Sues Exxon over Climate Change: Wildfires, Droughts and Water Are a Few Reasons Why," *Inside Climate News*, April 18, 2018.

338 As of 2021, at least 1,500: Tom Wilson, "Lawyer Who Defeated Shell Predicts 'Avalanche' of Climate Cases," *Financial Times*, December 30, 2021.

338 "The government accepts": John Schwartz, "Court Quashes Youth Climate Change Case Against Government," *New York Times*, January 17, 2020.

339 Climate change, they wrote: Patrick Greenfield and Jonathan Watts, "JP Morgan Economists Warn Climate Crisis Is Threat to Human Race," *Guardian*, February 21, 2020.

339 Like Mark Carney's predictions: Ibid. Between 2016 and 2018 alone, the top thirty-three banks loaned $1.9 trillion to the fossil fuel industry.

339 In fact, the world's sixty biggest banks: "Banking on Climate Chaos 2021: Fossil Fuel Finance Report," *Oil Change International*, March 24, 2021.

341 "Imagine Shell decided": Ben van Beurden, "The Spirit of Shell Will Rise to the Challenge," LinkedIn, June 9, 2021.

341 As bitumen royalties plummeted: Ainslie Cruikshank, "Alberta's Deficit Is Set to Reach Historic Levels. A Collapse in Oil Revenue Is a Big Reason Why," *Narwhal*, August 27, 2020.

341 Lawyers for Alberta's bitumen: "Opinion in Reference Re: Impact Assessment Act," p. 8, albertacourts.ca/.

342 Meanwhile, in Calgary: Geoffrey Morgan, "'Mental Degradation': A Fresh Wave of Layoffs Is Pushing Albertans to the Edge—And in Danger of Losing Their Homes," *Edmonton Journal*, October 26, 2020.

342 There hasn't been much: "Total Takes $7-Billion Writedown on Oilsands Projects, Labels Fort Hills, Surmont 'Stranded' Assets," *Bloomberg*, August 4, 2020.

342 That same month: "Keystone XL Is Dead, and Albertans Are on the Hook for $1.3B," *CBC News*, June 9, 2021.

342 In mid-July 2021: Sebastien Malo, "Maine Pipeline Co Drops Lawsuit Over City Law That Blocked Oil Export," *Reuters*, July 16, 2021.

342 A week earlier: Tony Seskus, "University of Calgary Hits Pause on Bachelor's Program in Oil and Gas Engineering," *CBC News*, July 8, 2021.

343 Van Beurden himself: @ClimatePower video, 8:31 a.m., May 5, 2022.

CHAPTER 25

344 "Maybe it wasn't as profitable": Personal communication, August 24, 2022.

344 "Virtue, then, is anything that moves you": "Joyfulness in Everything: A Conversation with George Saunders," February 15, 2017, oxfordexchange.com /blogs/.

344 On May 26, in a stunning conclusion: Maxine Joselow, "Historic Dutch Ruling Targets Corporate Emissions," *E&E News*, May 27, 2021.

344 It was, all in all: Ketan Joshi, "The Surprise Court Ruling That Cut Through Shell's Greenwashing Facade," *New Republic*, May 28, 2021. (This article is noteworthy for its thorough explanation—and dismantling—of common greenwashing tropes.)

345   (*Boulder v. Suncor Exxon*): *Boulder v. Suncor Exxon* court documents.

345   "This is the first time": "The Australian Government Has a Duty of Care to Protect Children from Climate Harm, Court Rules," *SBS* (*Special Broadcasting Service*), May 27, 2021.

345   The case, based on: A similar case, based on citizens' constitutional rights, is making its way through the Canadian courts. "Youth Climate Case Forges Ahead After Court Affirms Historic Decision," press release, March 26, 2021, ecojustice.ca/.

345   With the cataclysmic fires: Lisa Cox, "'Unprecedented' Globally: More Than 20% of Australia's Forests Burnt in Bushfires," *Guardian*, February 24, 2020.

345   "It is difficult to characterise": "*Sharma v. Minister for Environment*," Equity Generation Lawyers, equitygenerationlawyers.com/.

346   In January 2022: Lisa Friedman, "Court Revokes Oil and Gas Leases, Citing Climate Change," *New York Times*, January 27, 2022.

347   "Yeah, we knew": Justin Worland, "The Reason Fossil Fuel Companies Are Finally Reckoning with Climate Change," *Time*, January 16, 2020.

347   "the deliberate slowing": John Elkington, "Alex Steffen on Predatory Delay," April 30, 2016, johnelkington.com/2016/04/alex-steffen-on-predatory-delay/.

347   "The evidence that had been gathered": Hiroko Tabuchi and Lisa Friedman, "Oil Executives to Face Congress on Climate Disinformation," *New York Times*, October 27, 2021.

348   "the greatest threat": Worland, "The Reason Fossil Fuel Companies Are Finally Reckoning with Climate Change."

348   A fossil fuel divestment campaign: "Swarthmore Environmental Studies—Divestment Debates," swarthmore.edu/environmental-studies/.

348   University of California's $80 billion endowment: Umair Irfan, "The University of California System Is Ending Its Investment in Fossil Fuels," *Vox*, September 18, 2019.

348   As of mid-2021: Global Fossil Fuel Divestment Commitments Database, gofossilfree.org/. See also Bill McKibben, "This Movement Is Taking Money Away from Fossil Fuels, and It's Working," *New York Times*, October 26, 2021.

348   In 2019, Shell acknowledged: Bill McKibben, "Money Is the Oxygen on Which the Fire of Global Warming Burns," *New Yorker*, September 17, 2019.

348   "Removing fossil fuel securities": "Largest Federal Employee Union Applauds Biden Push to Remove Fossil Fuel Securities from Retirement Funds," press release, afge.org/publication/. See also "Blackstone, Inc., Once a Major Player in Shale Patches, Is Telling Clients Its Private Equity Arm Will No Longer Invest in the Exploration and Production of Oil and Gas," *Bloomberg*, February 22, 2022.

349   For generations, Big Oil: James Gustave Speth, "They Knew: How the U.S. Government Helped Cause the Climate Crisis," *Yale Environment 360*, September 15, 2021.

349   it is carrying debt loads: ExxonMobil: 2005 Summary Annual Report, p. 4

349   Meanwhile, dividends: Benji Jones, "Exxon Is Slashing Workers and Cutting Costs, and Employee Morale Has Collapsed. Here's Everything We Know," *Business Insider*, March 3, 2021.

349   The biggest companies: Clara Vondrich, "Big Oil and Investors Knew a Crash Was Coming: COVID-19 Just Sped Up the Clock," *Front Page Live*, May 11, 2020.

349   Home Depot was worth more: Kevin Crowley and Bryan Gruley, "The Humbling of Exxon," *Bloomberg*, April 30, 2020.

349   In 1980, the oil and gas sector: "IEEFA Update: ExxonMobil's Slide from the Top Ten of the S&P 500—Historic Turning Point for the Company," Institute for Energy Economics and Financial Analysis (IEEFA), August 30, 2019, ieefa .org/.

349   ExxonMobil, which was the most valuable: Clare Duffy, "Major Shakeup for the Dow Jones Industrial Average Index: 3 New Stocks Join," *CNN*, August 4, 2020.

349   "We have an obligation": Attracta Mooney, "Aviva Will Use Its 'Ultimate Sanction' to Force Action on Global Warming," *Financial Times*, January 30, 2021.

349   Truly unnerving to anyone: Mitchell Beer, "After Big Oil's Very Bad Week, the Message for Alberta Is Clear," *Policy Options*, June 2, 2021.

349   In their "World Energy Outlook": "World Energy Outlook 2020," International Energy Agency, October 13, 2020. See also "Renewable Power Generation Costs in 2020," International Renewable Energy Agency, 2021.

350   The threat is serious enough: Justin Worland, "The Reason Fossil Fuel Companies Are Finally Reckoning with Climate Change," *Time*, January 16, 2020.

350   by 2030 the European Union: Kate Abnett, "Climate 'Law of Laws' Gets European Parliament's Green Light," *Reuters*, June 24, 2021.

350   due largely to the bitumen industry: Barry Saxifrage, "Remember the Copenhagen Accord's 2020 Targets? Here's How Canada and Many of Its Peers Did," *National Observer*, May 24, 2022.

350   emissions have more than doubled: According to the Pembina Institute, the oil and gas sector accounted for 26 percent of Canada's GHG emissions in 2019. Emissions from the oil sands rose by 137 percent between 2005 and 2019. See also Barry Saxifrage, "Climate Snapshot: Bay du Nord," *National Observer*, February 23, 2022.

350   In a statement: Alex Ballingall, "'We Recognize the Problem': Canada's New Ministers for the Environment and Natural Resources Have the Oil and Gas Sector in Their Sights," *Toronto Star*, October 30, 2021.

350   In this way, fire resembles: James Mackintosh, "Shareholders Reign Supreme Despite CEO Promises to Society," *Wall Street Journal*, February 10, 2022. See also Kyle Bakx, "Banned for Decades, Releasing Oilsands Tailings Water Is Now on the Horizon," *CBC News*, December 6, 2021.

350   It has taken decades: Geoffrey Supran and Naomi Oreskes, "Assessing Exxon-Mobil's Climate Change Communications (1977–2014)," *Institute of Physics*, August 23, 2017. See also Max Binks-Collier, "For Decades, Alberta's Energy Regulator Massively Downplayed Crude Oil and Saline Water Spills," *National Observer*, February 16, 2022; Jeffrey Pierre and Scott Newman, "How Decades of Disinformation About Fossil Fuels Halted U.S. Climate Policy," *All Things Considered*, October 27, 2021; Rachel Siegel, "Brainard Questioned on Inflation, Climate Risk Issues as Part of Nomination to Become Fed's Second-In-Command," *Washington Post*, January 13, 2022 (in which Senator Pat Toomey

(R-PA) said, "There is no reason to believe that global warming poses a systemic risk to the financial system"); Hiroki Tabuchi, "House Panel Expands Inquiry into Climate Disinformation by Oil Giants," *New York Times*, October 28, 2021; Owen Walker, "HSBC Suspends Banker over Climate Change Comments," *Financial Times*, May 22, 2022 (regarding HSBC executive Stuart Kirk who gave a paper entitled "Why investors need not worry about climate risk"); and Zia Weise, "Shell Consultant Quits, Says Company Causes 'Extreme Harm' to Planet," *Politico*, May 23, 2022.

351 By a nearly two-thirds vote: "Chevron Investors Back Proposal for More Emissions Cuts," *Reuters*, May 26, 2021.

351 On the same day: Pippa Stevens, "Activist Firm Engine No. 1 Claims Third Exxon Board Seat," *CNBC*, June 2, 2021.

351 issued a landmark "flagship report": "Net Zero By 2050," International Energy Agency, May 2021, iea.org/reports/net-zero-by-2050.

CHAPTER 26

352 "it is a fire which consumes me": Jorge Luis Borges, "A New Refutation of Time," *Labyrinths* (New York: New Directions, 1962).

352 in recorded history (prior to 2020): "2020 Tied for Warmest Year on Record, NASA Analysis Shows," January 14, 2021, climate.nasa.gov/.

353 "We could have been looking": Mike Hager, "Will Wildfires Get Too Intense to Fight?," *Globe and Mail*, July 14, 2017.

353 "We had the radio on": Name withheld.

354 Since 2016, people all over the world: John Muyskens et al., "1 in 6 Americans Live in Areas with Significant Wildfire Risk," map, *Washington Post*, May 17, 2022. See also Bill Gabbert, "Wildfire Risk Rating Now Available for 145 Million Properties in the United States," *Wildfire Today*, May 16, 2022.

EPILOGUE

357 "How we get there": Via Harsha Walia ("How you get there is where you are"), personal communication, January 8, 13, 2020. See also Philip Booth's line, "How you get there is where / you'll arrive" from his poem "Heading Out."

BIBLIOGRAPHY

Asher, Damien (with Omar Mouailem). *Inside the Inferno: A Firefighter's Story of the Brotherhood That Saved Fort McMurray.* Toronto: Simon & Schuster, 2017.

Clark, Karl A. *The Bituminous Sands of Alberta.* Edmonton: W. D. McLean, 1929.

Davis, Lance. *The Pursuit of Leviathan: Technology, Institutions, Productivity and Profits in American Whaling, 1806–1906.* Chicago: University of Chicago Press, 1997.

Davis, Mike. "Let Malibu Burn: A Political History of the Fire Coast," *L.A. Weekly.* 1996.

Dembicki, Geoff. *Are We Screwed?: How a New Generation Is Fighting to Survive Climate Change.* New York: Bloomsbury, 2017.

The Derrick's Hand-Book of Petroleum. Oil City, PA: Derrick Publishing Co., 1898.

Ells, Sydney C. *Reflections of the Development of the Athabasca Oil Sands.* Ottawa: Department of Mines and Technical Surveys, 1962.

Forbes, R. J. *Bitumen and Petroleum in Antiquity.* London: Institute of Petroleum, 1931.

Gale, Thomas A. *Rock Oil: The Wonder of the Nineteenth Century.* Erie, PA: Sloan & Griffith, 1860.

George, Rick. *Sun Rise: Suncor, the Oil Sands and the Future of Energy.* Toronto: Harper Collins, 2012.

Giddens, Paul H. *The Birth of the Oil Industry.* New York: Macmillan, 1938.

Gough, Zachary. *The Elusive Mr. Pond: The Soldier, Fur Trader and Explorer Who Opened the Northwest.* Madeira Park, BC: Douglas & McIntyre, 2014.

Hawley, Jerron et al. *Into the Fire: The Fight to Save Fort McMurray.* Toronto: McClelland & Stewart, 2017.

Heming, Arthur. *The Drama of the Forests.* Toronto: Doubleday, Page and Co., 1922.

Henry, James D. *History and Romance of the Petroleum Industry.* London: Bradbury, Agnew & Co., 1914.

Huberman, Irwin. *The Place We Call Home: A History of Fort McMurray as Its People Remember.* Fort McMurray: Fort McMurray Historical Society, 2001.

Hudson's Bay Company Archives, under auspices of Archives of Manitoba. See hbca@gov.mb.ca, manitoba.ca/archives.

Hunt, Joyce E. *Local Push—Global Pull: The Untold Story of the Athabaska Oil Sands, 1900–1930.* Calgary: PushPull, Ltd., 2011.

Jackson, Joe. *A World on Fire: A Heretic, An Aristocrat, and the Race to Discover Oxygen.* New York: Viking, 2005.

Jean, Francis K. *More Than Oil: Trappers, Traders & Settlers in Northern Alberta.* Fort McMurray: City Centre Group, 2012.

Johnson, Steven. *The Invention of Air: A Story of Science, Faith, Revolution, and the Birth of America.* New York: Riverhead, 2008.

Keith, Lloyd, ed. *North of Athabasca: Slave Lake and Mackenzie River Documents of the North West Company, 1800–1821.* Montreal: McGill-Queen's University Press, 2001.

Koldeway, R. *The Excavations at Babylon.* London: Macmillan & Co., 1914.

Lane, Nick. *Oxygen: The Molecule That Made the World.* Oxford: Oxford University Press, 2002.

Mackenzie, Alexander. *Voyages from Montreal on the River St. Laurence . . .* London: T. Cadell Jr. and W. Davies, 1801.

McLaurin, John J. *Sketches in Crude-Oil.* Harrisburg, PA: Self-published, 1898.

Miller, Carl F., ed. "Fire Fighting Operations in Hamburg, Germany During World War II," Washington, D.C.: Civil Defense Preparedness Agency, 1971.

M'Lean, John. *Notes of a Twenty-Five Years' Service in the Hudson's Bay Territory.* London: Richard Bentley, 1849.

Morse, Eric. *Fur Trade Canoe Routes of Canada.* Ottawa: Queen's Printer, 1969.

Murphy, Peter J. *History of Forest and Prairie Fire Control Policy in Alberta.* Edmonton: Alberta Energy and Natural Resources, 1985.

Mystic Seaport Museum. "Whalemen's Shipping List and Merchants' Transcript— 1843–1914." See research.mysticseaport.org/.

Newman, Peter C. *Company of Adventurers.* Markham, ON: Viking, 1985.

Nichols, Peter. *Oil and Ice: A Story of Arctic Disaster and the Rise and Fall of America's Last Dynasty.* London: Penguin, 2010.

Nikiforuk, Andrew. *Tar Sands: Dirty Oil and the Future of a Continent.* Vancouver: Greystone, 2009.

Priestley, Joseph. "Observations on Different Kinds of Air," *Philosophical Transactions.* London: Royal Society, January 1, 1772.

Pyne, Stephen J. *Fire: A Brief History.* Seattle: University of Washington Press, 2001.

Scott, Andrew C., et al. *Fire on Earth: An Introduction.* Oxford: John Wiley & Sons, 2014.

Smil, Vaclav. *Oil: A Beginner's Guide.* Oxford: One World Press, 2008.

———. *Energy and Civilization: A History.* Cambridge: Cambridge University Press, 2017.

Struzik, Edward. *Firestorm: How Wildfire Will Shape Our Future.* Washington, DC: Island Press, 2017.

Syncrude Canada, Ltd. *The Syncrude Story.* Fort McMurray: Syncrude Canada, Ltd., 1990.

Tarbell, Ida M. *The History of the Standard Oil Company.* New York: Macmillan, 1904.

Tobin, Ashley, Ed. *93/88,000: Stories of Evacuation, Re-entry and the In-between.* Fort McMurray: The 88,000 Project, 2016.

———. *159 More/88,000: Stories of Evacuation, Re-entry and the In-between.* Fort McMurray: The 88,000 Project, 2017.

Turner, Chris. *The Patch: The People, Pipelines, and Politics of the Oil Sands.* Toronto: Simon & Schuster, 2018.

Tymstra, Cordy. *The Chinchaga Firestorm: When the Moon and Sun Turned Blue.* Edmonton: University of Alberta Press, 2015.

Yergin, Daniel. *The Prize: The Epic Quest for Oil, Money, & Power.* New York: Simon & Schuster, 1991.

INDEX

Page numbers in italics refer to illustrations.

IMAGE CREDITS

COLOR INSERT